State-of-the-Art Sensors Technology in Spain 2015

Volume 2

Special Issue Editor

Gonzalo Pajares Martinsanz

Special Issue Editor
Gonzalo Pajares Martinsanz
Department Software Engineering and Artificial Intelligence
Faculty of Informatics
University Complutense of Madrid
Spain

Editorial Office
MDPI AG
St. Alban-Anlage 66
Basel, Switzerland

This edition is a reprint of the Special Issue published online in the open access journal *Sensors* (ISSN 1424-8220) from 2015–2016 (available at: http://www.mdpi.com/journal/sensors/special_issues/state-of-the-art-spain-2015).

For citation purposes, cite each article independently as indicated on the article page online and as indicated below:

Author 1; Author 2; Author 3 etc. Article title. *Journal Name*. **Year**. Article number/page range.

Vol 2 ISBN 978-3-03842-372-0 (Pbk)   Vol 1-2 ISBN 978-3-03842-312-6 (Pbk)
Vol 2 ISBN 978-3-03842-373-7 (PDF)   Vol 1-2 ISBN 978-3-03842-313-3 (PDF)

Table of Contents

About the Guest Editor

Gonzalo Pajares received his Ph.D. degree in Physics from the Distance University, Spain, in 1995, for a thesis on stereovision. Since 1988 he has worked at Indra in critical real-time software development. He has also worked at Indra Space and INTA in advanced image processing for remote sensing. He joined the University Complutense of Madrid in 1995 on the Faculty of Informatics (Computer Science) at the Department of Software Engineering and Artificial intelligence. His current research interests include computer and machine visual perception, artificial intelligence, decision-making, robotics and simulation and has written many publications, including several books, on these topics. He is the co-director of the ISCAR Research Group. He is an Associated Editor for the indexed online journal *Remote Sensing* and serves as a member of the Editorial Board in the following journals: *Sensors, EURASIP Journal of Image and Video Processing, Pattern Analysis and Applications.* He is also the Editor-in-Chief of the *Journal of Imaging*

Preface to "State-of-the-Art Sensors Technology in Spain 2015"

Since 2009, three Special Issues have been published on sensors and technologies in Spain, where researchers have presented their successful progress. Thirty-one high quality papers demonstrating significant achievements have been collected and reproduced in this book.

They are self-contained works addressing different sensor-based technologies, procedures and applications in several areas, including measurement devices, wireless sensor networks, robotics, imaging, optical systems or electrical/electronic devices among others.

Readers will find an excellent source of resources for the development of research, teaching or industrial activity.

Although the book is focused on sensors and technologies in Spain, it describes worldwide developments and references on the covered topics. Some works have been or come from international collaborations.

Our society demands new technologies for data acquisition, processing and transmission for immediate actuation or knowledge, and with important impact on one's welfare when required.

The international, scientific and industrial communities worldwide will also be an indirect beneficiary of these works. Indeed, the book provides insights and solutions for the varied problems covered. Also, it lays the foundation for future advances toward new challenges and progress in many areas. In this regard, new sensors will contribute to the solution of existing problems, and, where the need arises for the development of new technologies or procedures, this book paves the way.

We are grateful to all the people involved in the preparation of this book. Without the invaluable contributions of the authors together with the excellent help of reviewers, this book would not have reached fruition. More than 120 authors have contributed to this book.

Thanks also to the *Sensors* journal editorial team for their invaluable support and encouragement.

Gonzalo Pajares Martinsanz
Guest Editor

Article

Merge Fuzzy Visual Servoing and GPS-Based Planning to Obtain a Proper Navigation Behavior for a Small Crop-Inspection Robot

José M. Bengochea-Guevara, Jesus Conesa-Muñoz, Dionisio Andújar and Angela Ribeiro *

Center for Automation and Robotics, CSIC-UPM, Arganda del Rey, Madrid 28500, Spain;
jose.bengochea@csic.es (J.M.B.-G.); jesus.conesa@csic.es (J.C.-M.); dionisioandujar@hotmail.com (D.A.)
* Correspondence: angela.ribeiro@csic.es; Tel.: +34-918-711-900; Fax: +34-918-717-050

Academic Editor: Gonzalo Pajares Martinsanz
Received: 21 December 2015; Accepted: 19 February 2016; Published: 24 February 2016

Abstract: The concept of precision agriculture, which proposes farming management adapted to crop variability, has emerged in recent years. To effectively implement precision agriculture, data must be gathered from the field in an automated manner at minimal cost. In this study, a small autonomous field inspection vehicle was developed to minimise the impact of the scouting on the crop and soil compaction. The proposed approach integrates a camera with a GPS receiver to obtain a set of basic behaviours required of an autonomous mobile robot to inspect a crop field with full coverage. A path planner considered the field contour and the crop type to determine the best inspection route. An image-processing method capable of extracting the central crop row under uncontrolled lighting conditions in real time from images acquired with a reflex camera positioned on the front of the robot was developed. Two fuzzy controllers were also designed and developed to achieve vision-guided navigation. A method for detecting the end of a crop row using camera-acquired images was developed. In addition, manoeuvres necessary for the robot to change rows were established. These manoeuvres enabled the robot to autonomously cover the entire crop by following a previously established plan and without stepping on the crop row, which is an essential behaviour for covering crops such as maize without damaging them.

Keywords: generation of autonomous behaviour; crop inspection; visual servoing; fuzzy control; precision agriculture; GPS

1. Introduction

Farming practices have traditionally focused on uniform management of the field and ignored spatial and temporal crop variability. This approach has two main negative outcomes: a) air and soil pollution, with consequent pollution of groundwater, and b) increased production costs [1]. Moreover, agricultural production must double in the next 25 years to sustain the increasing global population while utilising less soil and water. In this context, technology will become an essential aspect of minimising production costs while crops and environment are properly managed [2–4].

The development of technologies such as global positioning systems (GPS), crop sensors, humidity or soil fertility sensors, multispectral sensors, remote sensing, geographic information systems (GIS) and decision support systems (DSS) have led to the emergence of the concept of precision agriculture (PA), which proposes the adaptation of farming management to crop variability. Particularly important within PA are techniques aimed at selective treatment of weeds (site-specific management) by restricting herbicide use to infested crop areas and even varying the amount of treatment applied according to the density and/or type of weeds, in contrast to traditional weed control methods.

Selective herbicide application requires estimations of the herbicide needed for each crop unit [5]. First, data must be acquired in the field to determine the location and estimated density of the weeds (perception stage). Using this information, the optimal action for the crop is selected (decision-making stage). Finally, the field operations corresponding to the decision made in the previous stage must be performed to achieve the selective treatment of weeds (action stage). At the ground level, data collection can be accomplished by sampling on foot or using mobile platforms. Sampling on foot is highly time-consuming and requires many skilled workers to cover large treatment areas. Discrete data are collected from pre-defined points throughout an area using sampling grids, and interpolation is employed to estimate the densities of the intermediary areas [6].

In continuous sampling, data are collected over the entire sample area. Continuous data enable a qualitative description of abundance (e.g., presence or absence; zero, low, medium, or high) rather than the quantitative plant counts that are usually generated by discrete sampling [7].

To effectively implement PA, the perception stage should be substantially automated to minimise its cost and to increase the quality of the gathered information. Among the various means of collecting well-structured information with reasonably priced autonomous, vehicles that are equipped with on board sensing elements are considered to be one of the most promising technologies in the medium-term. However, the use of mobile robots in agricultural environments remains challenging as navigation in agricultural environments presents difficulties due to the variability and nature of the terrain and vegetation [8,9].

Research in navigation systems for agricultural applications has focused on guidance methods that employ global or local information. Guidance systems that use global information attempt to direct the vehicle along a previously calculated route based on a terrain map and the position of the vehicle relative to an absolute reference. In this case, global navigation satellite systems (GNSS), such as GPS, are usually employed. Guidance systems that utilise local information attempt to direct a vehicle based on the detection of local landmarks, such as planting patterns and intervals between crop rows.

The precision of the absolute positions that are derived from a GNSS can be enhanced by real-time kinematics (RTK), which is a differential GNSS technique that employs measurements of the phase of the signal carrier wave and relies on a single reference station or interpolated virtual station to provide real-time corrections and centimetre-level accuracy. The use of a RTK-GPS receiver as the only positioning sensor for the automatic steering system of agricultural vehicles has been examined in several previous studies [10,11]. Furthermore, recent studies [12,13] evaluate the use of low-cost GPS receivers for the autonomous guidance of agricultural tractors along straight trajectories.

Regardless of the type of GNSS used, this navigational technology has some limitations when the GNSS serves as the only position sensor for autonomous navigation of mobile robots. For this reason, RTK-GNSS is frequently combined with other sensors, such as inertial measurement units (IMUs) [14,15] or fibre-optic gyroscopes (FOGs) [16–18]. When a GNSS, even a RTK-GNSS, is employed as the only sensor for navigating across a crop without stepping on plants, an essential behaviour in crops such as maize, it is an indispensable requirement to perfectly know the layout of the crop rows and therefore, crops must be sowed with an RTK-GNSS-guided planting system or mapped using a georeferenced mapping technique. This approach is expensive and may not be feasible, which reduces the scope of the navigation systems that are based in GNSS. In this context, the proposed approach integrates a GNSS sensor with a camera (vision system) to obtain a robot's behaviour, which enables it to autonomously cover an entire crop by following a previously established plan without stepping on the crop rows to avoiding damage to the plants. As discussed in the next section, the plan only considers the field contour and the crop type, which is readily available.

Vision sensors have been extensively utilised in mobile robot navigation guidance [19–26] due to their cost-effectiveness and ability to provide large amounts of information, which can also be employed in generating steering control signals for mobile robots. In addition, diverse approaches have been proposed for crop row detection. In previous studies [25,27], a segmentation step is applied to a colour image to obtain a binary image, in which white pixels symbolise the vegetation cover. Then, the binary image is divided into horizontal strips to address the perspective of the camera. For each strip, they review all columns of pixels. Columns with more white pixels than black pixels are labelled as potential crop rows, and all pixels in the column are set to white; otherwise, they are set to black. To determine the points that define crop rows, the geometric centre of the block with the largest number of adjacent white columns of the image is selected. Then, the method estimates the line defined by these points, which is based on the average values of their coordinates. Other approaches [28] transform an RGB colour image to grayscale and divide it into horizontal strips, where maximum grey values indicate the presence of a candidate row. Each maximum defines a row segment, and the centres of gravity of the segments are joined via a similar method to the centre of gravity that is utilised in the Hough transform or by applying linear regression. In [29], the original RGB image is transformed to a grayscale image and divided into horizontal strips. They construct a bandpass filter that is based on the finding that the intensity of the pixels across these strips exhibits a periodic variation due to the parallel crop rows. Sometimes, detection of the row is difficult as crops and weeds form a unique patch. The Hough transform [30] has been employed for the automatic guidance of agricultural vehicles [23,31–33]. Depending on the crop densities, several lines are feasible, and a posterior merging process is applied to lines with similar parameters [34–36]. When weeds are present and irregularly distributed, this process may cause failure detection. In [26,37], the authors employ stereo-images for crop row tracking to create an elevation map. However, stereo-based methods are only adequate when appreciable differences exist between the heights of crops and the heights of weeds, which is usually not the case when an herbicide treatment is performed in the field. In [38–40], crop rows are mapped under perspective projection onto an image that shows some behaviours in the frequency domain. In maize fields, where the experiments were performed, crops did not show a manifest frequency content in the Fourier space. In [41], authors analyse images that were captured from the perspective from a vision system that is installed onboard a vehicle and consider that the camera is being submitted to vibrations and undesired movements, which are produced as a result of vehicle movements on uneven ground. They propose a fuzzy clustering process to obtain a threshold to separate green plants or pixels (crops and weeds) from the remaining items in the field (soil and stones). Similar to other approaches, crop row detection applies a method that is based on image perspective projection, which searches for the maximum accumulation of segmented green pixels along straight alignments.

Regarding the type of vehicle, most studies of the autonomous guidance of agricultural vehicles have focused on tractors or heavy vehicles [10,12,13,15,17,20–24,26]. Moreover these large vehicles can autonomously navigate along the crop rows, but they are unable to autonomously cover an entire crop by performing the necessary manoeuvres for switching between crop rows. In other cases, the autonomous navigation is related to fleets of medium sized tractors able to carry weed control implements [42]. In PA, the use of small robots for inspecting an entire crop is a suitable choice over large machines to minimise the soil compaction. In this context, this study was conducted to support the use of small autonomous vehicles for inspection, with improved economic efficiency and reduced the impact on crops and soil compaction, which integrate both global location and vision sensors for obtaining a navigation system that enables the covering of an entire field without damage to the crop. Inspection based on small autonomous vehicles can be very useful for early pest detection by gathering geo-referenced information that is needed to construct accurate risk maps. More than one sampling is performed throughout the year, due to minimal crop impact, primarily if they can navigate across a field by following the crop rows, and soil compaction.

2. Materials and Methods

The robot used in this project is a commercial model (mBase-MR7 built by MoviRobotics, Albacete, Spain). It has four wheels with differential locomotion, no steering wheel and can rotate on its vertical axis. This work considered that the robot carries out the manoeuvres as if it could not rotate on its vertical axis, like other vehicles used within the agricultural fields (tractors, all-terrain vehicles, *etc.*). The reason is that often, in our experiments, the vehicle's wheels have dug up the land and the robot has gotten stuck. The on-board camera is a digital single-lens reflex camera (EOS 7D, Canon, Tokyo, Japan). The camera is located 80 cm aboveground at a pitch angle of 18° and is connected to an on-board computer (a Toughbook CF-19 laptop, Panasonic, Osaka, Japan equipped with an Intel Core i5 processor and 2 GB of DDR3 RAM) via a USB connector. The camera supplies approximately five frames per second, and each frame has a resolution of 1056 × 704 pixels. Other camera locations were studied, such as placing it facing down, focusing directly on the soil. However, this case only covered a small portion of terrain, and therefore the detection of the crop row was more vulnerable to local changes as sowing errors, weed patches, *etc.* Furthermore, the covered terrain was the area immediately in front of the robot, so when the robot needed to react to what it was present in the image, part of it had been left behind. Another analysed option was to place it ahead of the robot, in a forward position using a steel mast. However, this caused more vibrations on the camera during robot navigation, deteriorating the system operation significantly.

The vehicle equipment is complemented with a R220 GPS receiver (Hemisphere, Scottsdale, AZ, USA) with RTK correction for geo-referencing the gathered data and determining whether the robot has reached a field edge. A laptop (a tablet) is used to remotely control the robot. The architecture of the developed system is illustrated in Figure 1a, and photographs of the vehicle and maize crop rows in a field are shown in Figure 1b.

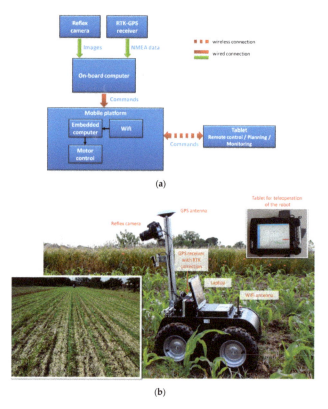

(a)

(b)

Figure 1. (**a**) The system architecture. (**b**) Devices integrated in the mBase-MR7 robot (right) and a maize crop field (left).

The plan to be followed by the robot is generated by a path planner. Path planning in agricultural fields is a complex task. Generally, it can be formulized as the well-known Capacited Vehicle Routing Problem (CVRP), as stated in [43]. Basically, the problem consists of determining the best inspection route that provides complete coverage of the field considering features such as the field shape, the crop row direction, the type of crop and some characteristics of the vehicles, such as the turning radii. The vehicles must completely traverse each row exactly once; therefore, the planner determines the order for performing the rows in such a manner that some optimisation criterion is minimal. Given a field contour, the planner can deduce the layout of the rows and the inter-row distance required by the plants, due to it assumed that the sowing was carried out by a mechanical tool that kept that distance in a reasonably precise way. In addition, a high precision is not required since the proposed platform and the proposed method exclusively use the trajectory points as guiding references to enter and leave the field rows.

The planner employed in this work is described in [44] and uses a simulated annealing algorithm to address a simplified case of the general path planning problem with only one vehicle and considering the travelled distance as the optimisation criterion. Figure 2 shows the route that is generated by the planner for the crop field in which the experiments were performed. The field size was approximately 7 m × 60 m, which represents a total of ten crop rows in a maize planting schema with 0.7 m of distance between rows. In this case, the optimal trajectory was to sequentially explore the field, beginning at one edge and always going to an adjacent row as the vehicle is very small and has a turning radius that enables movement between adjacent rows despite the very small distances between rows. It is important to note that the system proposed in this study can work with any planner able to return the path to be followed as an ordered sequence of the number of pairs of GPS points as crop rows must be travelled, where the first pair of points represents the input point to the field to scout a row and the second pair point defines the field output point to go to the next row to be inspected.

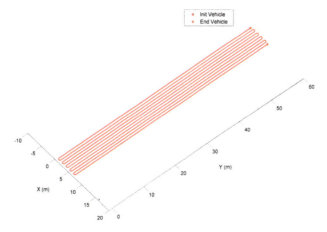

Figure 2. Path taken by the planner traversed all rows of the crop field in which the experiments were performed.

To inspect a crop row, the robot is positioned at the beginning of the row, *i.e.*, in the approximate point established by the plan, with the row between its two front wheels. The robot advances, tracking the crop row using its on-board camera, until it reaches the end of the row. Once it has inspected a row, the robot executes the necessary manoeuvres to position itself at the head of the next row to be inspected; the approximate position of this row is also established in the plan. The process is repeated until all rows in the field have been inspected or when the plan has been completely executed. To achieve field inspection with complete coverage, a set of individual behaviours is required, including:

(1) tracking of a crop row, (2) detection of the end of a crop row, and (3) transition to the head of the next row to be inspected (note that the plan provides only an approximate point). The proposed approaches to generate these behaviours in the robot are discussed in the following sections.

2.1. Crop Row Tracking Behaviour

An image-processing method capable of extracting the layout of the crop rows in real time from images acquired with the camera positioned on the front of the robot was designed to allow the robot to use images of the crop rows to navigate. The purpose of the image processing is to obtain the vehicle's position with respect to the crop row (see Figure 3), *i.e.*, the motion direction angle (α) and displacement or offset (d) between the robot centre and the closest point along the line defining the crop row.

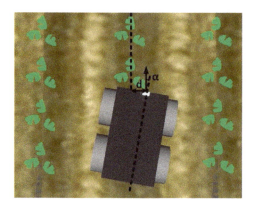

Figure 3. Pose of the robot with respect to the central crop row.

The values of the vehicle's offset (d) and angle (α) are provided to two fuzzy controllers, one for angular speed and one for linear speed, which determine the correction values for the steering to generate crop row tracking behaviour in the robot.

2.1.1. Image Processing

In the row recognition process, the main problem is the identification of accurate features that are stable in different environmental conditions. The row detection process is accompanied by some difficulties, such as incomplete rows, missing plants, and irregular plant shapes and sizes within the row. In addition, the presence of weeds along the row may distort row recognition by adding noise to the row structure. The majority of the studies have focused on large agricultural vehicles, in which the displacement is more uniform than the displacement of small vehicles. In this study, the challenge is to robustly detect a crop row in the presence of weeds, despite the vibrations and variations in the camera, which are caused by the movement of a vehicle in the field. The majority of all methods for vegetation detection usually consider that all pixels that are associated with vegetation have a strong green component [45–52]. To take advantage of this characteristic, the utilisation of digital cameras in the visible spectrum and the use of the RGB colour model is frequent when working at the ground level [27–29,33,35,40,41,45,49–52]. The proposed row detection approach takes advantage of our previous study (refer to introduction) for designing and developing a real-time technique that properly works with RGB images that are acquired in varying environmental conditions.

A typical image acquired by the camera of the robot is shown in Figure 4a. In the upper corners of the image, the crop rows are difficult to distinguish due to the perspective in the image. To avoid these effects, the image was divided in half, and the upper half was discarded (Figure 4b). Thus, the image processing presented below only utilises the bottom half of each frame. Figure 5 shows a flowchart diagram of the image processing phase.

(a)   (b)

Figure 4. (**a**) A typical image acquired by the camera of the robot. (**b**) The working area of the image is delimited in red.

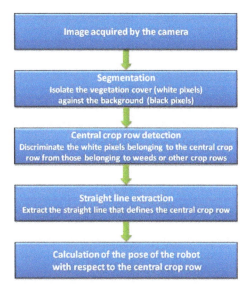

Figure 5. Diagram of the image processing phase.

The objective of the first processing stage (segmentation stage) is to isolate the vegetation cover against the background, *i.e.*, to convert the input RGB image into a black-and-white image in which the white pixels represent the vegetation cover (weeds and crop) and the black pixels represent the remaining elements in the image (soil, stones, debris, straws, *etc.*).

Segmentation exploits the strong green components of the pixels representing vegetation. The coloured image can be transformed into a greyscale image by a linear combination of the red, green and blue planes, as shown in Equation (1):

$$\forall i \in rows\_image \wedge \forall j \in colums\_image :$$
$$Grey\,(i,j) = r * input_{red(i,j)} + g * input_{green(i,j)} + b * input\_blue\,(i,j)$$

(1)

where i varies from 0 to 352, j from 0 to 1056, the *input_red(i, j)*, *input_green(i, j)*, *input_blue(i, j)* values are the non-normalised red, green, and blue intensities (0–255), respectively, at pixel (i, j) and r, g, b are the set of real coefficients that determine how the monochrome image is constructed. These values are crucial in the segmentation of vegetation against non-vegetation, and their selection is discussed in detail in [46,51]. In the proposed approach, the constant values were established to a set of values ($r = -0.884$, $g = 1.262$, and $b = -0.311$) that previously showed good results for similar images [53] compared with other well-known indices, such as *ExG* ($r = -1$, $g = 2$, $b = -1$) [46].

In the next step, a threshold is used to convert the monochrome greyscale image into a binary image in which the white pixels represent vegetation and the black pixels non-vegetation. The threshold depends on the lighting conditions. Therefore, the threshold is not fixed in the approach outlined here but is instead calculated for each analysed image as the mean value of the grey intensities in the image. The results of the segmentation stage are illustrated in Figure 6.

(a)   (b)

Figure 6. (a) Original image with marked weed presence. (b) Segmented image.

The goal of the next stage (central crop row detection), which processes the binary images obtained in the previous stage, is to discriminate the white pixels belonging to the central crop row from those belonging to weeds or other crop rows. To achieve this goal, the method developed based on [54] first performs a morphological opening operation (erosion followed by dilation) of the binary image to eliminate isolated white pixels and highlight areas with a high density of white pixels. One of the aims of this operation is to eliminate the small groups of black pixels that appear inside the crops. The structural element used for the dilation and erosion is a 3×3 square. The borders of the resulting image are then extracted using the Sobel operator such that all pixels in the transitions (white to black and vice versa) are marked. The image is then divided into three horizontal strips to deal with perspective. Each strip is processed independently using the following methods.

The potential centre of the central crop row is the column of the strip with the greatest number of white pixels within a search window. To identify this column, a vector is built using the same number of components as the size of the window, where each component stores the number of white pixels (vegetation) of the associated column. The perspective of the original images is also considered when defining this window; thus, the size of the window varies depending on the proximity of the camera to the analysed strip (Thales' intercept theorem). The search window is centred in the middle of the image in the first frame, but due to overlap between subsequent frames (the robot advances 6 cm between frames at its highest speed), the possible centre of the row in the next frame is searched around the central position identified in the previous frame.

After the potential centre of the row is identified, the algorithm begins to identify the edges delimiting the crop row, searching to the right and to the left from the centre found. To confirm that a crop edge has been reached, the method uses three pixel labels: white, black and border. When the pixel encountered is white, it is marked as belonging to the crop row, and the algorithm continues with the next pixel. When a border pixel is located, the exploration has reached either a crop edge or a group of black pixels inside the crop. The distance to the next border pixel can be used to distinguish between these two cases. The distance to the next crop row or to a weed between crop rows is greater in the former case than in the latter, *i.e.*, inside the crop row. In fact, two distance thresholds are established,

$D_1 \leqslant D_2$, such that if the computed distance is greater than threshold D_2, the exploration has reached a crop edge, whereas if the distance is less than D_1, a group of black pixels inside the crop has been reached. If the distance is between D_1 and D_2, the method uses the previously generated vector to locate the centre of the row and proceeds as follows. The percentage of white pixels in each column is calculated for the range of components of the vector between the current position of the pixel and the position of the edge. If this percentage is higher than a threshold called *min_proportion*, the algorithm indicates that it has reached a group of black pixels inside the crop because the column to which this group of black pixels belongs has a large number of white pixels because it is part of a crop row. If it is lower than this threshold, the algorithm indicates that it has reached the edge of the crop row because the number of black pixels in the columns that separate it from the next crop row or weed is large.

This procedure is formally set out in Table 1, where p is the pixel currently being explored, n is the next (not black) pixel in the processing order at a distance d, and D_1, D_2 and *min_proportion* are the three parameters of the method.

Table 1. Crop row detection method.

Type of Current Pixel	Distance d (in pixels) until next non-Black Pixel		
	$d \leqslant D_1$	$D_1 < d \leqslant D_2$	$d > D_2$
White	Mark all pixels from p to n and jump to n $(p \leftarrow n)$	Mark all pixels from p to n and jump to n $(p \leftarrow n)$	Stops
Border	Mark all pixels from p to n and jump to n $(p \leftarrow n)$	IF White pixels(input$(1 \dots N), p \dots n)) > min\_proportion$ THEN Mark all pixels from p to n and jump to n $(p = n)$ ELSE Stops	Stops
Black	Jump to n $(p \leftarrow n)$	Jump to n $(p \leftarrow n)$	Stops

Due to the effects of perspective in the image, the width of the crop rows varies depending on proximity to the camera. This phenomenon is considered in the method. The parameter D_1 varies between 5 and 10, starting at 5 when the farthest strip from the camera is analysed and reaching 10 when analysing the closest strip. Likewise, the parameter D_2 varies between 10 and 20. The value of the threshold *min_proportion* is 0.6. Using this process, the central crop row can be detected from the binary image in the presence or absence of weeds (see Figure 7).

(a)   (b)

Figure 7. (a) Segmented image. (b) Central crop row detected by applying the proposed detection method to Figure 6a.

After the central crop row is detected, the purpose of the next stage is to extract the straight line that defines the central crop row from the image resulting from the last stage. To address the slight perspective of the camera, the image is divided into three horizontal strips, which are processed to obtain three points to define the central crop row. The previous stage assumed that the potential centre of the central crop row was the column with the greatest number of white pixels. However, this may not be the case when groups of black pixels are present inside the crop. In the previous stage, such an occurrence does not affect the algorithm, that is, it does not matter whether the algorithm starts in the exact centre of the crop row as long as it is located inside the row. However, the extraction of the straight line that defines the crop row requires that the centre of the crop row be located as accurately as possible. Thus, a search window is defined in the same way as in the previous stage, and a vector is obtained with as many components as the window size, where each component stores the number of white pixels of the associated column. Next, the maximum value of the vector is computed, and all columns in the image whose vector component is greater than 80% of this maximum value are converted to white, whereas the rest are converted to black (Figure 8a). To determine the points that define the central crop row, in each strip, the geometric centre of the block with the largest number of white columns together is chosen. The algorithm then estimates the straight line defining the three centres identified (one for each strip) using the least squares method (Figure 8b). If less than two centres are located (*i.e.*, sowing errors), the algorithm employs the straight line that is obtained in the previous frame. After obtaining the straight line that defines the crop row, the angle (α) between the direction of motion of the robot and the centre line of the crop row and the displacement (d) between the centre of the vehicle and the centre of the crop row are calculated (refer to Figure 3).

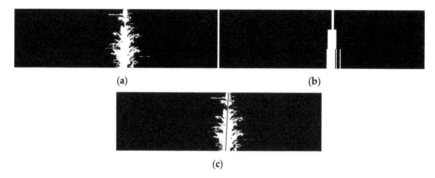

(a)   (b)

(c)

Figure 8. (**a**) Central crop row obtained in the previous stage. (**b**) White and black columns defined by the algorithm from Figure 8a. (**c**) Straight line defining the crop row centre in Figure 8a.

2.1.2. Navigation Control

Many approaches exist that address the actuator control of a car. Conventional control methods produce reasonable results at the expense of high computational and design costs because obtaining a mathematical model of the vehicle becomes extremely expensive [55,56], since wheeled mobile robots are characterised by nonlinear dynamics and are affected by an important number of disturbances, such as turning and static friction or variations in the amount of cargo. Alternatively, we can approach human behaviour for speed and steering control using artificial intelligence techniques, such as neural networks [57]. However, the technique that provides a better approximation to human reasoning and gives a more intuitive control structure is the fuzzy logic [58,59]. Some authors have proposed solutions that are based in fuzzy logic for autonomous navigation [60–64], which demonstrates their robustness. In [60], fuzzy control is employed in a real car to perform trajectory tracking and obstacle avoidance in real outdoor and partially known environments. In [61], fuzzy controllers are implemented in a real car to conduct experiments on real roads within a private circuit. Their results show that the fuzzy controllers perfectly mimic human driving behaviour in driving and route tracking, as well as complex, multiple-vehicle manoeuvres, such as adaptive cruise control or overtaking. In [62], the navigation of multiple mobile robots in the presence of static and moving obstacles that employ different fuzzy controllers is discussed. Their experiments demonstrate that robots are capable of avoiding obstacles and negotiating dead ends, as well as efficiently attaining targets. In [63], authors develop and implement fuzzy controllers for the steering and speed control of an autonomous guided vehicle. Their results indicate that the proposed controllers are insensitive to parametric uncertainty and load fluctuations and outperformed conventional proportional-integral-derivative (PID) controllers, particularly in tracking accuracy, steady-state error, control chatter and robustness. In [64], an unknown path-tracking approach is addressed, based on a fuzzy-logic set of rules, which emulates the behaviour of a human driver. The method applies approximate knowledge about the curvature of the path ahead of the vehicle and the distance between the vehicle and the next turn to attain the maximum value of the linear velocity that is required by the vehicle to safely drive on the path.

The proposed navigation control that enables a robot to follow crop rows comprises two fuzzy controllers (Figure 9): one for angular speed and one for linear speed. Both controllers are fuzzy and therefore imitate the behaviour of a skilled driver; for example, if the vehicle is moved to one side of the crop row to be tracked, the robot must correct its position in the other direction such that it navigates with the crop row between its wheels.

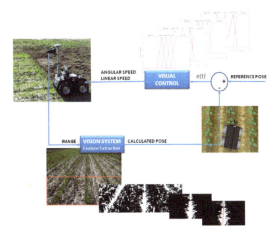

Figure 9. Visual scheme of the developed control.

The inputs of the controller acting on the angular speed of the robot are the displacement of the centre of the vehicle from the midpoint of the crop row (d in Figure 3) and the angle of orientation of the robot (α in Figure 3). The controller produces the angular speed of the vehicle as an output. The rules used in the controller take the following form:

If (Offset is Negative Big) and (Angle is Positive Small) then (Angular Speed is Positive Small)
These rules are summarised in Table 2.

Table 2. Fuzzy control rules for angular speed. The fuzzy labels in this table correspond to the fuzzy sets shown in Figure 10. Furthermore, Figure 11 shows the fuzzy set for linear speed used in Table 3 whereas fuzzy sets for the output variables are illustrated in Figure 12.

Offset d / Angle α	Negative Big	Negative Small	Zero	Positive Small	Positive Big
Negative Big	Positive Big	Positive Big	Positive Big	Positive Small	Zero
Negative Small	Positive Big	Positive Small	Positive Small	Zero	Negative Small
Zero	Positive Big	Positive Small	Zero	Negative Small	Negative Big
Positive Small	Positive Small	Zero	Negative Small	Negative Small	Negative Big
Positive Big	Zero	Negative Small	Negative Big	Negative Big	Negative Big

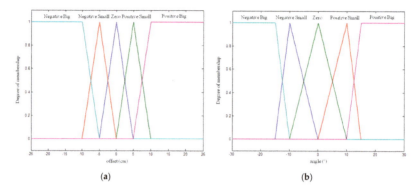

Figure 10. Angular speed controller. Fuzzy sets of input variables: (**a**) offset and (**b**) angle.

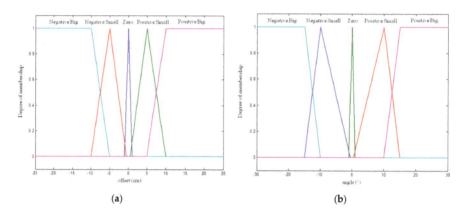

Figure 11. Linear speed controller. Fuzzy sets of input variables: (**a**) offset and (**b**) angle.

Table 3. Fuzzy control rules for linear speed. The fuzzy labels in this table correspond to the fuzzy sets shown in Figures 11 and 12b.

Angle α \ Offset d	Negative Big	Negative Small	Zero	Positive Small	Positive Big
Negative Big	Minimum	Minimum	Minimum	Minimum	Minimum
Negative Small	Minimum	Minimum	Medium	Medium	Medium
Zero	Minimum	Medium	Maximum	Medium	Minimum
Positive Small	Medium	Medium	Medium	Minimum	Minimum
Positive Big	Minimum	Minimum	Minimum	Minimum	Minimum

Negative_Big, Negative_Small, Zero, Positive_Small and Positive_Big are the fuzzy sets shown in Figure 10 and are determined in consonance with the features of the robot. The value ranges of each set in the input corresponding to the offset of the robot regarding the midpoint of the crop row (d) were selected based on the distance between the wheels of the robot (41 cm). In the case of the angle of orientation (α), to select the value ranges of each set, it was assumed that the robot would crush the crop row if it turned at an angle of 30° or −30°.

The Takagi-Sugeno implication is compatible with applications that require on-time responses [58] and is therefore used in this work. The output of the controller is the singleton-type membership functions shown in Figure 12a. The range of angular speeds allowed by the robot is −90 °/s to 90 °/s, and the fuzzy sets are chosen to cover the speed range necessary to control the robot smoothly. The input variables in the controller of the linear speed of the robot are the same as the previous controller, *i.e.*, offset and angle, and the fuzzy rules defined are summarised in Table 3.

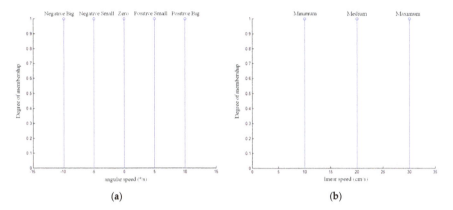

(a)   (b)

Figure 12. Fuzzy sets for the output variables: (**a**) angular speed and (**b**) linear speed.

The value ranges of the fuzzy sets (Figure 11) for both displacement and angle are selected so that the vehicle moves at maximum speed (30 cm/s) provided that it is correctly positioned; otherwise, it slows to avoid treading on the crop row and crushing the crop.

The controller output is the linear speed applied to the vehicle and is characterised by three fuzzy sets (MIN: minimum, MED: medium, MAX: maximum), whose values are selected to cover the robot's range of allowed speeds (Figure 12).

2.2. End of Crop Row Detection

To complete tracking of the inspected crop row, the robot must detect the end of the crop row (Figure 13). For this purpose, the number of pixels in the image belonging to vegetation is used. When the robot begins to track a crop row, the number of pixels belonging to vegetation in the first image of the tracking is stored as a reference value. As the robot advances, it detects the number of pixels that are associated with the vegetation in each image. When this number is less than 80% (threshold was determined using the trial and error method) of the reference value, the robot assumes that it has reached the end of the crop row and checks with the on board GPS to ensure that this position is consistent with the output point given in the plan (with a margin of error). If they are consistent, the robot stops; otherwise, it continues tracking the crop row using the direction of the central row that is obtained in the previous frame. The loss of pixels that are associated with vegetation is consistent with a possible sowing error.

(a)   (b)

Figure 13. (**a**) Image acquired by the robot of the end of a crop row. (**b**) Segmented image of Figure 13a. The number of pixels that is associated with vegetation (white pixels) is less than 80% of the number of pixels in the reference image (Figure 6), and the robot location is very close to the output point that is defined in the plan for the row that is being scouted. Thus, the robot determines that it has reached the end of the crop row.

2.3. Row Change Behaviour

After detecting the end of a crop row, the row change behaviour is performed. In this behaviour, the robot performs the necessary manoeuvres to position itself at the head of the next row to be inspected. This set of manoeuvres consists of a combination of straight-line and circular-arc movements with the help of the on board GPS. Specifically, it is a sequence of the following five moves (Figure 14): (1) straight-line forward movement to position itself outside the crop, in the header of the crop, to avoid crushing the crop in subsequent manoeuvres; (2) circular-arc movement towards the corresponding side of the next crop row to be inspected to position itself perpendicular to the crop rows; (3) straight-line forward movement; (4) reverse circular-arc movement away from the crop row to position itself parallel to the direction of the crop rows; and (5) straight-line forward movement to place itself at the head of the next crop row to be inspected according to the plan.

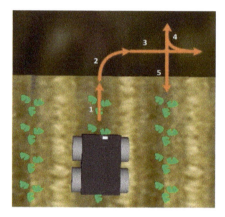

Figure 14. Manoeuvres defined for crop row change.

3. Results and Discussion

To evaluate the performance and robustness of the proposed approach for central row detection, a set of 500 images, which were acquired by the robot working in remote control mode in different maize fields and distinct days at Arganda del Rey (Madrid, Spain), were utilised. The images were captured in different conditions of illumination, growth stages, weed densities and camera orientations; the robot operates in these conditions when it autonomously navigates. Figure 15 illustrates several examples of the employed images.

The proposed approach was compared with a detection method that is based on the Hough transform [30], which is a strategy that was successfully integrated in some of our previous studies [36] in terms of effectiveness and processing time. The last aspect is really important in cases, such as this case, in which real-time detection is required. Both methods analysed the bottom half of each image, *i.e.*, 1056×352 pixels. The effectiveness was measured based on an expert criterion, in which the line that was detected was considered to be correct when it matched the real direction of the crop row. Table 4 shows the results from processing the 500 images with both approaches. The performance of the proposed approach exceeds the performance of a Hough-transform based strategy by 10%. The processing time that is required by the proposed approach is approximately four times less than the processing time required for the Hough transform, which indicates that the proposed approach can process approximately 14 frames in the best case compared with the Hough transform approach. The Hough transform approach can process three frames per second, which is less than the five frames per second that the EOS 7D camera provides (refer to Section 2 of this paper).

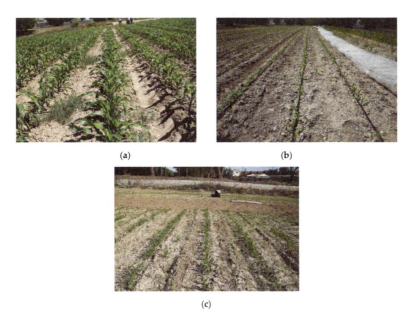

(a) (b)

(c)

Figure 15. (a)–(c) are examples of different images that were acquired by the camera of the robot. (a) shows the maize crops; (b) shows the crop rows where the irrigation system used is visible; (c) shows the kind of image to be taken when a crop border is being reaching.

Table 4. Detection of the central crop row. Performance of the proposed approach and an approach that is based on the Hough transform.

Approach	Effectiveness (%)	Mean Processing Time (seconds)
Proposed approach	96.4	0.069
Hough-transform-based approach	88.4	0.258

To test the different behaviours developed for the robot, a test environment was established. Green lines were painted in an outside soil plane to simulate crop rows. The lines were 30 m in length and spaced 70 cm apart, as illustrated in Figure 16a. To make the environment more realistic, weeds and sowing errors were introduced along the lines.

(a) (b)

Figure 16. (a) First test environment. (b) Crop employed in the experiments.

In this test environment, several experiments were performed to verify the proper functioning of the three autonomous robot behaviours described above: tracking, detection of the end of the crop row and row change. During the experiments, the robot followed the lines without headings, detected the row end, and changed rows, continuing the inspection, without human intervention. Figure 17 shows the evolution of the linear and angular speeds of the robot during the inspection of a line and the position of the robot with respect to the line (offset and angle) extracted by the image algorithm. Table 5 shows the mean, standard deviation, minimum and maximum of the linear speed, angular speed, offset and angle. The robot adjusted its linear speed depending on the error in its position; the speed was highest when the error in its position was 0 and decreased as the error increased, confirming the proper operation of the fuzzy controller of the linear speed. Variations in the angular speed were minimal because the error in the position was very small (offset and angle). The slight corrections to the left in the angular speed (negative values) were due to the tendency of the robot to veer towards the right when moving forward in its working operation (remote control mode). The design of the controller allowed these corrections to be made softly and imperceptibly during robot navigation and yielded movement in a straight line, which is impossible in remote control mode.

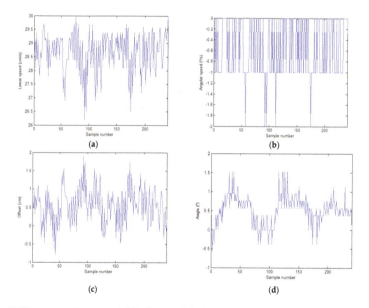

Figure 17. First test environment: (**a**) Evolution of the linear speed of the robot. (**b**) Evolution of the angular speed of the robot. (**c**) Evolution of the offset of the robot relative to the line. (**d**) Evolution of the angle of the robot relative to the line.

Table 5. Results obtained in the test environment.

Test environment	Mean	Std. dev.	Minimum	Maximum
Linear speed (cm/s)	28.66	0.68	26.20	30.00
Angular speed (°/s)	−0.57	0.53	−2.00	0.00
Offset (cm)	0.55	0.47	−0.78	1.88
Angle (°)	0.52	0.38	−0.56	1.51

After confirming the performance of the robot in the test environment, several experiments were conducted in a real field at Arganda del Rey (Madrid, Spain). The field was a cereal field with crop row spacing of 70 cm, which resembles a maize sowing schema. The rows were not perfectly straight and were characterised by strong high weed presence, as illustrated in Figure 16b. The robot was tested in different crop rows that covered the entire field. In this environment, the robot again followed the crop rows without crushing the crops (Figure 18), detected the row end, and performed the necessary manoeuvres to change rows (Figure 19) to continue the inspection without human intervention.

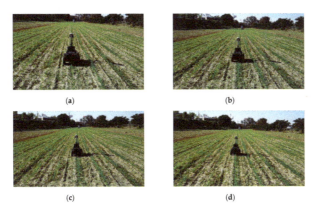

Figure 18. Crop row tracking behaviour. Images from (**a**) to (**d**) show a sequence of different frames of the video available in [65].

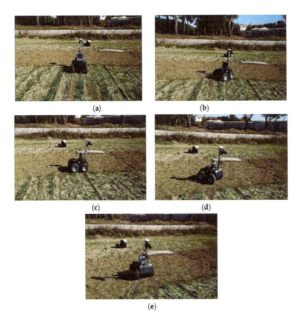

Figure 19. Sequence of manoeuvres performed by the robot to change rows. (**a**) corresponds with manoeuvre 1 in Figure 14, *i.e.*, the straight-line forward movement; (**b**) with the circular-arc movement, manoeuvre 2; (**c**) with straight-line forward movement, manoeuvre 3; (**d**) with reverse circular-arc movement, manoeuvre 4; (**e**) with straight-line forward movement, manoeuvre 5. Video can be found in [65].

However, the crop row tracking performance was worse in the real environment compared to the first test environment. Figure 20 shows the evolution of the linear and angular speeds of the robot during the inspection of a line and the position of the robot with respect to the line (offset and angle) extracted by the image algorithm. Table 6 presents the values (mean, standard deviation, minimum and maximum) obtained for linear speed, angular speed, offset and angle. The difference in performance relative to the first environment is evident. The results were a consequence of the estimated position of the robot relative to the crop row (offset and angle). The image-processing algorithm correctly extracted the straight line defining the crop row. In contrast to the first test environment, the rows were not perfectly straight, which affected the performance of the robot. However, the difference in performance between the environments was primarily due to the differences between the flat soil of the first environment and the abrupt soil of the field. The irregularities and roughness of the field, which hindered the movement of the robot, and the features of the vehicle, which did not provide any damping, caused vibrations and swinging in the pitch, yaw and roll angles of the camera during robot navigation, particularly on the most rugged soil. These effects increased the variation in the estimated position of the robot as it moved in the field, which decreased the performance of the robot, slowed row tracking, and induced headings in stretches where the wheels of the robot had poor traction.

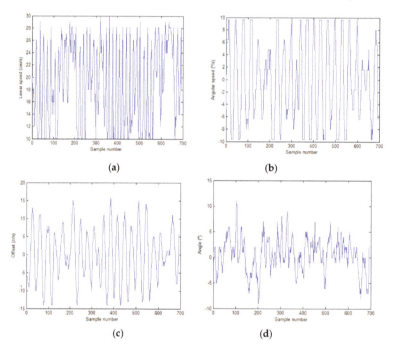

Figure 20. Tests in a real field: (**a**) Evolution of the linear speed of the robot. (**b**) Evolution of the angular speed of the robot. (**c**) Evolution of the offset of the robot relative to the line. (**d**) Evolution of the angle of the robot relative to the line.

Table 6. Results obtained in a real field.

Real Field	Mean	Std. dev.	Minimum	Maximum
Linear speed (cm/s)	18.87	6.16	10.04	29.94
Angular speed (°/s)	−0.14	6.40	−9.60	9.60
Offset (cm)	0.05	7.33	−14.00	16.00
Angle (°)	0.53	3.32	−9.00	11.00

The observed variations are one of the challenges of using small field inspection vehicles instead of tractors or large machines; for the latter, the displacement is more uniform, with fewer vibrations and variations in the camera. As observed in the experiments, although the robot was able to navigate along the crop rows without crushing them, to improve its performance, the mechanical features of the robot must be improved, or a more suitable vehicle must be adopted to navigate the crop field. Note that the proposed navigation system can be easily adapted to control other types of vehicles as it is based on driver behaviour rather than the vehicle model.

4. Conclusions

Crop inspection is a very important task in agriculture and PA, particularly when information about weed distribution is essential for generating proper risk maps that aid in site-specific treatment tasks. Among the various means of collecting field information, the use of small, autonomous vehicles with on-board sensing elements to minimise the impact on crop and soil compaction is quite promising. An autonomous field inspection vehicle, which autonomously navigates by visually tracking crop rows and following a previously established plan that guarantees the entire coverage of a field, was presented in this study.

A set of basic behaviours necessary for an autonomous mobile robot to inspect a crop field with full coverage was established and implemented, including the tracking of a crop row, the detection of the end of a crop row, and the correct positioning of the robot at the head of the next row.

An image-processing method capable of extracting the central crop row in the presence of weeds under uncontrolled lighting conditions in real time from images acquired with a reflex camera positioned at the front of the robot was developed. Two fuzzy controllers were also designed to achieve visual servoing for robot navigation. A method for detecting the end of the crop row using the images acquired by the camera was developed. In addition, the manoeuvres required to change rows were implemented.

Several experiments were conducted to test the performance of the proposed and developed behaviours. These behaviours were performed in a test environment with plane soil. The robot was able to follow the lines, detect the row end, and change lines to continue the inspection without human intervention. The good results are due to the very small errors in the estimated position of the robot relative to the crop row.

After these tests, a set of trials was performed on a real crop field. The robot was able to perform its behaviours correctly, although with reduced performance compared to the first test environment. This decreased performance is due to the variations in the estimated position of the robot relative to the crop row. The irregularities and roughness of the field hindered the movement of the robot, which lacked damping features, resulting in camera vibrations and swinging in its pitch, yaw and roll angles during robot navigation, particularly in the most rugged soil. These effects slowed the tracking of the crop row and induced headings in some areas where the wheels of the robot had poor traction.

These variations were the main disadvantage we encountered when working with this small vehicle in the field. To improve the performance of the developed behaviours, the mechanical features of the robot should be modified, or an alternative vehicle better suited to navigating a crop field should be adopted. In the latter case, the approaches that are proposed in this paper can be integrated in the new vehicle just adapting slightly the value ranges of the fuzzy sets of the navigation control to the new vehicle.

Acknowledgments: The Spanish Government has provided full and continuing support for this research work through projects AGL2011-30442-C02–02 and AGL2014-52465-C4-3-R. The authors wish to thank Pedro Hernáiz and his team (ICA-CSIC) for their invaluable help in the field trials. The authors wish to acknowledge the invaluable technical support of Damian Rodriguez.

Author Contributions: The work was developed as a collaboration among all authors. J.M. Bengochea-Guevara and A Ribeiro designed the study. J.M. Bengochea-Guevara carried out the system integration and programming. D. Andújar posed the field experiments. J. Conesa-Muñoz mainly contributed to the development of the planner and provided support in the field tests with D. Andújar. A. Ribeiro directed the research, collaborating in the testing and the discussion of the results. The manuscript was mainly drafted by J.M. Bengochea-Guevara and A. Ribeiro and was revised and corrected by all co-authors. All authors have read and approved the final manuscript.

Conflicts of Interest: The authors declare no conflicts of interest.

Abbreviations

The following abbreviations are used in this manuscript:

CVRP	Capacited Vehicle Routing Problem
DSS	Decision Support System
FOG	Fibre Optic Gyroscope
GIS	Geographical Information System
GNSS	Global Navigation Satellite System
IMU	Inertial Measurement Unit
PA	Precision Agriculture
PID	Proportional Integral Derivative
RTK	Real Time Kinematic
ExG	Excess green index
b	coefficient of the blue plane
d	distance
g	coefficient of the green plane
$min\_proportion$	Percentage threshold
r	coefficient of the red plane
α	motion direction angle
D_1 :	Lower distance threshold
D_2	Upper distance threshold

References

1. Gebbers, R.; Adamchuk, V.I. Precision agriculture and food security. *Science* **2010**, *327*, 828–831. [CrossRef] [PubMed]
2. Srinivasan, A. *Handbook of Precision Agriculture: Principles and Applications*; Food Products Press: New York, NY, USA, 2006.
3. Stafford, J.V. Implementing precision agriculture in the 21st century. *J. Agric. Eng. Res.* **2000**, *76*, 267–275. [CrossRef]
4. Reid, J.F. The impact of mechanization on agriculture. *Bridge Agric. Inf. Technol.* **2011**, *41*, 22–29.
5. Senay, G.B.; Ward, A.D.; Lyon, J.G.; Fausey, N.R.; Nokes, S.E. Manipulation of high spatial resolution aircraft remote sensing data for use in site-specific farming. *Trans. ASAE* **1998**, *41*, 489–495. [CrossRef]
6. Rew, L.J.; Cousens, R.D. Spatial distribution of weeds in arable crops: are current sampling and analytical methods appropriate? *Weed Res.* **2001**, *41*, 1–18. [CrossRef]
7. Marshall, E.J.P. Field-scale estimates of grass weed populations in arable land. *Weed Res.* **1988**, *28*, 191–198. [CrossRef]
8. Slaughter, D.C.; Giles, D.K.; Downey, D. Autonomous robotic weed control systems: A review. *Comput. Electron. Agric.* **2008**, *61*, 63–78. [CrossRef]
9. Li, M.; Imou, K.; Wakabayashi, K.; Yokoyama, S. Review of research on agricultural vehicle autonomous guidance. *Int. J. Agric. Biol. Eng.* **2009**, *2*, 1–16.
10. Stoll, A.; Kutzbach, H.D. Guidance of a forage harvester with GPS. *Precis. Agric.* **2000**, *2*, 281–291. [CrossRef]
11. Blackmore, B.S.; Griepentrog, H.W.; Nielsen, H.; Nørremark, M.; Resting-Jeppesen, J. Development of a deterministic autonomous tractor. In Proceedings of the 2004 CIGR Olympics of Agricultural Engineering, Beijing, China, 11–14 October 2004.
12. Gomez-Gil, J.; Alonso-Garcia, S.; Gómez-Gil, F.J.; Stombaugh, T. A simple method to improve autonomous GPS positioning for tractors. *Sensors* **2011**, *11*, 5630–5644. [CrossRef] [PubMed]

13. Alonso-Garcia, S.; Gomez-Gil, J.; Arribas, J.I. Evaluation of the use of low-cost GPS receivers in the autonomous guidance of agricultural tractors. *Spanish J. Agric. Res.* **2011**, *9*, 377–388. [CrossRef]
14. Eaton, R.; Katupitiya, J.; Siew, K.W.; Howarth, B. Autonomous farming: Modelling and control of agricultural machinery in a unified framework. *Int. J. Intel. Syst. Technol. Appl.* **2010**, *8*, 444–457. [CrossRef]
15. Kise, M.; Noguchi, N.; Ishii, K.; Terao, H. The development of the autonomous tractor with steering controller applied by optimal control. In Proceeding of the 2002 Automation Technology for Off-Road Equipment, Chicago, IL, USA, 26–27 July 2002; pp. 367–373.
16. Nagasaka, Y.; Umeda, N.; Kanetai, Y.; Taniwaki, K.; Sasaki, Y. Autonomous guidance for rice transplanting using global positioning and gyroscopes. *Comput. Electron. Agric.* **2004**, *43*, 223–234. [CrossRef]
17. Noguchi, N.; Kise, M.; Ishii, K.; Terao, H. Field automation using robot tractor. In Proceeding of the 2002 Automation Technology for Off-Road Equipment, Chicago, IL, USA, 26–27 July 2002; pp. 239–245.
18. Bak, T.; Jakobsen, H. Agricultural robotic platform with four wheel steering for weed detection. *Biosyst. Eng.* **2004**, *87*, 125–136. [CrossRef]
19. Marchant, J.A.; Hague, T.; Tillett, N.D. Row-following accuracy of an autonomous vision-guided agricultural vehicle. *Comput. Electron. Agric.* **1997**, *16*, 165–175. [CrossRef]
20. Billingsley, J.; Schoenfisch, M. The successful development of a vision guidance system for agriculture. *Comput. Electron. Agric.* **1997**, *16*, 147–163. [CrossRef]
21. Slaughter, D.C.; Chen, P.; Curley, R.G. Vision guided precision cultivation. *Precis. Agric.* **1999**, *1*, 199–217. [CrossRef]
22. Tillett, N.D.; Hague, T. Computer-vision-based hoe guidance for cereals—An initial trial. *J. Agric. Eng. Res.* **1999**, *74*, 225–236. [CrossRef]
23. Åstrand, B.; Baerveldt, A.-J. A vision based row-following system for agricultural field machinery. *Mechatronics* **2005**, *15*, 251–269. [CrossRef]
24. Benson, E.R.; Reid, J.F.; Zhang, Q. Machine vision-based guidance system for agricultural grain harvesters using cut-edge detection. *Biosyst. Eng.* **2003**, *86*, 389–398. [CrossRef]
25. Gottschalk, R.; Burgos-Artizzu, X.P.; Ribeiro, A.; Pajares, G. Real-time image processing for the guidance of a small agricultural field inspection vehicle. *Int. J. Intel. Syst. Technol. Appl.* **2010**, *8*, 434–443. [CrossRef]
26. Kise, M.; Zhang, Q.; Rovira Más, F. A stereovision-based crop row detection method for tractor-automated guidance. *Biosyst. Eng.* **2005**, *90*, 357–367. [CrossRef]
27. Sainz-Costa, N.; Ribeiro, A.; Burgos-Artizzu, X.P.; Guijarro, M.; Pajares, G. Mapping wide row crops with video sequences acquired from a tractor moving at treatment speed. *Sensors* **2011**, *11*, 7095–7109. [CrossRef] [PubMed]
28. Søgaard, H.T.; Olsen, H.J. Determination of crop rows by image analysis without segmentation. *Comput. Electron. Agric.* **2003**, *38*, 141–158. [CrossRef]
29. Hague, T.; Tillett, N.D.; Wheeler, H. Automated crop and weed monitoring in widely spaced cereals. *Precis. Agric.* **2006**, *7*, 21–32. [CrossRef]
30. Hough, P.V. Method and Means for Recognizing Complex Patterns. US Patent US3069654 A, December 1962.
31. Marchant, J.A. Tracking of row structure in three crops using image analysis. *Comput. Electron. Agric.* **1996**, *15*, 161–179. [CrossRef]
32. Hague, T.; Marchant, J.A.; Tillett, D. A system for plant scale husbandry. *Precis. Agric.* **1997**, *2*, 635–642.
33. Leemans, V.; Destain, M.-F. Application of the Hough transform for seed row localisation using machine vision. *Biosyst. Eng.* **2006**, *94*, 325–336. [CrossRef]
34. Tellaeche, A.; Burgos-Artizzu, X.P.; Pajares, G.; Ribeiro, A. A vision-based method for weeds identification through the Bayesian decision theory. *Pattern Recognit.* **2008**, *41*, 521–530. [CrossRef]
35. Tellaeche, A.; BurgosArtizzu, X.P.; Pajares, G.; Ribeiro, A.; Fernández-Quintanilla, C. A new vision-based approach to differential spraying in precision agriculture. *Comput. Electron. Agric.* **2008**, *60*, 144–155. [CrossRef]
36. Tellaeche, A.; Pajares, G.; Burgos-Artizzu, X.P.; Ribeiro, A. A computer vision approach for weeds identification through Support Vector Machines. *Appl. Soft Comput.* **2011**, *11*, 908–915. [CrossRef]
37. Kise, M.; Zhang, Q. Development of a stereovision sensing system for 3D crop row structure mapping and tractor guidance. *Biosyst. Eng.* **2008**, *101*, 191–198. [CrossRef]
38. Vioix, J.-B.; Douzals, J.-P.; Truchetet, F.; Assémat, L.; Guillemin, J.-P. Spatial and spectral methods for weed detection and localization. *EURASIP J. Appl. Sign. Process.* **2002**, *2002*, 679–685. [CrossRef]

39. Bossu, J.; Gée, C.; Guillemin, J.P.; Truchetet, F. Development of methods based on double Hough transform and Gabor filtering to discriminate crop and weeds in agronomic images. In Proceedings of the SPIE 18th Annual Symposium Electronic Imaging Science and Technology, San Jose, CA, USA, 15–19 January 2006.

40. Bossu, J.; Gée, C.; Jones, G.; Truchetet, F. Wavelet transform to discriminate between crop and weed in perspective agronomic images. *Comput. Electron. Agric.* **2009**, *65*, 133–143. [CrossRef]

41. Romeo, J.; Pajares, G.; Montalvo, M.; Guerrero, J.M.; Guijarro, M.; Ribeiro, A. Crop row detection in maize fields inspired on the human visual perception. *Sci. World J.* **2012**, *2012*. [CrossRef] [PubMed]

42. Conesa-Muñoz, J.; Gonzalez-de-Soto, M.; Gonzalez-de-Santos, P.; Ribeiro, A. Distributed multi-level supervision to effectively monitor the operations of a fleet of autonomous vehicles in agricultural tasks. *Sensors* **2015**, *15*, 5402–5428. [CrossRef] [PubMed]

43. Bochtis, D.D.; Sørensen, C.G. The vehicle routing problem in field logistics part I. *Biosyst. Eng.* **2009**, *104*, 447–457. [CrossRef]

44. Conesa-Munoz, J.; Bengochea-Guevara, J.M.; Andujar, D.; Ribeiro, A. Efficient Distribution of a Fleet of Heterogeneous Vehicles in Agriculture: A Practical Approach to Multi-path Planning. In Proceedings of the 2015 IEEE International Conference on Autonomous Robot Systems and Competitions (ICARSC), Vila Real, Portugal, 8–10 April 2015; pp. 56–61.

45. Meyer, G.E.; Neto, J.C. Verification of color vegetation indices for automated crop imaging applications. *Comput. Electron. Agric.* **2008**, *63*, 282–293. [CrossRef]

46. Woebbecke, D.M.; Meyer, G.E.; Von Bargen, K.; Mortensen, D.A. Color indices for weed identification under various soil, residue, and lighting conditions. *Trans. ASAE* **1995**, *38*, 259–269. [CrossRef]

47. Andreasen, C.; Rudemo, M.; Sevestre, S. Assessment of weed density at an early stage by use of image processing. *Weed Res.* **1997**, *37*, 5–18. [CrossRef]

48. Perez, A.J.; Lopez, F.; Benlloch, J.V.; Christensen, S. Colour and shape analysis techniques for weed detection in cereal fields. *Comput. Electron. Agric.* **2000**, *25*, 197–212. [CrossRef]

49. Aitkenhead, M.J.; Dalgetty, I.A.; Mullins, C.E.; McDonald, A.J.S.; Strachan, N.J.C. Weed and crop discrimination using image analysis and artificial intelligence methods. *Comput. Electron. Agric.* **2003**, *39*, 157–171. [CrossRef]

50. Yang, C.-C.; Prasher, S.O.; Landry, J.-A.; Ramaswamy, H.S. Development of an image processing system and a fuzzy algorithm for site-specific herbicide applications. *Precis. Agric.* **2003**, *4*, 5–18. [CrossRef]

51. Ribeiro, A.; Fernández-Quintanilla, C.; Barroso, J.; García-Alegre, M.C.; Stafford, J.V. Development of an image analysis system for estimation of weed pressure. In Proceedings of 5th European Conference on Precision Agriculture, Uppsala, Sweden, 9–12 June 2005; pp. 169–174.

52. Van Evert, F.K.; Van Der Heijden, G.W.; Lotz, L.A.; Polder, G.; Lamaker, A.; De Jong, A.; Kuyper, M.C.; Groendijk, E.J.; Neeteson, J.J.; Van der Zalm, T. A Mobile Field Robot with Vision-Based Detection of Volunteer Potato Plants in a Corn Crop. *Weed Technol.* **2006**, *20*, 853–861. [CrossRef]

53. Burgos-Artizzu, X.P.; Ribeiro, A.; Tellaeche, A.; Pajares, G.; Fernández-Quintanilla, C. Analysis of natural images processing for the extraction of agricultural elements. *Image Vision Comput.* **2010**, *28*, 138–149. [CrossRef]

54. Bengochea-Guevara, J.M.; Burgos Artizzu, X.P.; Ribeiro, A. Real-time image processing for crop/weed discrimination in wide-row crops. In Proceedings of RHEA, Madrid, Spain, 21–23 May 2014; pp. 477–488.

55. Sheikholeslam, S.; Desoer, C. Design of decentralized adaptive controllers for a class of interconnected nonlinear dynamical systems. In Proceedings of the 31st IEEE Conference on Decision and Control, Tucson, AZ, USA, 16–18 December 1992; pp. 284–288.

56. Rossetter, E.J.; Gerdes, J.C. Performance guarantees for hazard based lateral vehicle control. In Proceedings of the ASME 2002 International Mechanical Engineering Congress and Exposition, New Orleans, LA, USA, 17–22 November 2002; pp. 731–738.

57. Pomerleau, D.A. *Alvinn: An Autonomous Land Vehicle in a Neural Network*; Technical Report CMU-CS-89-107; Carnegie Mellon University: Pittsburgh, PA, USA, 1989.

58. Sugeno, M. On stability of fuzzy systems expressed by fuzzy rules with singleton consequents. *IEEE Trans. Fuzzy Syst.* **1999**, *7*, 201–224. [CrossRef]

59. Zadeh, L.A. Fuzzy sets. *Inf. Control* **1965**, *8*, 338–353. [CrossRef]

60. Fraichard, T.; Garnier, P. Fuzzy control to drive car-like vehicles. *Rob. Autom. Syst.* **2001**, *34*, 1–22. [CrossRef]

61. Naranjo, J.E.; Sotelo, M.; Gonzalez, C.; Garcia, R.; Sotelo, M.A. Using fuzzy logic in automated vehicle control. *IEEE Intell. Syst.* **2007**, *22*, 36–45. [CrossRef]
62. Pradhan, S.K.; Parhi, D.R.; Panda, A.K. Fuzzy logic techniques for navigation of several mobile robots. *Appl. Soft Comput.* **2009**, *9*, 290–304. [CrossRef]
63. Kodagoda, K.R.S.; Wijesoma, W.S.; Teoh, E.K. Fuzzy speed and steering control of an AGV. *IEEE Trans. Control Syst. Technol.* **2002**, *10*, 112–120. [CrossRef]
64. Antonelli, G.; Chiaverini, S.; Fusco, G. A fuzzy-logic-based approach for mobile robot path tracking. *IEEE Trans. Fuzzy Syst.* **2007**, *15*, 211–221. [CrossRef]
65. Digital.CSIC. Avalaible online: http://digital.csic.es/handle/10261/110162 (accessed on 11 December 2015).

Article

Assistant Personal Robot (APR): Conception and Application of a Tele-Operated Assisted Living Robot

Eduard Clotet, Dani Martínez, Javier Moreno, Marcel Tresanchez and Jordi Palacín *

Department of Computer Science and Industrial Engineering, Universitat de Lleida, Jaume II, 69, 25001 Lleida, Spain; eclotet@diei.udl.cat (E.C.); dmartinez@diei.udl.cat (D.M.); jmoreno@diei.udl.cat (J.M.); mtresanchez@diei.udl.cat (M.T.)
* Correspondence: palacin@diei.udl.cat; Tel.: +34-973-702-760; Fax: +34-973-702-702

Academic Editor: Gonzalo Pajares Martinsanz
Received: 31 December 2015; Accepted: 24 April 2016; Published: 28 April 2016

Abstract: This paper presents the technical description, mechanical design, electronic components, software implementation and possible applications of a tele-operated mobile robot designed as an assisted living tool. This robotic concept has been named Assistant Personal Robot (or APR for short) and has been designed as a remotely telecontrolled robotic platform built to provide social and assistive services to elderly people and those with impaired mobility. The APR features a fast high-mobility motion system adapted for tele-operation in plain indoor areas, which incorporates a high-priority collision avoidance procedure. This paper presents the mechanical architecture, electrical fundaments and software implementation required in order to develop the main functionalities of an assistive robot. The APR uses a tablet in order to implement the basic peer-to-peer videoconference and tele-operation control combined with a tactile graphic user interface. The paper also presents the development of some applications proposed in the framework of an assisted living robot.

Keywords: Assistant Personal Robot; telecontrol; mobile robot; peer-to-peer video; assisted living

---

1. Introduction

The Assistant Personal Robot (or APR) is proposed as a remotely telecontrolled mobile robotic platform with videoconference capabilities. The APR is designed to provide telepresence services [1] primarily for the elderly and those with mobility impairments. Reports from diverse institutions, such as the United Nations [2] and World Health Organization [3], postulate that the proportion of people aged 60 or over is expected to rise from 12% to 21% during the next 35 years as a result of a clear increase of human life expectancy. According to these predictions, elderly care is one of the fields where new technologies have to be applied in the form of passive monitoring systems [4,5] and also as autonomous and tele-operated robots. Such robots can be used to provide assistance to the elderly and people with mobility problems in homes and institutions [6].

The development of assistive robots can provide benefits through the development of optimized solutions for some specific problems that appear in typical household domains [7,8]. A review of the technical description of some mobile robots proposed for telepresence applications is available in [8]. As examples of these applications: in [9] a robotic assistant was used in order to help the residents of a geriatric institution perform their daily routine; in [10] an assistive robot was proposed to help elderly people with mild cognitive impairments; in [7] a telepresence robot showed positive results in domestic elder care assistance; in [11] a telepresence robot was used to improve communication and interaction between people with dementia and their relatives; finally, in [12] a home care monitoring system was proposed, based on a fixed sensor network infrastructure and some physiological sensors, offering the possibility of a virtual visit via a tele-operated mobile robot.

In the context of this paper, a typical example for the application of a household robot is the development of functionalities focused on detecting human inactivity, falls or other accidental situations. In this case, one of the goals is the reduction of the reaction time for external assistance. This aspect is very important because approximately one third of the population aged over 65 falls every year [13] and, in most of the cases, are unable to recover by their own means or ask for assistance [14]. The consequences for elderly people are not limited to physical injuries because a loss of self-confidence also affects mobility drastically [15]. Therefore, telecontrolled mobile robots can be used for social interaction by decreasing isolation and promoting positive emotions and psychological reinforcement.

The APR is designed to provide remote general assistance and help to elderly people and those with impaired mobility in conventional unstructured environments. Figure 1 shows an image of the APR next to some volunteers to indicate its size. Mobility is provided by omnidirectional wheels that can move the APR in any direction without performing intermediate trajectory maneuvers. The head of the APR is based on an Android Tablet, can be rotated horizontally and vertically, and implements the videoconference capabilities for telecontrol and telepresence. The APR includes onboard Light Detection and Ranging (LIDAR) sensors which detect the distance to the surrounding objects on a two-dimensional plane by measuring the time-of-flight of a rotating laser beam. The main electronic board of the APR uses the information of the LIDAR sensors directly to implement a high-priority collision avoidance procedure which is able to stop the mobile robot. The APR also has two mobile arms with one degree of freedom and which can be rotated clockwise and counterclockwise through 360° and placed at any desired angular orientation. The arms can be used for gestural interaction or as a soft handle as a walking support.

Figure 1. Comparative image of the APR.

The new contribution of this paper is the complete technical description of a tele-operated mobile robot designed for assisted living. Comparing the capabilities of the APR with the 16 commercial and prototype mobile robots reviewed in [8], the APR includes common devices and functionalities available in other mobile robots designed for telepresence. The APR is not designed with adjustable height, a capability included in four of the 16 robots reviewed. The APR includes an omnidirectional motion system, which is included in two of the 16 robots reviewed. The APR includes ambient sensors, a capability not included in any of the mobile robots reviewed. Finally, the APR includes soft arms that can be used as a support for walking, a feature which is not usually available in telepresence mobile robots.

2. Mechanical Design

The mechanical design of the APR is steered by the aim to provide personal assistance services in households or institutions without interfering with the inhabitants. The overall weight of the APR is 30 kg. All the heavy elements in the APR are in the base, close to ground level to ensure a lower center of gravity and stable displacement. The definitive height of the APR was determined by performing an ergonomic test with volunteers of different ages. In this test, the volunteers agreed on a robot height of 1.70 m to ensure a good view of the screen included in the head of the APR. The width was limited to 40 cm to simplify the remote telecontrol of the mobile robot when passing through doorways. The head of the APR contains a panoramic screen that can be moved with two degrees of freedom (up/down and left/right). The torso of the APR has two shoulders with one degree of freedom in order to move the arms forward and backward. The motion of the APR is accomplished with three omnidirectional wheels. The physical design of the APR is inspired in and includes several resemblances to humans in order to operate, maneuver and move the head and the arms in a similar way in a typical household or institutional scenario. In [1], a low velocity was considered a drawback when moving long distances with a telecontrolled mobile robot. This drawback could potentially deter people from utilizing this kind of robots. The APR has a maximum forward velocity of 1.3 m/s, which is comparable with the speed of a person and considered adequate for long indoor displacements, such as along a corridor.

The mechanical structure of the base of the APR (Figure 2a) is made of stainless steel for durability and resistance. This mechanical structure supports three battery shafts for three 12Ah sealed lead acid batteries, three P205 DC motors from Micromotor (Micromotor, Verderio, Italy) and a sloping surface to hold the LIDAR sensors, the main electronic control board and the battery charger system. The APR has three omnidirectional wheels (Figures 1 and 2b) shifted 120° and attached to the motors by conical mechanical connectors. This motion system maintains the engine axle, provides proper torque transmission, and simplifies the assembly of other internal components.

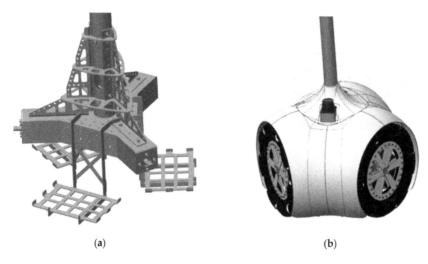

(a)   (b)

Figure 2. CAD design of the base of the APR: (**a**) mechanical structure and (**b**) plastic cover.

The omnidirectional wheels contain multiple rollers on the external edge of each wheel creating a holonomic mobile platform where each wheel has two degrees of freedom, allowing wheel spin and perpendicular displacements from the wheel's forward direction and thus direct displacement in any desired direction (Figure 3). The external white cover of the base is made of ABS plastic (Figure 2b) using a fast prototyping 3D printer. This provides a cheap method of fabrication, a smooth-finished appearance and easy replacement of damaged parts. The circular outline of the base of the robot provides an external space-efficient shape and the absence of protruding and potentially harmful elements. This circular design also minimizes the probability of it becoming accidentally hooked on such furnishings as mats, curtains or clothing.

Figure 3. Representation of the direct motion capabilities of the APR.

The chest and shoulders of the APR are located at a height of approximately 1.3 m, slightly lower than the average for human shoulders. This position was chosen to enable elderly people to hold the arms of the robot directly. The shoulders of the APR contain two heavy Micromotor DC motors, which are connected to two soft arms with a 35 cm separation between them. The arms are 55 cm long, just for aesthetical reasons, and can be used as a support by elder people when walking or for gesture interaction. The arms are moved periodically to mimic the natural movements of humans when the APR moves forward.

The head of the APR (Figure 4) is designed as the main interface between the tele-operator, the mobile robot and the user. The head is based on the Cheesecake 10.1″ XL QUAD tablet (Approx Iberia, Gelves, Spain). This has four processors, a ten-finger touch interface, network connectivity, and multimedia capabilities in order to create and reproduce audio and video in videoconference communication between the APR and a remote tele-operator. The head of the APR has two small Micromotor DC motors (BS138F-2s.12.608) that provide two degrees of freedom. The head can rotate 120° (60° to each side) and also tilt from 0° to 90°: 0° is with the camera pointing at the ground to offer a zenithal view of people lying down and the contour of the mobile robot, while 90° is the case with the camera pointed to the front. The APR is complemented with a low-cost 180° fish eye lens magnetically attached to the camera of the tablet that can provide a panoramic view of the surroundings.

(a)   (b)

Figure 4. Detail of the head of the APR: (**a**) lateral view, (**b**) frontal view.

3. Electronic Components

The electronic components of the APR are placed at the base and are composed of a motor control board (MCB) and three battery charger modules. The MCB (Figure 5) has an ARM (Cambridge, England) Cortex-M4 based Microcontroller Unit (MCU) with a STM32F407VGT6 processor from STMicroelectronics (Geneva, Switzerland), operating at 168 MHz. This MCB was selected because it is a low power device that is able to control up to seven DC motors with dual-channel encoders while including support for different serial interfaces. This MCU provides a Full Speed (12 Mb/s) USB 2.0 On-The-Go (OTG) wired interface with the tablet device located at the head of the APR. The motors are controlled by four H Bridges on a single Quad-Dual-Bridge Driver integrated circuit (TB6555FLG from Toshiba, Tokyo, Japan), two H Bridges to control the robot's arms (HIP4020IBZ from Intersil, Milpitas, CA, United States), and three H Bridges to control the motors that drive the omnidirectional wheels (VNH2SP30 from STMicroelectronics). The MCU implements all the specifications of the Android Accessory Development Kit (ADK) so as to be recognized by any Android Tablet or Smartphone through a Full-speed USB wired connection. The MCB communicates directly with different LIDAR sensors (such as the small and compact URG and UTM models, Hokuyo, Osaka, Japan), laser range sensors that provide a two-dimensional planar description of a predefined planar-space around the robot. The URG model obtains new raw data at a sample rate of approximately 100 milliseconds and provides an angle of view of 240° with a maximum operative range of 5.6 m and a maximum current consumption of 0.5 A at 5 V. The UTM model provides a higher sampling rate and operative range but, in this case, the maximum current consumption is 1.0 A at 12 V.

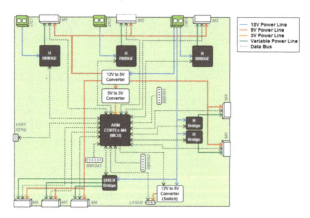

Figure 5. Schematic representation of the motor control board (MCB) of the APR.

Finally, the MCB is connected to three complementary battery charger boards that provide information about the state of the internal batteries, allowing remote battery management without user intervention. The batteries are charged when the APR is manually plugged in to recharge. Additionally, the tablet of the APR has his own internal battery that is always fully charged in normal APR operation. Thanks to this internal battery, the tablet (and then the audio and video communication capabilities of the APR) remain active and available for calls for several hours after full APR discharge. Table 1 shows the description of the connectors on the MCB (Figure 5).

Table 1. Description of the MCB connectors.

TAG	Description	TAG	Description
CN1	Connector for battery 1.	M6	Connector for the Head Pan motor.
CN2	Connector for battery 2.	M7	Connector for the Head Tilt motor.
CN3	Connector for battery 3.	M8	Unused.
M1	Connector for the front left wheel motor.	USB1	Micro-USB "On the Go" connector.
M2	Connector for the front right wheel motor.	LASER	Connector for a Hokuyo LIDAR device
M3	Connector for the back wheel motor.	BMON1	Connector to monitor battery 1.
M4	Connector for the Left Arm motor.	BMON2	Connector to monitor battery 2.
M5	Connector for the Right Arm motor.	BMON3	Connector to monitor battery 3.

The base of the APR (Figure 6) has three high-capacity batteries and three battery charger boards based on an UC3906 Linear Lead-Acid Battery Charger combined with a MJD2955T4 PNP bipolar transistor that regulates the charging process. The three battery charge boards are complemented with a main AC/DC switching adapter (220 V AC/18 V DC) for battery charge and a secondary AC/DC switching adapter (220 V AC/12 V DC) for simultaneous powering the MCB during recharging of the internal batteries.

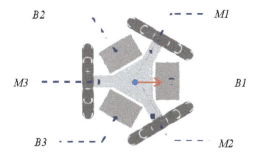

Figure 6. Batteries (B1, B2 and B3) and main DC motors (M1, M2 and M3) at the base of the APR.

Table 2 shows the electrical current provided by the batteries of the APR for a set of predefined robot motions. In this prototype, the batteries operate independently in order to power different parts of the APR directly. Battery 1 (Figure 6-B1) is exclusively dedicated to powering motor 1 (Figure 6-M1) and battery 2 (Figure 6-B2), to powering motor 2 (Figure 6-M2). These two motors are used exclusively for forward (and backward) displacement, which is usually performed at full speed and for long periods of time. Battery 3 (Figure 6-B3) is used to power motor 3 (Figure 6-M3), the MCB, the LIDAR sensors, and the Tablet, which also has its own batteries. Table 2 shows that the maximum energy consumption is reached when the APR goes forward or backward at full speed, the rotation and the transversal displacement of the APR is performed with the three motors in simultaneous operation. Most of the current provided by Battery 3 is dedicated to powering the LIDAR sensors. These can be electrically disconnected in order to reduce power consumption when the APR is in standby operation. The APR has been tested in periods of up to four hours but there is currently still no estimation of battery life as this will largely depend on usage.

Table 2. Average electrical current of the batteries when performing different predefined motions.

Motion	Battery 1 (mA)	Battery 2 (mA)	Battery 3 (mA)	Total (mA)
Standby	5	5	480	490
Stopped	5	5	890	900
Go Forward	700	803	1202	2705
Go Backward	934	951	880	2765
Rotate (left/right)	120	97	976	1193
Move right	176	194	1582	1952
Move left	154	181	1447	1782

4. Software Implementation

The APR is defined as a remotely operated platform that has to be compatible with environments designed for humans rather than for robots. The APR has to offer a reliable remote control operation with a stable remote videoconference system in order to ensure its application as a tele-operated assisted-living robot. The transmission of audio, video and control orders is carried out by using direct peer-to-peer (P2P) or client-to-client communication to reduce delay in the remote control system.

4.1. Transmission of Motion Control Orders

The remote control of the APR is performed with an Android application (APP) running in a remote smartphone or tablet. Figure 7 shows an image of the remote control interface. This is based on the touch-screen capabilities of such mobile devices. The telecontrol APP is composed of an on-screen joystick that is used to control the position of the robot, two sliders to control the pan and tilt of the robot head, two sliders to control the robot arms (left and right arm), and an extra slider used to rotate the APR over its vertical axis. The telecontrol APP can be used with one or two fingers by trained and non-trained people. The remote control of the mobile robot is not a challenging task because the high-mobility capabilities of the APR do not usually require intermediate trajectory maneuvers. Future implementations of the telecontrol APP will include such additional features as direct control of the trajectory of the APR with small head displacements [16] in order to allow remote robot control by people with impaired mobility.

Figure 7. Android APP used for the remote control of the APR.

Table 3 shows the structure of the commands sent in a single seven-byte user datagram protocol (UDP) packet where each byte has a specific interpretation. The main on-screen joystick codifies the desired input motion orders in two bytes: robot angle (Angle) and movement velocity (Module). Other controls, such as the rotation of the APR, the orientation of the head, and the rotation of the arms, are coded and submitted simultaneously. The complete control of the mobile robot is performed with network messages containing only seven bytes.

Table 3. Structure of an UDP motors control packet.

ID	Head Pan	Head Tilt	Left Arm	Right Arm	Module	Angle	Rotation
Byte	0	1	2	3	4	5	6

The UDP protocol provides fast transmission rates at the expense of omitting error detection procedures and also with a lack of knowledge about the status of the communication. The use of the UDP protocol in a real-time remote control system is justified since there is no interest on recovering lost or delayed packets containing outdated information. In this implementation, a control command is sent at least every 100 ms according to the operator's tactile-gestures. Finally, the APR automatically stops after a time-lapse of 300 ms with no new control commands as a way to avoid the execution of unsupervised movements. At this moment, the APR automatically stops in case of network connection failure and remains inactive waiting for a new connection. In the future, this state will be complemented with the possibility of returning to the starting point or another predefined point.

The tablet located in the head of the APR receives the UDP control packets and generates specific low-level orders for the MCB through the Fast USB wired connection. Table 4 show the basic instruction set orders interpreted by the MCB. There are specific motion orders (starting with MF and MB) that are used for long straight forward and backward displacements. These displacements require the application of an electric brake to the unused (and unpowered) M3 motor to prevent free rotation and the continuous rectification of the angle of the wheel.

Table 4. Instruction set orders interpreted by the MCB.

Command	Parameters	Description
AL $d°$	d: from -45 to $45°$	Moves the left arm to the specified position in degrees.
AR $d°$	d: from -45 to $45°$	Moves the right arm to the specified position in degrees.
HP $d°$	d: from -60 to $60°$	Rotates the head of the robot to the specified position in degrees.
HT $d°$	d: from 0 to $90°$	Tilts the head of the robot to the specified position in degrees.
M $S1S2S3°$	$S1$: PWM from 0 to 60% $S2$: PWM from 0 to 20% $S3$: PWM from 0 to 20%	Fixes the Pulse Width Modulation (PWM) applied to motors M1, M2 and M3 of the APR.
MF $S°$	S: PWM from 0 to 60%	Applies electric braking to M3 and fixes the same PWM to M1 and M2 to generate a forward displacement.
MB $S°$	S: PWM from 0 to 60%	Applies electric braking to M3 and fixes the same PWM to M1 and M2 to generate a backward displacement.
TL $S°$	S: PWM from 0 to 20%	Fixes the same PWM to M1, M2 and M3 to rotate the robot to the left.
TR $S°$	S: PWM from 0 to 20%	Fixes the same PWM to M1, M2 and M3 to rotate the robot to the right.

4.2. Transmission of Audio and Video Streaming

The bidirectional transmission of audio and video data streaming is a fundamental task in a mobile robot designed for tele-operation and remote interaction. This section presents a detailed description of the generic bidirectional audio and video transmission system developed for the APR.

This system is specifically designed to connect two Android devices: the tablet that controls the APR and the remote operator's smartphone or tablet.

4.2.1. Videoconference Architecture

The videoconference system defined in an Android device is controlled from the upper-class *CommunicationsManager*. This class contains all the methods required to configure and initiate a new videoconference with a remote device. The videoconference system is made up of four different classes: *UdpRxImages*, *CameraSurfaceView*, *Recorder*, and *UdpRxAudio*. Each of these classes is executed in an independent thread and an independent network socket to limit the possibility of blocking the device in case of unexpected network problems when performing long operations. Figure 8 shows the hierarchy of the videoconference system.

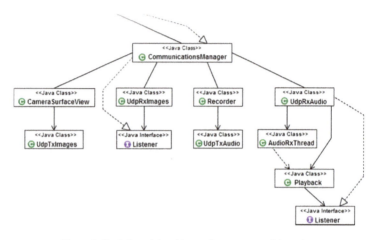

Figure 8. Depiction of the videoconference system hierarchy.

The exchange of data between classes is performed though *custom event listeners* (from now on, listeners) and by creating callback functions that are triggered by specific events and notified to all the classes registered to that listener. The main reason that justifies the use of listeners instead of handlers is that listeners are immediately executed once received, while handler callbacks are queued for future execution.

4.2.2. Network Communications Protocol

The audio and video data is exchanged by using UDP packets based on sending single packets to a specific address and port instead of creating a dedicated communication channel. Since UDP does not perform any control on the packets, it offers a time efficient communication protocol. This lack of control makes UDP one of the best transmission methods for communications where delivery time is more important than data integrity. The main reason of using UDP in audio/video communications is the prevention of communication delays caused by errors produced in a single packet. Then, the lost data packets are ignored and communication is restored when the next packet is received. Unlike the video and audio data, the control parameters for the communication are simultaneously transmitted by using the transmission control protocol (TCP). The use of TCP sockets provide more reliable communication channels, which are used to exchange essential configuration data from videoconference parameters that each device has to know in order to successfully decode the video and audio data received from the remote device.

4.2.3. Video Communications

Video transmission/reception is an indispensable feature for a tele-operated assisted living robot. Video communication is a computationally expensive application that needs to be efficient, reliable, robust, and with low-delay in the transmissions.

Currently, one of the most commonly used encoders to perform live streams is the H.264/MPEG-4 due to its high-compression capabilities. This encoder is designed to generate stable and continuous transmissions with low-bandwidth requirements but has several drawbacks. On one hand, this encoder can have a variable communication delay from 3 to 5 s and high-motion scenes can appear blurred because of the dependence between consecutive frames [17]. On the other hand, this encoder requires highly optimized libraries which are difficult to integrate into a custom application. The bidirectional video communication method implemented in the APR overcomes these drawbacks by transmitting video as a sequence of compressed images in JPEG format instead of using a common video streaming format. This approach simplifies the implementation and control of the communication. Therefore, the bidirectional video communication implemented in the APR has to acquire, compress and submit surrounding images while receiving, decompressing and showing the images submitted by the remote device.

The procedure for image acquisition in the Android environment starts in the *CameraSurfaceView* class, which extends the Android class *SurfaceView*. The *SurfaceView* class is used as a canvas providing a dedicated drawing area where the preview of the camera is displayed. Once the camera has been initialized, the *onPreviewFrame(byte[] data, Camera camera)* callback is triggered each time a new image is obtained, retrieving an array containing the YUV image data encoded with the NV21 Android native format for camera preview as well as the camera instance that has captured the image. The images are then compressed to the standard JPEG format (which reduces the file size at the expense of quality and computing time) to be submitted as an image packet. Table 5 shows an example of typical common image resolutions with the relationship between size, quality and time required for the JPEG compression of the image, and the theoretical maximum rate of frames per second (fps) that can be obtained in communication with the Tablet of the APR. Video communication can use any resolution available in the telecontrol device. Under normal conditions, the time required to compress one image is similar to the time needed to decompress the image. The large size reduction achieved with the JPEG compression (Table 5) is because the original YUV-NV21 image format is already a compressed image format with averaged and grouped U and V color planes so this image has less spatial variability than a conventional raw RGB color image and can be described with less information and size after compression.

The implementation of a videoconference system requires image acquisition, image compression, transmission, reception, and image decompression in a bidirectional way. The real frame rate of the image acquisition devices is usually 15 fps. Some images may be lost during this process: (a) because a new image is available while the previous image is still being compressed so this must be discarded; (b) because of UDP transmission on the network; and (c) because a new image has been received while the previous image is still being decompressed so this must be discarded. However, current smartphones and tablets usually include multiple CPUs and use different isolated threads for transmission and reception so they can achieve higher videoconference frame rates than when only one CPU is available for processing.

Figure 9 shows a visual example of the impact of the quality parameter on a JPEG compression and decompression. Figure 9 shows the face the remote operator, acquired with a smartphone, compressed as JPEG, submitted to the network, received by the tablet of the APR, decompressed, and shown on the screen of the APR for visual interaction. In this example with small images, the difference between 100% and 60% qualities is almost imperceptible to the human eye while the image size is 12 times smaller. There are also small differences between 60% and 30% qualities in terms of compressed image size, processing time and expected frames per second. The APR uses a video transmission procedure configured to operate with a starting default resolution of 320 × 240 pixels with a JPEG quality factor of 60%. Then, the expected average image submitted to the network is 56 times smaller than the original YUV-NV21 images acquired by the device, and 84 times smaller than the RGB version of the original image. This image resolution can be changed manually during the videoconference or adapted dynamically to the network bandwidth capabilities.

Table 5. Relationship between image size and JPEG quality.

Color Image			JPEG Compression			
Resolution (Pixels W × H)	RGB Size (Bytes)	YUV-NV21 Size (Bytes)	JPEG Quality (%)	JPEG Size (Bytes)	Compression Time (ms)	Fps (max)
176 × 144	76,032	50,688	30	1233	12.46	80
176 × 144	76,032	50,688	60	1391	11.74	85
176 × 144	76,032	50,688	100	11,911	16.50	60
320 × 240	230,400	153,600	30	2280	19.96	50
320 × 240	230,400	153,600	60	2718	19.51	51
320 × 240	230,400	153,600	100	33,617	24.74	40
640 × 480	921,600	614,400	30	6351	29.90	33
640 × 480	921,600	614,400	60	7526	30.26	33
640 × 480	921,600	614,400	100	115,491	55.14	18
720 × 480	1,036,800	691,200	30	7618	32.54	30
720 × 480	1,036,800	691,200	60	9316	31.86	31
720 × 480	1,036,800	691,200	100	126,430	61.68	16
1280 × 720	2,764,800	1,843,200	30	17,810	72.47	13
1280 × 720	2,764,800	1,843,200	60	21,230	72.87	13
1280 × 720	2,764,800	1,843,200	100	311,409	131.72	7
1280 × 960	3,686,400	2,457,600	30	23,699	96.58	10
1280 × 960	3,686,400	2,457,600	60	28,647	98.02	10
1280 × 960	3,686,400	2,457,600	100	426,045	177.25	5

Figure 9. Sample 320 × 240 color image showing the small differences obtained after a JPEG compression and decompression procedure with quality settings of (**a**) 100%; (**b**) 60%; (**c**) 30%.

The compressed image is sent to the *UdpTxImages* thread, which creates a new UDP packet containing the image. Once an image has been received at the *UdpTxImages* thread, a flag is activated to notify the *CameraSurfaceView* class that there is a transmission in progress, so newly generated images are not transmitted. This flag is used to indicate to the *CameraSurfaceView* that there is no need to compress and send new images to the *UdpTxImages* thread until the current transmission is completed. The main goal of this procedure is to avoid wasting CPU cycles by compressing images that will not be sent. However, this method slightly increases the video delay since the next image to be transmitted is not compressed until the last transmission ends. In general, a video transmission of compressed JPEG images of 320 × 240 pixels obtained at a frame rate of 15 fps and compressed with a quality factor of 60% with an average size of 2720 bytes requires a network bandwidth of 0.318 Mb/s per video streaming (0.881 Mb/s for a color image of 640 × 480 pixels). This image-based video streaming gives shorter delays in communication but may require higher network bandwidth than alternative buffered-based streaming systems. In general, typical domestic networks have enough upload and download bandwidth for two-directional video streaming communication with color images 640 × 480-pixels processed at 15 fps.

Video reception is carried out by the *UdpRxImages* execution thread. Once initialized, this thread starts receiving UDP packets containing the frames transmitted by the remote device. Each time an image packet is received, the data is decoded into a bitmap using the Android method *BitmapFactory.decodeByteArray()*. The new bitmap is then sent to the application main activity using the *OnImageReceived()* listener. Once the new image has been received at the main activity, the interface view is updated by the User Interface (UI) thread. In order to improve the memory usage at the reception side of the application, the *inMutable* flag from the *BitmapFactory* class is set to true, forcing the *BitmapFactory* class to always use the same memory space to store new decoded images by avoiding the creation of a new bitmap each time a new image is received. When using the *inMutable* flag the bitmap used to decode the received image must have the same resolution. If an image with a different resolution is received, the bitmap memory used by the *BitmapFactory* class is reallocated using the new image dimensions, so the only situation where a new bitmap is created is when the image resolution of the images is changed.

There is no a specific optimum resolution suitable for all situations and networks. The resolution of the local and remote images used in the video communication can be changed dynamically. If required, the video communication sends a TCP packet to the remote device to request a list containing the resolutions at which the remote camera can operate. In order to obtain such information, the *getSupportedRes()* function from the *CameraSurfaceView* class is called. This function uses the *getParameters()* method from the *camera* class that contains the instance of the camera that is being used. The *getSupportedPreviewSizes()* method from camera instance parameters retrieves a list of the supported resolutions for live video preview. This list is then sent back to the device that initially made the query and all the resolutions are shown on the screen. This system allows a fast change or selection of the best image resolution required for each specific situation.

Once the image resolution has been changed, a new TCP packet containing the new image width and height is sent to the remote device. When this packet is received, the remote device calls the *changeResolution()* method from the *CameraSurfaceView* class. This method stores the new resolution and enables a flag that notifies the *onPreviewFrame()* callback to change the configuration of the camera. In this case, the camera is stopped and the method *setNewConfiguration()* is called in order to change the video configuration and restarts the video transmission.

4.2.4. Audio Communications

The transmission of audio during a remote interaction with the APR is a natural, direct and efficient communication channel between the person in charge of the telecontrol of the robot and the person assisted by the robot. The main problem that arises when establishing bidirectional audio communication is the delay between audio transmission and reproduction at the remote device when using buffered-based libraries such as VoIP/SIP. This problem is produced because the Android VoIP requirements force the application to ensure that a minimum amount of data is received (and buffered) before starting the audio playback. This audio transmission is similar as the P2P procedure used for video transmission. This implementation uses the Android *AudioRecord* and *AudioTrack* classes to record and reproduce audio data in both devices respectively (telecontrol device and APR head device). Although *AudioRecord* is not designed to perform audio streaming, the functionality of this class has been modified to generate small audio chunks that can be sent through the network immediately and reproduced at the remote device.

Firstly an instance of the *AudioTrack* class is initialized with the following specifications: Sample rate of 8000 Hz, encoding format PCM 16bit, 1 channel (mono). Once initialized, the *AudioRecorder* instance stores all the data obtained by the device microphone in an internal buffer. When the internal buffer is filled with 3840 bytes of voice data, the buffer is sent to the *UdpAudioTx* thread, wrapped inside an UDP packet, and transmitted to the remote device. The UDP packet is received at the

UdpAudioRx execution thread of the remote device and sent to the *AudioTrack* class running inside the *Playback* thread. This system has been initialized with the same audio parameters as the *AudioRecord* instance and it queues the received audio data for reproduction. In order to avoid an increasing delay on the reception side of the application, packets received with a time difference greater than 50 ms, or empty audio streams, are dropped thus ensuring that the delay of a single packet does not accumulate into a whole communication delay.

4.3. External Network Server

The communication of the APR with a remote telecontrol device not located in the same local area network (LAN) is a specific problem that requires the development of an external network server. The main function of this server is to provide a common control space where telecontrol clients and robots are managed in order to establish a P2P communication between two devices. In this common space, the server is continuously waiting for a new incoming connection. This is then associated with an event listener executed in an independent execution thread. Once a new thread has been started, the server requests the mandatory information related to the connection itself (Table 6). The communication can be initiated by the APR or by a remote telecontrol device according to a predefined policy usage.

Table 6. Required data to connect with the external server.

Tag	Type	Provided by	Description
Role	String	Device APP	A string containing the role of the device connected to the server (service or robot)
Name	String	Configuration	A string containing the name of the robot or service. If no name is specified, the string "Default" is used as a name.
NAT Port	Integer	Network protocol	Network Address Translation (NAT) port assigned to the communication
IP Address	InetAddress	Network protocol	The IP address of the device that requested the connection (external IP)
Local IP Address	String	Device APP	The IP address of the device that requested the connection (internal IP)

The connection data is analyzed to determine if the device is connected under the same network domain as the server. This procedure is realized by checking the domain of the IP address provided during the connection process. If the IP address provided exists inside the private address ranges (Table 7), the device is then under the same network domain as the server and thus the internal IP address will be used to send packets to the device. If the device fails at providing the requested information the communication will be closed, otherwise, the communication will be accepted and registered to the server. Once a device is connected, the specified role determines the accessible functions for the device. Therefore, when an operator wants to initiate the control of a specific APR, a packet containing the robot name is sent to the server and, once received; the server sends a request to both devices to fire multiple UDP packets to the server through specific ports. The objective of these UDP packets is to open a direct path between the client and the robot on both network firewalls by using the UDP hole punching technique [18,19]. This protocol allows direct P2P communication between different registered APRs and different registered telecontrol devices using the ports opened when starting the communication with the external server. When P2P communication starts, the server closes the UDP communications and uses the TCP socket to send and receive specific packets from both devices. As a final task, the external server keeps an updated list of on-line APRs ready for new connections.

In general, the use of an external server allows long-range internet communications between a client and a robot regardless of whether they are connected at the same network domain or not.

However, when using UDP packets this method can fail to engage a communication channel if any of the devices is connected to the internet using mobile networks (such as 3G or 4G) in the case of staying under multiple levels of the network address translation (NAT) protocol, see [19] for additional details. This problem can be solved by using TCP packets instead of UDP packets or by changing the configuration of the communication channel in order to accept UDP packets.

Table 7. Ranges of private IP addresses.

RFC1918 Name	First Available IP	Last Available IP
24-bit block	10.0.0.0	10.255.255.255
20-bit block	172.16.0.0	172.31.255.255
16-bit block	192.168.0.0	192.168.255.255

5. Tests

This section presents the preliminary usability tests performed with the APR and also the specific test of the high-priority collision system implemented in the mobile robot. Future works will center on developing usability tests focused on the experience of the users such as the one proposed in [1] and on improving the human-robot interaction experience according to the guidelines proposed in [20].

5.1. Preliminary Usability Test

The preliminary usability tests were performed with 12 volunteers aged from 12 to 50 years during prototype development in order to evaluate the degree of satisfaction of the different features implemented in the APR according to the different classification: "poor", "acceptable" and "good". Future exhaustive tests will use additional classification levels such as proposed in [1]. The results obtained in these preliminary tests have been used to refine the design and implementation of the APR.

The first test consisted of introducing the robot to each of the 12 volunteers, who were then invited to rate the appearance of the APR. The volunteers gave an average score of "good", showing that the participants liked the APR design. Future in depth usability tests will be performed with target users in order to evaluate the usability of the APR, but currently, the focus was to obtain the subjective impression caused by the design of the mobile robot. The assumption was that a good impression is required to facilitate the continuous use a mobile robot at home.

The second test consisted of evaluating the videoconference capabilities implemented in the APR and the remote telecontrol APP under static conditions. During this test, the 12 volunteers were grouped into pairs and placed in different rooms in the same building and network. Each volunteer participating in the experiment played both roles: as a robot tele-operator and as a passive user, performing a total of 12 experiments. The volunteers rated this static videoconference as "acceptable" and "good". Some of the volunteers complained about a small but sometimes noticeable desynchronization between the audio and the images in the video communication. The cause of this desynchronization is that the audio and images of the video are sent in different network UDP packets that may not be received with the same cadence as they are submitted at. This problem was partially improved by including synchronization information into the audio packets to allow automatic discarding of delayed audio packets with no audio information.

The third experiment consisted of evaluating the maneuverability of the APR. In this test each of the 12 volunteers has to control the APR as a remote tele-operator. Each volunteer was trained for ten minutes in the use of the remote telecontrol APP and then the APR had to be displaced between different starting points and destinations. The volunteers rated this telecontrol as "good", expressing increasing confidence during the experiment.

The last test consisted of evaluating the videoconference capabilities implemented in the APR and the remote telecontrol APP under dynamic conditions. During this test, the 12 volunteers were grouped into pairs. Each volunteer participating in the experiment played both roles: as a static robot tele-operator and as a passive user who has to walk along a corridor with the mobile robot, performing a total of 12 experiments. This test was performed in the same building and network. The volunteers rated this dynamic videoconference as "poor" and "acceptable". The cause of this low score was a loud background noise from a mechanical vibration generated in the APR during displacements. This problem was addressed and partially solved by adding rubber bands to the structure that supports the tablet to minimize the transmission of mechanical vibrations to the microphone embedded in the tablet. Then, this dynamic experiment was repeated with these improvements and the volunteers rated the videoconference as "acceptable".

5.2. High-Priority Collision Avoidance System

The APR is a tele-operated mobile robot designed to operate in domestic environments with limited space and its operation must be compatible with the presence of static objects such as walls and furniture, and also dynamic objects such as people and doors [21]. In general, control of a tele-operated robot is strongly conditioned by external factors such as network stability and the operator's experience.

The APR has a maximum forward velocity of 1.3 m/s and, for this reason, includes a high-priority collision avoidance system implemented in the MCB. This high-priority procedure processes the information from the LIDARs directly in order to stop the mobile robot automatically without delay when a collision situation is estimated or detected regardless of any other consideration. Figure 10 depicts the planar information gathered by the main LIDAR placed parallel to the ground at a height of 38 cm. The high-priority procedure is based on the delimitation of different semicircular safety areas around the APR [22]. Currently, other alternatives, such as an automatic change in the trajectory [23], are not implemented as they may cause confusion to the operator in charge of the telecontrol of the mobile robot.

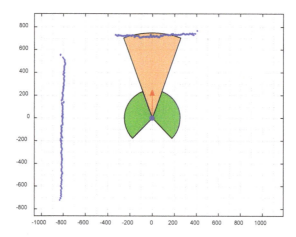

Figure 10. Depiction of the APR position and direction (red arrow), distance data points detected with the LIDAR (blue points) and depiction of the frontal (brownish) and lateral (greenish) safety areas around the mobile robot.

The collision avoidance system automatically stops the APR if the LIDAR detects anything inside one of the different safety areas. In this emergency situation, the tele-operator of the APR can rotate the robot or perform backward or transversal displacements. The radius and orientation of the safety areas around the APR are adjusted dynamically according to the trajectory and velocity of the mobile robot. For example, the APR stops if it tries to pass through a doorway at maximum speed while it will not stop when passing at a reduced speed.

A set of different experiments were carried out to verify the dynamic definition of the safety areas around the APR. Figure 11 shows a sequence of images while conducting one experiment where the APR goes forward at full speed in the direction of an obstacle. The depiction in Figure 10 corresponds approximately to sequence 3 in Figure 11 in which an object is detected in front of the mobile robot. The results of different collision avoidance experiments have shown that the APR can avoid collisions except in the case of objects with transparent glasses or mirrors because these objects are not properly detected by laser LIDAR sensors.

Figure 11. Sequence of images (1–4) obtained in a collision avoidance experiment with the APR moving forward.

Table 8 summarizes the collision avoidance experiments performed. In each experiment, the APR is configured to go forward at a fixed speed. The first column in Table 8 shows the forward speed relative to the maximum forward speed of 1.3 m/s. The second column in Table 8 shows the radius of the frontal safety area (see Figure 10) which is fixed according to the forward speed. The last column shows the final distance APR-obstacle measured when the robot has been effectively stopped. Table 8 shows the proposed calibration between the robot velocity and radius of the safety frontal distance, which has a value of 120 cm when the APR reaches the maximum forward velocity of 1.3 m/s. The minimum gap-distance between the APR and the object after braking at full speed was 22.5 cm. The procedure for stopping the mobile robot is based on using the motors as electrical brakes by forcing a shortcut through the H-bridges.

Table 8. Collision avoidance test results.

Relative APR Forward Speed (%)	Radius of the Frontal Safety Area (cm)	Distance APR-Obstacle When Stopped (cm)
100	120.0	22.5
88	108.6	23.0
75	96.4	26.3
66	86.7	25.5
50	72.3	30.8
33	56.3	29.0
25	48.9	30.2

In the future, this high-priority collision avoidance system will be improved in order to estimate the trajectory of dynamic obstacles around the mobile robot and reduce the braking distance.

6. Applications of the APR

This section proposes some sample applications for the APR to take advantage of its capabilities as a tele-operated assisted living robot.

6.1. Mobile Videoconference Service

The APR has been designed as a telecontrolled mobile videoconference service prepared to provide social and assistive services to elderly people and persons with impaired mobility. In the case of this application, the APR is located at home and the mobile robot can be configured to automatically to accept calls from registered remote operators or services that will take the control of the mobile robot during a mobile videoconference. The APR can be turned on by pressing a single button (Figure 12) so as to make it easy for users with low technological skills to use.

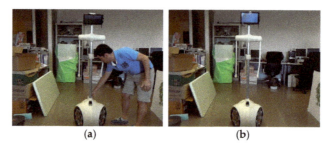

(a)   (b)

Figure 12. (**a**) Starting the APR; (**b**) APR initializing the onboard sensors and initializing the Tablet of the head.

This unique button starts the MCB which configures its main USB as a host interface. The tablet is also configured to start automatically when activity is detected in a USB host interface. The tablet of the APR always operates as a slave device connected to the USB host interface of the MCB. Once started, the tablet of the APR automatically launches the client APR application, which also connects to a global server to register the robot as an available remotely operated APR ready for a new connection (Figure 12). The APR only requires specific operational configuration the first time it is started (global server IP and robot name), and after that, the robot remembers the parameters used for the configuration and uses them as its default parameters.

On one hand, the APR operates like a basic phone or videoconference device that can use the information of the onboard LIDAR sensors automatically to generate a call or an alarm for relatives and preconfigured registered services. This call can also be generated simply by touching the screen of the tablet. On the other hand, the people or services registered can use the telecontrol APR APP to initiate a remote communication with the APR as audio only, audio and video, or audio and video combined with remote telecontrol operation. The remote tele-operation APP establishes a connection with a global external server and shows a list of all the authorized and available APRs registered for this service.

Figure 13 shows the screen of the tablet of the APR with a small window that previews the local image acquired by the APR and submitted during the videoconference, the image received during the videoconference and four main buttons used to control the communication. The buttons enable or disable the submission of video, the submission of audio, the reception of audio, and also change the resolution of the images used in the videoconference. Currently, the complete videoconference starts automatically but the automatic response of the APR can be configured depending on the authorization level of the service in charge of the call.

One of the objectives of the APR is to enable direct bidirectional communication with relatives and technical or medical services to be established. On one hand, this communication can be used to generate social interaction with elderly people. On the other hand, the sensors on the APR can detect problems or inactivity and automatically generate alarms. In such a context, the APR can be used by a remote tele-operated service to explore a typical household scenario visually, evaluate the situation and reduce the reaction time if external assistance is needed.

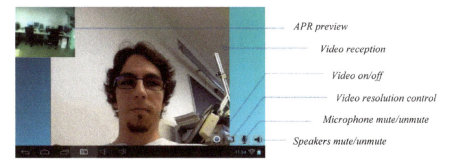

APR preview

Video reception

Video on/off

Video resolution control

Microphone mute/unmute

Speakers mute/unmute

Figure 13. Screenshot of the Tablet of the APR showing the face of the person calling.

6.2. Mobile Telepresence Service

Alternatively, the concept of telepresence [8] implemented in the APR can be used directly by people with impaired mobility [24] located in their home in order to connect with another place. In the case of this application, the APR is located at one facility and the mobile robot is configured to accept calls from registered users who will take control of the mobile robot as telepresence service. Figure 14 shows a simulated example of the APR used as a telepresence platform during an informal talk between three colleagues. In this simulated case, the APR is only used as a tool for social interaction. This is an alternative application demanded by people with impaired mobility and which will be deployed in future work.

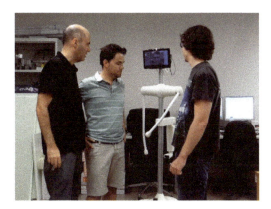

Figure 14. Depiction of the telepresence capabilities of the APR.

This application will be installed in a typical industry, office or institution to offer telepresence contact between employees. Telepresence is achieved by offering a physical and mobile depiction of the person who is controlling the APR. The telecontrol application of the APR enables direct control of the robot's arms and head as a way of simulating the creation of non-verbal communication between groups of people who usually collaborate with each other. In the future, this non-verbal communication will be enhanced with the inclusion of configurable buttons with predefined and personalized combination of small arm and head movements.

Future work based on this application will focus on deploying the APR as a telepresence service for the industry as an informal way of maintaining contact between groups of employees in case of someone having impaired mobility that impedes their presence and interaction with the group. The hypothesis is that frequent and informal telepresence contact can improve productivity while contributing to reducing the effort spent on unpleasant, long or difficult journeys. In this context, the APR will be tested as a telepresence tool that may contribute to enhancing the employability of people with impaired mobility.

Figure 15 shows a screenshot of the telecontrol application of the APR. This application is used by a trained operator and the interface is operated with one or two fingers. The interface includes a circular motion joystick where the red circle depicts the current direction of the APR and a rotation joystick used specifically to rotate the mobile robot. The external circle of the motion joystick is also used to show the relative location of obstacle warnings detected by the APR. There are additional sliders to control the head and the arms of the robot. This screen shows the status of the battery and wireless communication as text messages at top center of the screen but this information will be replaced with small graphic indicators in the future.

Figure 15. Screenshot of the remote telecontrol application.

6.3. Walking Assistant Tool

The APR can be used as a walking assistant tool [25] and to avoid the use of additional devices in limited spaces [26]. Figure 16 shows a sample image of a person using the soft arms of the APR as a walking support. In this mode, the velocity of the APR is automatically limited while the arms are set at an angle of 35° to the back. The functionality of the mobile robot as a walking support can be remotely supervised by using the front or rear camera of the tablet and by rotating the head. The orientation of the motors of the arms has a closed control loop and any disturbance (force applied by the user) [26] can be used to refine the displacement of the robot automatically. Finally, in this mode, the information gathered by the LIDAR sensors is used automatically to center the robot when passing through doorways or along a narrow corridor.

Figure 16. Applications (a–d) of the APR as a support for walking.

6.4. Scheduling Tool

The APR can be used to help elderly people to remember their daily routines [27,28]. This task is performed by means of individual programmable alarms that show a visual just-in-time representation of a daily routine on the screen of the APR, combined with additional acoustic signals or recorded advices. In this application, the tactile screen of the tablet of APR is used for direct feedback. Figure 17 shows a simulation of the APR operating as a scheduling tool. In our region, there is a public service for elderly people. This involves a daily call to remind them of all the daily routines but the hypothesis is that a visual indication at the scheduled moment of, for example, taking a medicine will be more effective than a daily phone-call.

Figure 17. Simulation of the APR providing a daily or regular routine notification.

6.5. Fall Detection Tool

In this context, the last APR application proposed in this paper is the detection of falls or inactivity [4,5]. This functionality requires a sequential analysis of the raw data provided by the onboard LIDAR sensors, the detection of the background objects and abnormal user movements. Figure 18 shows a simulation of the APR used as a fall detector that generates a warning when the volume of the user detected in the information provided by the LIDARs decreases or increases suddenly. Then, an automatic call is generated if the user does not press a button that appears on the screen of the tablet on the APR.

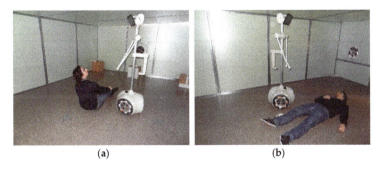

(a) (b)

Figure 18. Simulation of the APR used as a fall detector (**a**) with activity and (**b**) without activity.

6.6. Mobile Ambient Monitoring Platform

The APR can be used as a mobile ambient monitoring platform that gathers information from the environment. For example, in [29], a mobile robot with specialized onboard sensors was used to measure such environmental parameters as air velocity and gas concentrations in the air. Other ambient parameters, such as temperature, humidity, and luminance, were also measured using low-cost sensors. These environmental values are directly correlated with living conditions and were used to detected areas or rooms with abnormal conditions automatically. Alternatively, ambient information can be gathered with a fixed-sensor network infrastructure. For example, the GiraffPlus project [12] allowed remote examination of body temperature, blood pressure and electrical usage based on a sensor network infrastructure and some physiological sensors. This system offers the possibility of a virtual visit via a tele-operated robot to discuss the physiological data and activities occurring over a period of time and the automatic generation of alarms in case of detecting instantaneous problems or warnings based on long-term trend analysis.

Figure 19 shows the ambient information gathered by the APR during a telecontrol experiment based on the methodologies presented in [29,30] with a common sampling sensory time of 1 s. Figure 19 show the temperature registered by the APR with extreme values of 26 °C in one room and 12 °C in another room with an open window (in winter). The humidity reached abnormal values when the APR was in the room with an open window. The luminance level was poor with the lowest values in an indoor corridor. The air velocity was always low and probably originated by the displacement of the APR. Finally, the measurement of volatile gas concentrations in the air can be performed by using one sensor per gas type or by the use of a versatile sensor [29], which can be configured to measure different gasses. In this case, a photo-ionization sensor (PID) was configured to measure acetone because this was a substance used in the area explored [30]. The results showed a low concentration of acetone in the air during the experiment. The APR can be used as a supporting tool to implement a mobile ambient monitoring platform and develop new assisted-living applications.

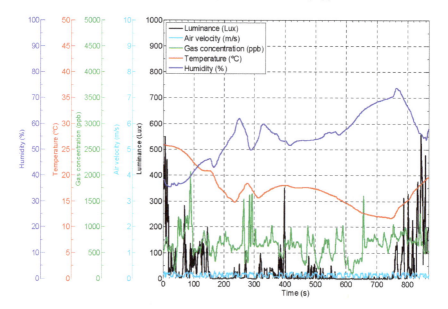

Figure 19. Example of ambient information automatically gathered by the APR.

7. Conclusions

This paper presents the conception and application of a tele-operated mobile robot as an assisted living tool. The proposed Assistant Personal Robot (APR) is described in terms of mechanical design, electronics and software implementation. The APR uses three omnidirectional wheels in order to offer high-mobility capabilities for effective indoor navigation. The APR has a maximum speed of 1.3 m/s and implements a high-priority collision avoidance system in the electronic board that controls the motors of the mobile robot, which can stop the APR regardless of the current motion orders. The APR is described in terms of applications proposed to develop the concept of a tele-operated assisted living tool mainly focused on assistive applications for elderly people and those with impaired mobility. The preliminary experiments performed in this paper have been used to refine the design of the APR as a tele-operated assisted living robot. Future work will focus on improving the design of the APR and the development of the assistive services proposed in this paper.

Acknowledgments: This work is partially founded by Indra, the University of Lleida, and the RecerCaixa 2013 grant. The authors would like to thank the collaboration of Marius David Runcan and Mercè Teixidó in the initial stage of mobile robot development. The authors would like to thank the anonymous reviewers for their helpful and constructive comments that greatly contributed to improving the final version of the paper.

Author Contributions: Eduard Clotet and Dani Martínez designed and developed the software applications and edited most parts of this manuscript. Javier Moreno designed and implemented the mechanical adaptation of the mobile robot. Marcel Tresanchez designed and implemented the electronic boards. Eduard Clotet and Dani Martínez also designed and performed the experimental tests described in the paper. Jordi Palacín coordinated the design and implementation of the software, the electronics and the mechanical design of the APR and also coordinated the final edition of the manuscript.

Conflicts of Interest: The authors declare no conflict of interest.

References

1. Desai, M.; Tsui, K.M.; Yanco, H.A.; Uhlik, C. Essential features of telepresence robots. In Proceedings of the 2011 IEEE Conference on Technologies for Practical Robot Applications, Woburn, MA, USA, 11–12 April 2011; pp. 15–20.
2. United Nations. World Population Ageing 2013. Available online: http://www.un.org/en/development/desa/population/publications/pdf/ageing/WorldPopulationAgeing2013.pdf (accessed on 15 December 2015).
3. World Health Organization. World Report on Ageing and Health. 2015. Available online: http://apps.who.int/iris/bitstream/10665/186463/1/9789240694811_eng.pdf (accessed on 18 February 2016).
4. Delahoz, Y.S.; Labrador, M.A. Survey on Fall Detection and Fall Prevention Using Wearable and External Sensors. *Sensors* **2014**, *14*, 19806–19842. [CrossRef] [PubMed]
5. Palmerini, L.; Bagalà, F.; Zanetti, A.; Klenk, J.; Becker, C.; Cappello, A. A Wavelet-Based Approach to Fall Detection. *Sensors* **2015**, *15*, 11575–11586. [CrossRef] [PubMed]
6. Vermeersch, P.; Sampsel, D.D.; Kelman, C. Acceptability and usability of a telepresence robot for geriatric primary care: A pilot. *Geriatr. Nurs.* **2015**, *36*, 234–238. [CrossRef] [PubMed]
7. Bevilacqua, R.; Cesta, A.; Cortellessa, G.; Macchione, A.; Orlandini, A.; Tiberio, L. Telepresence Robot at Home: A Long-Term Case Study. In *Ambient Assisted Living*; Springer International Publishing: New York, NY, USA, 2014; pp. 73–85.
8. Kristoffersson, A.; Coradeschi, S.; Loutfi, A. A Review of Mobile Robotic Telepresence. *Adv. Hum.-Comput. Interact.* **2013**. [CrossRef]
9. Pineau, J.; Montemerlo, M.; Pollack, M.; Roy, N.; Thrun, S. Towards robotic assistants in nursing homes: Challenges and results. *Robot. Auton. Syst.* **2003**, *42*, 271–281. [CrossRef]
10. Schroeter, C.; Muller, S.; Volkhardt, M.; Einhorn, E.; Bley, A.; Martin, C.; Langner, T.; Merten, M. Progress in developing a socially assistive mobile home robot companion for the elderly with mild cognitive impairment. In Proceedings of the IEEE/RSJ International Conference on Intelligent Robots and Systems (IROS), San Francisco, CA, USA, 25–30 September 2011; pp. 2430–2437.
11. Moyle, W.; Jones, C.; Cooke, M.; O'Dwyer, S.; Sung, B.; Drummond, S. Connecting the person with dementia and family: A feasibility study of a telepresence robot. *BMC Geriatr.* **2014**, *14*. [CrossRef] [PubMed]
12. Palumbo, F.; Ullberg, J.; Štimec, A.; Furfari, F.; Karlsson, L.; Coradeschi, S. Sensor Network Infrastructure for a Home Care Monitoring System. *Sensors* **2014**, *14*, 3833–3860. [CrossRef] [PubMed]
13. Tromp, A.M.; Pluijm, S.M.F.; Smit, J.H.; Deeg, D.J.H.; Boutera, L.M.; Lips, P. Fall-risk screening test: A prospective study on predictors for falls in community-dwelling elderly. *J. Clin. Epidemiol.* **2001**, *54*, 837–844. [CrossRef]

14. Sadasivam, R.S.; Luger, T.M.; Coley, H.L.; Taylor, B.B.; Padir, T.; Ritchie, C.S.; Houston, T.K. Robot-assisted home hazard assessment for fall prevention: A feasibility study. *J. Telemed. Telecare* **2014**, *20*, 3–10. [CrossRef] [PubMed]

15. Vellas, B.J.; Wayne, S.J.; ROMERO, L.J.; Baumgartner, R.N.; Garry, P.J. Fear of falling and restriction of mobility in elderly fallers. *Age Ageing* **1997**, *26*, 189–193. [CrossRef] [PubMed]

16. Palleja, T.; Guillamet, A.; Tresanchez, M.; Teixido, M.; Fernandez del Viso, A.; Rebate, C.; Palacin, J. Implementation of a robust absolute virtual head mouse combining face detection, template matching and optical flow algorithms. *Telecommun. Syst.* **2013**, *52*, 1479–1489. [CrossRef]

17. Khan, A.; Sun, L.; Ifeachor, E.; Fajardo, J.O.; Liberal, F. Video Quality Prediction Model for H.264 Video over UMTS Networks and Their Application in Mobile Video Streaming. In Proceedings of the IEEE International Conference on Communications (ICC), Cape Town, South Africa, 23–27 May 2010; pp. 1–5.

18. Ford, B.; Srisuresh, P.; Kegel, D. Peer-to-Peer Communication across Network Address Translators. In Proceedings of the 2005 USENIX Annual Technical Conference, Anaheim, CA, USA, 10–15 April 2005; pp. 179–192.

19. Halkes, G.; Pouwelse, J. UDP NAT and Firewall Puncturing in the Wild. In *Networking 2011*, Proceedings of the 10th international IFIP TC 6 conference on Networking, Valencia, Spain, 9–13 May 2011; pp. 1–12.

20. Tsui, K.M.; Dalphond, J.M.; Brooks, D.J.; Medvedev, M.S.; McCann, E.; Allspaw, J.; Kontak, D.; Yanco, H.A. Accessible Human-Robot Interaction for Telepresence Robots: A Case Study. *Paladyn J. Behav. Robot.* **2015**, *6*, 1–29. [CrossRef]

21. Pai, N.-S.; Hsieh, H.-H.; Lai, Y.-C. Implementation of Obstacle-Avoidance Control for an Autonomous Omni-Directional Mobile Robot Based on Extension Theory. *Sensors* **2012**, *12*, 13947–13963. [CrossRef] [PubMed]

22. Clotet, E.; Martínez, D.; Moreno, J.; Tresanchez, M.; Palacín, J. Collision Avoidance System with Deceleration Control Applied to an Assistant Personal Robot. *Trends Pract. Appl. Agents Multi-Agent Syst. Sustain.* **2015**, *372*, 227–228.

23. Almasri, M.; Elleithy, K.; Alajlan, A. Sensor Fusion Based Model for Collision Free Mobile Robot Navigation. *Sensors* **2016**, *16*. [CrossRef] [PubMed]

24. Tsui, K.M.; McCann, E.; McHugh, A.; Medvedev, M.; Yanco, H.A. Towards designing telepresence robot navigation for people with disabilities. *Int. J. Intell. Comput. Cybern.* **2014**, *7*, 307–344.

25. Wandosell, J.M.H.; Graf, B. Non-Holonomic Navigation System of a Walking-Aid Robot. In Proceedings of the 11th IEEE International Workshop on Robot and Human Interactive Communication, Berlin, Germany, 25–27 September 2002; pp. 518–523.

26. Shi, F.; Cao, Q.; Leng, C.; Tan, H. Based On Force Sensing-Controlled Human-Machine Interaction System for Walking Assistant Robot. In Proceedings of the 8th World Congress on Intelligent Control and Automation, Jinan, China, 7–9 July 2010; pp. 6528–6533.

27. Pollack, M.E.; Brown, L.; Colbry, D.; McCarthy, C.E.; Orosz, C.; Peintner, B.; Ramakrishnan, S.; Tsamardinos, I. Autominder: An intelligent cognitive orthotic system for people with memory impairment. *Robot. Auton. Syst.* **2003**, *44*, 273–282. [CrossRef]

28. De Benedictis, R.; Cesta, A.; Coraci, L.; Cortellessa, G.; Orlandini, A. Adaptive Reminders in an Ambient Assisted Living Environment. *Ambient Assist. Living Biosyst. Biorobot.* **2015**, *11*, 219–230.

29. Martinez, D.; Teixidó, M.; Font, D.; Moreno, J.; Tresanchez, M.; Marco, S.; Palacín, J. Ambient Intelligence Application Based on Environmental Measurements Performed with an Assistant Mobile Robot. *Sensors* **2014**, *14*, 6045–6055. [CrossRef] [PubMed]

30. Martinez, D.; Moreno, J.; Tresanchez, M.; Clotet, E.; Jiménez-Soto, J.M.; Magrans, R.; Pardo, A.; Marco, S.; Palacín, J. Measuring Gas Concentration and Wind Intensity in a Turbulent Wind Tunnel with a Mobile Robot. *J. Sens.* **2016**, *2016*. [CrossRef]

Article

3D Visual Data-Driven Spatiotemporal Deformations for Non-Rigid Object Grasping Using Robot Hands

Carlos M. Mateo [1], Pablo Gil [2],* and Fernando Torres [2]

[1] Computer Science Research Institute, University of Alicante, San Vicente del Raspeig, Alicante 03690, Spain; carlos.mateo@ua.es

[2] Physics, Systems Engineering and Signal Theory Department, University of Alicante, San Vicente del Raspeig, Alicante 03690, Spain; fernando.torres@ua.es

* Correspondence: pablo.gil@ua.es; Tel.: +34-965-90-3400 (ext. 2014)

Academic Editor: Gonzalo Pajares Martinsanz
Received: 22 December 2015; Accepted: 29 April 2016; Published: 5 May 2016

Abstract: Sensing techniques are important for solving problems of uncertainty inherent to intelligent grasping tasks. The main goal here is to present a visual sensing system based on range imaging technology for robot manipulation of non-rigid objects. Our proposal provides a suitable visual perception system of complex grasping tasks to support a robot controller when other sensor systems, such as tactile and force, are not able to obtain useful data relevant to the grasping manipulation task. In particular, a new visual approach based on RGBD data was implemented to help a robot controller carry out intelligent manipulation tasks with flexible objects. The proposed method supervises the interaction between the grasped object and the robot hand in order to avoid poor contact between the fingertips and an object when there is neither force nor pressure data. This new approach is also used to measure changes to the shape of an object's surfaces and so allows us to find deformations caused by inappropriate pressure being applied by the hand's fingers. Test was carried out for grasping tasks involving several flexible household objects with a multi-fingered robot hand working in real time. Our approach generates pulses from the deformation detection method and sends an event message to the robot controller when surface deformation is detected. In comparison with other methods, the obtained results reveal that our visual pipeline does not use deformations models of objects and materials, as well as the approach works well both planar and 3D household objects in real time. In addition, our method does not depend on the pose of the robot hand because the location of the reference system is computed from a recognition process of a pattern located place at the robot forearm. The presented experiments demonstrate that the proposed method accomplishes a good monitoring of grasping task with several objects and different grasping configurations in indoor environments.

Keywords: visual perception; vision algorithms for grasping; 3D-object recognition; sensing for robot manipulation

1. Introduction

Robot grasping and intelligent manipulation in unstructured environments require the planning of movement according to objects' properties and robot kinematics, as is discussed in [1] and a suitable perception of environment using sensing systems such as visual, tactile, force or combinations of them. In addition, working with the knowledge of the model's uncertainties can be useful when the objects and/or their properties are unknown [2]. In the past, most of the work in robot grasping was focused on providing movements and points to grasp. However, current methods are concerned with adapting the gripper or robot hand to objects and the environment [3]. Bohg *et al.* [4] give an overview of the methodologies and existing problems relating to object-grasp representation (local or global), prior-object knowledge (known and unknown) and its features (2D, 3D or multimodal information) and the type of hand used (gripper and multi-fingered). In other works, tactile sensors were used to provide information about an object's properties through physical contact. Thus, Chitta *et al.* [5] presented a tactile approach to classify soft and hard objects such as bottles, with or without liquids, according to texture measurements, as well as the hardness and flexibility properties of objects. Furthermore, Yousef *et al.* [6] provided a review of tactile sensing solutions based on resistive techniques, the predominant choice for grasping objects.

Although the detection of problems such as slippage and grip force can be controlled using hand kinematics and tactile sensors, the coordinated control among robot-hand joints and tactile sensors is not good enough to perform grasping complex tasks for deformable objects [7]. It is recommended to integrate other sensors that imitate human dexterity, that is, systems based on real-time visual inspection of the task. Therefore, this research involved the implementation of a new visual sensing system based on range imaging technology so as to increase the available sensing data and facilitate the completion of complex manipulation tasks using visual feedback. This work is motivated by the necessity of using other different sensors that tactile sensors for controlling the grasping process of elastic objects. In fact, empirical experiments prove as tactile sensors often supply pressure values close to zero in the contact points of elastic object surface when the object is being deformaded during a task grasping. In this case, the pressure is not adequate to control the grasping process. In general, the kinematics of robot systems and tactile feedback obtained directly from tactile sensors are both used to perform intelligent, dexterous manipulation but it is rare to find practical work in which visual sensors are used to check grasping tasks using robots. Nevertheless, a suitable sensing system, such as in [8], that combines tactile, force and visual sensors, allow us to adapt the grasping task in order to detect errors that can cause grasping failure . Furthermore, it evaluates the effectiveness of the manipulation process by improving the grasp of deformable objects. Similarly, Li *et al.* [9] proposed a controller based on visual and tactile feedback to perform robust manipulation, even in the presence of finger slippage, although flexible objects are not considered, only rigid ones. In our method, the physical model of the skeleton of multi-finger hand for the grasping process is estimated from the kinematics and spatial location of the forearm. Our method does not depend on the pose of the multi-finger hand because the location of the reference system is computed from a recognition process of a pattern located place at the robot forearm. The presented experiments demonstrate that the proposed method accomplishes a good monitoring of grasping tasks with several objects and different grasping configurations in indoor environments. Additionally, the features of the objects are changing in shape, size, surface reflectance and elasticity of the material of which they were made. Our research focuses on checking robot-grasping tasks using 3D visual data from a RGBD camera. Therefore, the approach presented here provides a new visual sensing system in which the grasped non-rigid object surface is supervised to prevent problems like slippage and lack of contact among the fingers, caused by irregular deformations of the object. The main goal of our work is to implement a visual perception system to achieve a robust robotic manipulation system by means of object surface analysis throughout the grasping process. Specific aspects relating to how the robot controller uses the data generated by our visual system and how those are combined with tactile or another information will be addressed in other works. The proposed method of 3D visual inspection for the manipulation of flexible objects has been tested in real experiments where flexible household objects are manipulated by a robot hand. These objects are made of different materials and have different size, shape and texture. Our method uses colour and geometrical information to detect an object in an environment and track its grasping, even if the object has unknown elasticity and flexibility properties. In comparison with other approaches, our method does not use deformations models of both objects and materials and, it also works well with both planar and 3d household objects in real time.

The paper is organized as follows: Section 2 is focused on the analysis of several related works with visual perception for attending to robotic grasping processes. Section 3 describes the proposed visual system for the surveillance of grasping tasks; details on the equipment and facilities used are presented here. Section 4 presents the specification, design and implementation of the recognition method for detection of an object and its surface, and also presents the theoretical principles and fundamentals for modelling grasping tasks, using the kinematics of the robot hand and the detected object which is being manipulated. Sections 5 and 6 present the novel method for measuring the deformation caused by pressure that occur during the grasping of flexible objects. Finally, Sections 7 and 8 describe the tests, outlines our approach and discuss the results of real grasping tasks using three objects with different physical properties, such as size, shape, material, texture and colour as well as reflectance properties due to the material of manufacturing. Additionally, an statistical analysis of experiments is realised to show the behaviour of our method.

2. Related Works

In the past, visual systems were used quite successfully for the manipulation of rigid objects, both for recognition [10] and for the location of an object [11]. This developments and applications of intelligent robot manipulation involve methods and approaches aimed at achieving an object classification and recognition goal. Nevertheless, 3D visual systems are combined with others to measure objects' shapes and the elasticity of deformable objects. We consider elasticity to be the ability of an object to recover its normal shape after being compressed. In [12] is shown an example of an embedded multisensor based on a CCD camera and a tactile membrane, which was designed to determine the contact area and forces' distribution in order to classify materials and their deformation properties.

Furthermore, other works are focused on detecting and tracking the deformation from the fusion of 2D images with force data [13,14]. Later, Khalil *et al.* [15] used stereoscopic vision to build a 3D surface mesh from contours and colour in order to discover the deformation of non-rigid objects, and Leeper *et al.* [16] used a low-cost stereo sensor mounted on the gripper to estimate grasp poses and to choose the best one according to a cost function based on points cloud features. More recently, Boonvisut *et al.* [17] proposed an algorithm for the identification of the boundaries of deformable tissues, and to use it for both offline and online planning in robot-manipulation tasks. Moreover, Calli *et al.* [18] presented a dataset to test method for manipulation tasks.

Continuing in the same line, Jia *et al.* [19] presented a grasping strategy for non-rigid planar objects using just two fingers. This research assumes a linear elasticity model of the object surface. Afterwards Lin, *et al.* [20] used this same idea applied to grasp and lift 3D objects measuring when a secure grip is achieved under contact friction. Unlike both works, our proposal does not use a deformation linear model and the detection of deformations is realised using 3D visual features over the entire object surface not only the contact regions close to fingers. Moreover, our system can work with nonplanar objects as 3D objects using more than two fingers; specifically, we use a multi-finger hand [21,22]. In our work, the secure grip is achieved under a measurement control to reach an appropriated deformation level. In line with our work, Navarro-Alarcon *et al.* [23,24] proposed a vision-based deformation controller for robot manipulators. In those works, the presented method estimated the deformation of an objects using visual-servoing to measure the Jacobian matrix of features on the object surface. The authors used a sponge with small marks located on the surface. In contrast, our proposal is based on the estimation of deformations to control the grasping tasks without artificial markers and without a previous known deformation model [25]. We also test our visual perception method with many different household objects, not only a sponge, as it was done in [24]. Recent research that use visual perception for manipulation tasks with non-rigid are included in works carried out by Alt *et al.* [26] and Sun *et al.* [27], although both the underlying idea and its applicability are very different to our proposal. Thus, on the one hand, in [26] are simultaneously combined both haptic and visual sensors for navigation and manipulation tasks. The authors used a deformable foam road mounted on the gripper to measure visually a 1d stress function when there was contact with deformation. On the other hand, in [27] a visual perception pipeline based on an active stereo robot head was implemented for autonomous flattening garments. It made a topology analysis of each wrinkle on clothes surface in order to identify deformations.

3. Visual Surveillance System for Robot Grasping Tasks of Objects

Modern approaches based on pressure data obtained from a tactile sensor may fail in complex grasping tasks (failure of the touch), in which the non-rigid objects change their shapes as their surfaces are deformed due to the forces applied by the fingers of a robot hand. In this case, the tactile sensors are often not able to obtain touch data correctly. Figure 1 shows the pressure of tactile sensors on grasping tasks of two different non-rigid objects. Figure 1a presents the pressure evolution for the grasping task of a brick. The pressure data retrieved from the tactile sensor are good enough due to that the pressures are greater than $0 \, \text{N/cm}^2$ for four fingers and the values are greater than $1.5 \, \text{N/cm}^2$ for three fingers. In contrast, Figure 1b presents the pressure evolution for the grasping task of a plastic glass. In this case the acquired pressure is not adequate. That is indicated by the fact the pressure values are close to zero for all fingers except the thumb ($0.5 \, \text{N/cm}^2$ is measured for thumb, but it is very low).

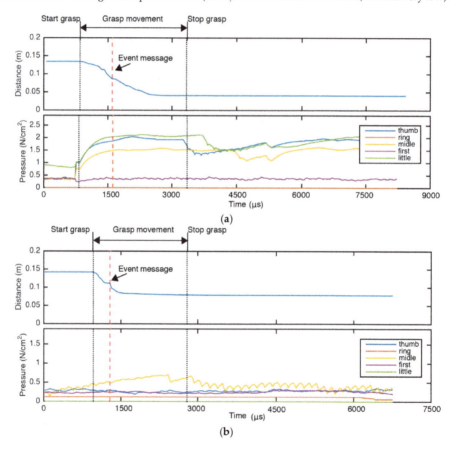

Figure 1. Comparison of different pressure for two real grasping tasks: (a) Pressure data allow to control the grasping; (b) Pressure is not valid to evaluate the grasping an object.

To overcome this problem, the novelty of our approach lies in that we use a visual sensing approach to obtain information about the object, such as shape, deformation and the interaction between the object and the robot hand. Therefore, our goal is to implement a visual system which serves as a surveillance module for the grasping control when there is no available tactile data, if, for example, the sensor is not working or fails due to its inability to measure pressure. In such scenarios, our visual system would be able validate if there is contact between the robot finger and the object.

Considering the recent advances and availability of depth cameras such as RGBDs and the advances of the 3D processing tools for point clouds, we have designed a strategy that considers RGBD images as input data for the visual inspection of grasping tasks. To obtain visual data, a Microsoft Kinect sensor is used in this work. The Kinect sensor consists of a visual camera (RGB) and a depth sensor (an infrared projector and a camera). The Kinect sensor operates at 30 Hz and can offer images of 640 × 480 pixels but it has some weaknesses, such as problems with depth resolution and accuracy. The depth resolution decreases quadratically with respect to the distance between the sensor and the scene. Thus, if the camera is the world reference frame, the point spacing in the z-axis (Figure 2a) is as large as 7 cm at the maximum range of 5 m. In addition, the error of depth measurements (or accuracy) increases quadratically, reaching 4 cm at the maximum range of 5 m [28]. Moreover, another known problem is the unmatched edges, which is caused by pixels near to the object's boundaries. In this case, wrong depth values are assigned by the Kinect sensor.

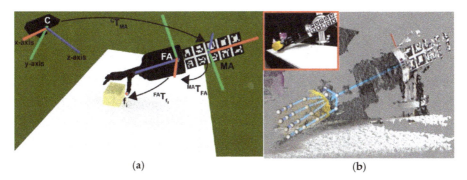

(a)   (b)

Figure 2. Workplace: (**a**) Scheme of the robot hand and RGBD sensor configuration; (**b**) RGB image and its point cloud associated with the robot hand and an object.

Our work assumes that the camera does not need to change its viewpoint because the robot hand is solely moved with according to the trajectory planning and grasp quality measure. Here, the sensor pose is static and is initially estimated according to the initial hand's pose. It is not the objective of this research to find the best viewpoint of the camera for grasping tasks using the robot hand. The RGBD sensor is positioned according to previous works of other authors such as [29] in order to maximize the visible area of the manipulated object.

The system was set up using ROS. The sensor is connected through a USB port to a computer which works as visual data server. Therefore, the visual system is ready to communicate with other systems such as the robot hand and tactile sensors. It should be noted that touch data from tactile sensors are not used as inputs or feedback of our visual system (Figure 3). It is the visual system which monitors the grasping tasks in which the tactile sensors can fail. Previously, Figure 1 has shown an example where the tactile sensor fails.

This work does not discuss how the control system works with the information received from the visual system, nor the readjustment of the fingers' position nor the equilibrium during the grasping task, Although the control system used (yellow block in Figure 3) was presented by Delgado *et al.* in [30], this one implements a control system based on the kinematic model of the robot hand but without using the physical model of the manipulated object. Another approach of implementing a control system is presented in [8] (Chapter 2), which uses the physical model of the manipulated object.

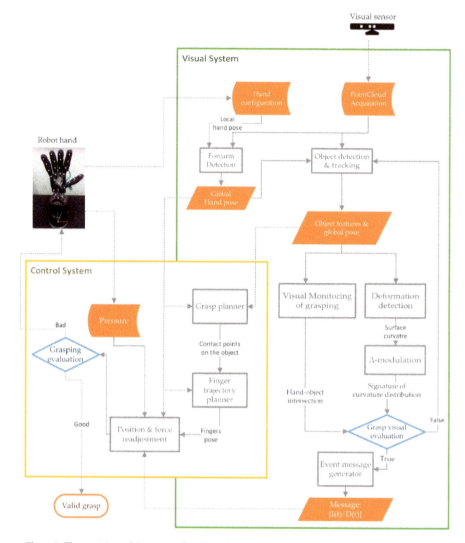

Figure 3. The overview of the system flowchart. The control system architecture (**yellow block**); and the proposed method for the visual system (**green block**).

On the one hand, the methods for determining contact points in a grasping process depend on whether the object is known or unknown. In the first case, fingertips contact positions are calculated as discussed in [31]. Here, Kragic *et al.* presented a real-time grasping planner to compute grasp points for known objects. To do this, is also required that the objects are detected and recognized by comparison of points cloud of object with models. We can use a surface descriptor-based method as is shown in our previous works [32] in order to describe the points cloud of object and to compare it with a surface model. In the second case, as is done in this work, the object is unknown then we extract the points cloud which represents the object located on a planar surface as a worktable by means of filtering of points cloud of scene and Random Sample Consensus (RANSAC). Thus, the points cloud of both object and table are separated. Afterwards, the centroid and the principal axis of points cloud of object are computed to approximate their position and orientation. The object normal vector is built using its centroid and the direction of the normal vector of the worktable. This way, the robot hand can be positioned and orientated for grasping. Later, the contact points for the thumb and middle fingertips are estimated by calculation of the orthogonal vector which intersects to the plane formed by the principal axis and the object normal vector.

On the other hand, the readjustment algorithm used in this work was presented in [33,34]. The algorithm receives as an input the finger joint trajectories and adapts them to the real contact pressure in order to guarantee that undesired slippage or contact-breaking is avoided throughout the manipulation task.

The contribution of this work is the visual system method and it is represented in the green block in Figure 3. This block presents a scheme of our method, based on data-driven 3D visual recognition. In particular, it shows the processing steps for the inspection of the grasping tasks where the robot hand starts moving forward until deformation is detected by the implemented method. The method is discussed in Sections 4–6. More specifically, our approach can be understood as a data-driven visual surveillance method that generates control events when potential grasping problems or anomalies are detected for the manipulated object—meaning significant deformations, slips or falls of the object that are caused by improper manipulation. Therefore, while the control system is working with the robot kinematics and tactile sensor, the visual system is working simultaneously, when it detects some of those anomalies in the grasping task, it sends an event message to the control system which determines a grasp quality measure and realizes the fingers movement planning (Figure 3). Figure 1 shows two examples were the visual system sends an event message before the control system finishes the grasping task.

The method is designed as a tracking loop with two pipelines (chains of processing blocks). The first one is composed by a processing block named *"visual monitoring of grasping"*. The second pipeline has two blocks named *"Deformation detection"* and *"Δ-modulation"*. The method is completed when the conditional block named *"Grasping evaluation"* determines whether there has been a problem or whether the grasping task is realised. At the end of the method, the tracking loop sends an event message. This message is encapsulated as a data packet by the processing block named *"Event message generato"*. The event message contains data about object deformation level and object-hand intersection level. These processes are well explained in Sections 4–6. The remaining part of this section is focused on providing detail on the block named *"Object detection & tracking"*.

3.1. Object Detection

To carry on a tracking process, it is essential to carry out an initial object detection process. Let $I_m(r, g, b, d)$ denote the input data, the system has a point cloud $P(x, y, z, r, g, b)$ by mapping the RGBD image from a calibration process of the visual sensor. The point cloud is partitioned into a set of disjointed regions. One of them represents the object to be manipulated, another is the robot hand and the rest is noise (data which are not considered). The green block of Figure 3 represents a complete overview of the proposed visual system in which the first time that process the *"object detection & tracking"* segments P into these regions. This process uses an object-recognition pipeline based on shape retrieval and interest region detections. There are several previous works that use these approaches, a case in point are [32,35], in which 3D objects are recognized and located on a worktable by using shape descriptors. Others, such as Aldoma *et al.* in [36], presented a review of current techniques of object recognition, comparing local and global mesh descriptors. Even though there are several works which use data sets of object mesh models for benchmarking in shape retrieval. These studies are not always helpful for manipulation experiments because they do not provide the physical properties of the objects, such as material stiffness or weight and how objects can change shape while they are being manipulated in real experiments. For this reason, our visual system (designed specifically for manipulation) is implemented considering that objects can be deformed over time.

The object recognition process is divided into the following stages:

1. Initialize the hand workspace
2. Get the object points
3. Select the object parts as super-voxels

The first stage initializes the hand workspace using the input data of two sets of sensors. They are internal sensors in the joints of the robot $S_1(E, t)$ and RGBD sensor $S_2(E, t)$. The goal of this stage is to obtain points in $P(t)$ belonging to a points cloud that could potentially be part of the object $P^*(t)$ (Figure 4a,b). Both sensors are dependent on an environment E and time t as:

$$S_1(E, t) = S_1(Q(t)) \tag{1}$$

where $Q = [f_1, ..., f_N]$ is the pose for both the robot's fingertips and palm, being $f_i = (q_1, ..., q_M)$, which defines the joint parameters for each finger and the palm. In addition, $S_2(E, t)$ is defined as:

$$S_2(E, t) = S_2(P(t)) \tag{2}$$

where $P(t)$ is a retrieved point cloud from the RGBD sensor at time t. From now on, P means $P(t)$ for simplicity. Likewise, they work cooperatively, combining data to obtain the pose of a grasped object in the environment, which means that S_1 helps to S_2 to determine region P^* from P. To do this, three sequential sub-processes are used: building the hand area; erasing noise; and sampling data. Firstly, a crop area placed in the geometric centre of the fingertips and palm kinematics is set; besides the radius of this area, this is two times the minimum distance between a fingertip and the palm position. Then, all points outside the crop area are removed. Secondly, the border, shadow and veil points are erased from the survivors' points in the previous sub-process. In addition, the points around the position of the robot-hand links are erased. Thirdly, the resulting set of points is sampled using a voxelized strategy, with a voxel size of 2% of the minimum distance between a fingertip and the palm position.

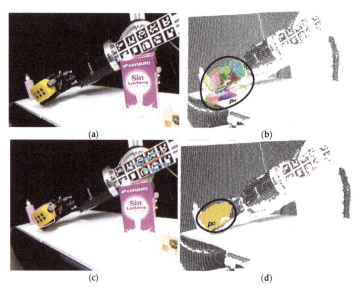

(a)   (b)

(c)   (d)

Figure 4. Samples of the object-detection process: (**a**) Original scene; (**b**) Colour and geometry segmentation process based on robot kinematics; (**c**) Clustering of regions to determine an object area; (**d**) Result of the 3D object detection.

The second stage obtains object points by means of a voxel cloud-connectivity segmentation method. This is used to determine better the object boundaries, as in Papon *et al.* [37]. Voxel Cloud Connectivity Segmentation (VCCS) is a variation of k-means clustering, with two important constraints: the seeding of super-voxel clusters is achieved by partitioning the 3D space and the iterative clustering algorithm enforces the strict spatial connectivity of occupied voxels. In short, VCCS efficiently generates and filters seeds according to how the neighbouring voxels are calculated. Each super-voxel cluster is represented with colours in Figure 4b. Finally, an iterative clustering algorithm enforces spatial connectivity. The statistical similarity test used here is Fisher's test [38] in contrast with [37].

The third stage details the selection of super-voxels or clusters that belong to the manipulated object. This stage involves three sub-processes: selecting the seed cluster; iterating them via the neighbouring cluster; and merging the connected clusters. Firstly, the nearest cluster to the geometrical centre of the fingertips' set and palm kinematics is selected as the seed cluster. Secondly, the remaining clusters are iterated to find newly connected clusters. Each new cluster is then labelled in line with how it should be merged. Thirdly, the labelled clusters are merged into the seed cluster, recovering the point cloud P^O which represents the object from P^* (Figure 4d).

Accuracy is increased in the segmentation process P^O from P^* due to the fact that S_1 helps S_2 to determine the region P^* from P, instead of determining P^O directly from P, as it would be done without the known kinematics of the robot hand. The reference frame for the robot kinematics is found using marker boards, as in [39]. This is done because the robot pose is unknown with respect to the world reference frame located in the RGBD sensor. Thus, once the forearm is located, we are able to obtain the pose of the robot's fingertips and its palm as:

$$^{C}T_{f_i} = {}^{C}T_{MA} \times {}^{MA}T_{FA} \times {}^{FA}T_{f_i} \tag{3}$$

where f_i is the frame of a robot's fingertips or palm, C is the camera frame, MA is the marker board frame and FA is the forearm frame. If ${}^{j}T_i$ denotes the transformation of the frame i with respect to the frame j, ${}^{C}T_{MA}$ is the transformation of the MA w.r.t. the C, ${}^{MA}T_{FA}$ is the the FA transformation with regards to the MA and the ${}^{FA}T_{f_i}$ is the transformation of a f_i with respect to the FA. Thus, the transformation of each robot's fingertips or palm frame w.r.t. the camera frame ${}^{C}T_{f_i}$ is given by the multiplication of these homogeneous transformations. Each homogeneous transformation represent rotations and translations.

3.2. Tracking Object Surface

The used tracking process is inspired by Fox's works [40,41]. The author presents a statistical approach in order to increase the efficiency of particle filters by adapting the size of sample sets on-the-fly. The key idea of the author in these works is to bound the approximation error introduced by the sample-based representation of the particle filter. The measure used to approximate the error was Kullback-Leibler distance. This approach chooses a small number of samples if the density is focused on a small part of the state space. In another case, if the state uncertainty is high, it chooses a large number of samples.

Unfortunately, this approach presents undesirable characteristics to track points of object surface while this object is being manipulated; our work proposes to use the change in the number of sample as a tracking driver. This is that the tracked target (the set of points on object surface) is replaced when the number of samples changes, significantly. The idea is that the particle filter tracker is efficiency tracking when the number of particles is small. Thereby, the goal is to find when the number of these particles has dramatically grown up. Consequently, the method perceives that the object surface is being changed. Thus, the method detects this change on object surface by mean of the statistical relationship among the number of particles in the sample set over time as:

$$\Gamma(t) = \mu(t) + \sigma(t), \quad \mu(t) = \frac{\sum_{i=1}^{t} \gamma(i)}{t}, \quad \sigma(t) = \sqrt{\frac{\sum_{i=1}^{t}(\gamma(i)-\mu(t))^2}{t}} \tag{4}$$

where t is the time and $\gamma(t)$ is the number of particles in the sample set at time t. $\mu(t)$ is the mean value and $\sigma(t)$ is the standard deviation of the number of particles from $t = 0$ until time $t > 0$. Thus $\Gamma(t)$ is the statistical relationship which determines when the target (the point cloud of surface object) must be replaced by the point cloud at time $t - 1$. This approach sets up $P^O(t - 1)$ as a new target when $\Gamma(t - 1) < \gamma(t)$. The efficiency of this strategy is dependent on how the method determines when the change on surface is significantly large to replace the tracked target with another point cloud more recent (at time $t - 1$).

4. Visual Monitoring of Grasping

4.1. Modelling the Grasping Task like a Regional Intersection

One way to monitor and supervise the grasping tasks is to measure the interaction between items in the environment, the robot hand and a manipulated object. Our proposal is based on the visual relationship between the regions $H(t)$ and $O(t)$. Figure 5 illustrates three cases where this spatial information provides an interaction relationship between an object and the hand. At first, the object is not caught and the hand is not positioned on the object. At this stage, the fingers have not yet started to move and the tactile pressure should be zero (Figure 5a). When the hand is positioned according to the "*Grasp planner*" and "*Finger trajectory planne*" (Figure 3), the fingers are moved and some of them can press the object but it is not enough to grasp and lift the object without sliding (Figure 5b). Later, at a given time, the contact and appropriate touch of the fingertips allow the robot hand to pick up the object correctly, or not. That is, sometimes, if the object has flexible properties, then it can change its position due to the finger pressure, causing a sliding of the object among the fingers (Figure 5c).

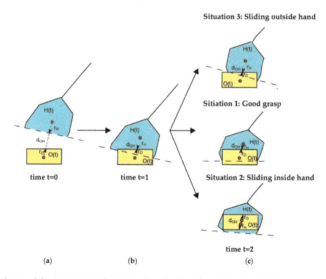

Figure 5. Scheme of the intersection between the robot hand and objects: (**a**) The hand volume has not yet intersected with the object volume; (**b**) The hand volume has started to intersect with the object volume; (**c**) Three final possible gasping situations.

4.2. Computing the Intersection between the Robot Hand and the Object

The proposed method of "*visual monitoring of grasping*" supervises the grasping process and is based on an estimation of the intersection region. In any real grasping task (Figure 6a), the manipulated flexible objects $O(t)$ change their volume and shape in relation to time t. Likewise, the robot hand $H(t)$ changes its pose according to the kinematics. From now on to simplify, H means $H(t)$ and O

means $O(t)$. Both the object and the hand are represented as 3D regions in this work. On the one hand, O is calculated using a segmentation process from the point cloud P of the scene acquired by the Kinect sensor (Figure 6), as discussed in Section 3. On the other hand, H is given by the hand pose and calculated from the kinematics of the robot (Figure 6a). We compute the 3D convex hull of the set of points of the object surface for O. Furthermore, we calculate a 3D virtual bounding hull that fits the points of fingertips and the palm area for the representation of H. The convex hull of a set of points is the smallest area which contains those points and its volume is calculated as:

$$\left\{ \sum_{i=1}^{|S|} \alpha_i p_i \Big| (\forall i : \alpha_i \geqslant 0) \wedge \sum_{i=1}^{|S|} \alpha_i = 1 \right\} \tag{5}$$

where S is either H or O set and p_i is a point of P^*. α_i is the weight assigned to p_i in such a way that is a non-negative value if the point belongs to S. Thus, on the one hand, we have for each value α_i^H when $S = H$ that:

$$\alpha_i^H (p_i) = \begin{cases} \frac{1}{|H|}, & p_i \in Q \times {}^C T_{f_i} \\ 0, & other\ case \end{cases} \tag{6}$$

where $p_i \in P^*$ is the point associated to α_i, $|H|$ is the number of points in H and $Q \times {}^C T_{f_i}$ is the set of points indicating the fingertip poses. On the other hand, each value α_i^O when $S = O$ is computed as:

$$\alpha_i^O (p_i) = \begin{cases} \frac{1}{|O|}, & p_i \in P^O \\ 0, & other\ case \end{cases} \tag{7}$$

The interaction between the robot hand and the object is estimated by the relationship between the object and the robot hand, and it is computed by the comparison of the 3D regions H and O. In order to do this, we obtain the overlap between the 3D convex hull, object, and the virtual bounding hull, robot hand (Figure 6b). This is done as follows:

$$I = H \cap O = \left\{ \sum_{i=1}^{|I|} \alpha_i^H \alpha_i^O p_i \Big| \left(\forall i : \alpha_i^H \alpha_i^O \geqslant 0 \right) \wedge \sum_{i=1}^{|S|} \alpha_i^H \alpha_i^O = 1 \right\} \tag{8}$$

where I is the intersection volume between H and O. Therefore point $p_i \in I$ when at the same time $p_i \in H$ and $p_i \in O$. Moreover, the hand grasps the object when the intersection is greater than $I(t) = O$ and smaller than $I(t) = \min(H(t), O(t))$. Note that, $I(t)$ is the generalization of I for the time t.

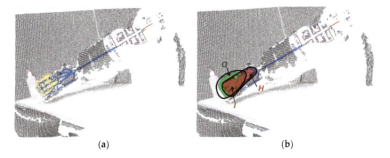

(a)   (b)

Figure 6. Visualization of the main steps of the visual monitoring algorithm for a grasped object scene: (**a**) point cloud of the scene and robot kinematics; (**b**) result of the detection of 3D regions of both the robot hand and the object.

Finally, the grasping task is evaluated by the "*Grasp visual evaluation*" (Figure 3). This conditional block determines whether the visual system must finish and it generates an event message by "*Event message generator"* block or continues with the supervision of grasping task. Thus, when $I(t) = 0$, and the object has fallen from the robot hand to the ground, *"Grasp visual evaluation"* finishes the tracking loop of the visual system and *"Event message generator"* block creates an event message encapsulated as a packet with the connect values of intersection and deformation. Hence this event message is sent to the control system.

5. Detection of Deformations by Means of Surface Curvatures

A novel contribution of this work is its presentation of an approach for the detection of deformation in flexible objects. The method *"detection of deformations"* is based on the general idea presented in [42], where the authors compute surface gradients for the modelling of surfaces as curvature levels, although the objective is much more ambitious in this work. Here, the issue goes further than a specific implementation of that general idea, insofar as we present a new method for detecting when a deformation is occurring in real time by means of differential deformation estimation among time points. This is done by analysing the curvatures of the object's surface through a timing sequence.

The aim is to know whether the flexible object is grasped properly. That is to say, the object is considered it has been well grasped if the deformation distribution measured as a surface variation undergoes meaningful changes in connecting with a reference frame. The ideal situation is given when the distribution of surface variation is constant throughout the grasping task. The comparison process between the surface at various time points is conducted by comparing variations in the surface in each point of a points cloud which contains the object.

The method locally analyses the points cloud which represents the object surface for extracting the surface variation as curvature values c_{p_i} at each point of the surface $P^O = \{p_i \in \mathbb{R}^3\}$. To analyse the surface variation at point p_i, the eigenvalues and the eigenvector of covariance C_P in a neighbourhood environment with a radius r matrix are extracted as in [42]. C_P is computed as:

$$C_P = PP^T = \begin{bmatrix} p_{i_1} - \overline{p} \\ \cdots \\ p_{i_k} - \overline{p} \end{bmatrix} \begin{bmatrix} p_{i_1} - \overline{p} \\ \cdots \\ p_{i_k} - \overline{p} \end{bmatrix}^T \tag{9}$$

where each p_{i_j} is a point of the neighbourhood environment N_j and $\overline{p}$ is the centroid of the path. k defines the number of points N_j.

Besides, the computation of eigenvalues λ_j and eigenvectors v_j of C_P is done by applying singular value decomposition (SVD) as:

$$C_P \cdot v_j = \lambda_j \cdot v_j \tag{10}$$

Once the eigenvalues are computed, the surface variation (curvature) is calculated as:

$$c_{p_i} = \frac{\lambda_0}{\lambda_0 + \lambda_1 + \lambda_2} \tag{11}$$

where $\lambda_0 \leqslant \lambda_1 \leqslant \lambda_2$ are the eigenvalues associated to the eigenvectors of a covariance matrix. Then, the points with similar curvature values c_{p_i} are clustered together in level curves. These level curves are defined as a function $S_P : \mathbb{R}^3 \to \mathbb{R}$, as follows:

$$S_P = \left\{ (x, y, z) \in \mathbb{R}^3 : \quad \Phi(x, y, z) = l \right\} \tag{12}$$

where l is a constant value and represents a level curve on the surface. The control rule used to determine whether a specific point p_i belongs to a specific level curve l_i is:

$$\left\{ p_i \in l_k \mid k = curvature(p_i) \times \frac{|P^*|}{|P|} \right\} \tag{13}$$

Consequently, each of the level curves is computed as a cluster, and it is represented with the same colour when the points of surface have a similar value of the curvature (Figure 6a). Therefore, two significant curves are highlighted, such as the level of maximum curvature (displayed in dark blue) and the level of boundary curvature (in yellow), which represents all points with the minimum curvature, without considering the zero value (displayed in red is the region with a value of zero). If the curvature is zero, then there is no deformation and all points of the surface lie on a flat.

Figure 7 represents P^O and its level curves S_P at two different time t; in this case, $t = 0$ and $t = 53$. It is later obtained via a signal which adjusts the original signature of the curvature distribution for P^O. This is done by using a least-squares fitting of the curvature values by means of a quadratic function $g(f, x)$ that allows the method to obtain a more representative signature of the curvature distribution as:

$$g(f, x) = w_1 x^n + w_2 x^{n-1} + \ldots + w_n x + w_{n+1} = \sum_{k=0}^{n} w_{k+1} x^{n-k} \cong f(x) \tag{14}$$

where n is the range of the quadratic function and w_n is a coefficient associated to the polynomial term x. Figure 7b represents S_P and its adjust $g(S_P, i)$ with a continuous curve line (the colour is orange in the first case, and blue in the second case) at two different time.

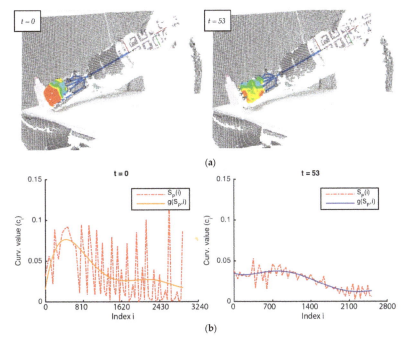

Figure 7. Two frames of the same manipulation sequence at two different time points: (**a**) The model of deformation computing the surface curvatures, grouped as level curves; (**b**) The signature of the curvature distribution.

6. Δ-Modulation and Grasping Evaluation

The detection of deformation is a huge problem that cannot be tackled without regard to the time variable t, *i.e.*, the signature of the curvature distribution of an object, as is shown in Figure 7, gives information regarding its surface shape but not its deformation. Therefore, in order to obtain deformation data, the way as changes the surface shape with respect to the time was studied using our method of 'Δ-*modulation*. Figure 7 is an example of how the signature of curvature distribution can be retrieved with regards to time, by comparing the curvature values between different time points. Figure 7, especially, shows the curvature and its Δ-modulation for the point clouds shown in Figure 6.

The method of Δ-modulation used to encode the signature of curvature distribution with regards to *t* in concrete is a sequence with two time values, for reducing its complexity. This method provides advantages for the comparison of signals over time. In Δ-modulation, the input signature is approximated by a step function where each sampling interval α increases or decreases one quantization level δ. Figure 7a,b both show an example in which the step function is overlapped with the original signature with regard to time. The main feature of the step function is that its behaviour is binary: at every α, the function moves up or down by the amount δ. Therefore, the Δ-modulation output can be represented by just one bit for each sample, and is represented in Figure 7a,b as a binary function. Consequently, Δ-modulation obtains a bit chain that approximates the derivative of the original signature. It is generated as 1 if the step function increases or 0 if the step function decreases.

Once Δ-modulation has encoded the signature of a curvature's distribution in both times, the outputs are compared by a subtraction logic operation as is shown in Figure 8c. A simple interpretation of this operation result is as follows:

- Case 1: The object is not deformed whether the output is always 0.
- Case 2: It is considered that the object is deformed.

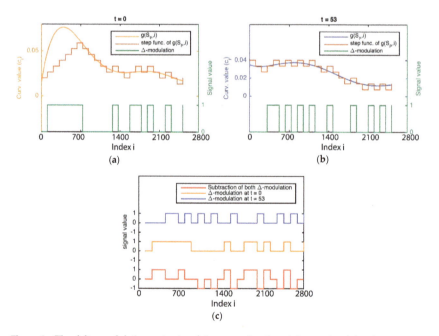

Figure 8. The delta modulation output and its approximation staircase signal for the curvature signature: (**a**) At time 0; (**b**) At time *t*; (**c**) The result of subtracting both modulated signals.

A disadvantage of this method is that it is dependent on the type of fitting used (here, a quadratic function), which could result in an over-fitting or under-fitting problem and then the function would not be sufficiently representative. To overcome this problem, we simplify the signature, making a histogram into a generating function that represents the signature of the curvature distribution in a simpler way that before (Figure 9a,b). The use of a histogram makes this method adjustable in terms of its sensitiveness. For instance, if we want a more reactive approach for deformation detection, it will be better the use of many classes. Therefore, we use this new representation of the signature, named curvature histogram H_P in order to apply the "Δ-*modulation*" method and so we can easily acquire a derivative function which is comparable itself between two time points (Figure 9c,d). The binary comparison of both derivative functions is a new pulse function, as shown in Figure 9e,f. This signal is created by adding 0 in the context of "case 1", or 1 in the context of "case 0".

Furthermore, the proposed method is able to determine when the manipulated object is being deformed, implementing it as a Finite-State Machine (FSM). Figure 10 represents this implementation. The FSM starts in the "undeformed" state and remains in this state until it receives the value "1" from the deformation signal; then the FSM changes to the state called "on hold". Once in the "on hold" state, the FSM returns to the previous state is named "undeformed", or changes to another state called "deformed", depending on whether the deformation signal represents a low or high level, respectively. This middle state is used to prevent the method from detecting false positives (a value of "1" that should be "0") according to the deformation signal.

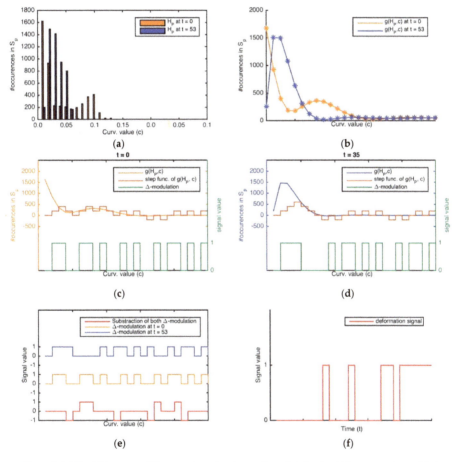

Figure 9. (a) The histograms of the curvature signatures of a point cloud, which represents an object, at two time points; (b) The signal of the histogram; (c,d) The delta modulation output and its approximation staircase signal for the signal of the histogram in both cases; (e) The result of subtracting both modulated signals; (f) The deformation signal measured as a time signal in a grasping task.

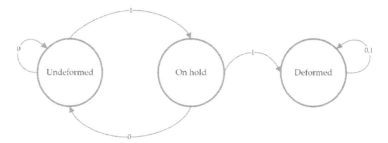

Figure 10. The Finite-State Machine for identifying from the deformation signal whether the object has been already deformed.

Aside from the evaluation of grasping by measuring of intersection *I* and deformation "*Grasp visual evaluation*" (Figure 3) is a conditional block to check which is the state of the FSM. Thus, "*Grasp visual evaluation*" finishes the tracking loop of the visual system and "*Event message generator*" block creates an event message, as it is commented in Section 4, if FSM is on "deformed" state.

7. Experiments and Results

In this section, we show the capabilities and effectiveness of our visual method. This method combines 3D visual processing from RGBD and the dexterous Shadow hand kinematics in order to accomplish the proper grasping of non-rigid objects. Our method implements the grasp adjustment from vision-based switching controller. In each step, the algorithm checks the interaction between the robot hand and the grasped object using visual information. The ability of our visual perception algorithm allows the robot system to correct the uncertainty/error caused by the lack of tactile/force data, or bad measurements when the grasped object has flexibility properties, such as in objects made from plastic polymers.

The next four subsections describe several relevant grasping experiments using flexible objects that were made from different materials, and that also have different shapes, sizes, textures and colours: a sponge, a brick, a plastic glass and a shoe insole. Each of those objects presents a challenge for the method. Some relevant frames of the manipulation task showing visual information computed by our system are presented for three of them, such as the result of the detected object tracker, the curvature data and the volume evolutions.

7.1. Experiment 1: Sponge

The sponge is ideal for testing the grasping tasks of flexible objects because it has a homogeneous distribution of elasticity and stiffness on its surface. These properties generate high levels of deformation close to attachment points where the fingers are located, but only slight deformations in the object's centre (Figure 11a). For this reason, the sponge is rapidly deformed close to the fingertips at the beginning of the grasping process, although the deformation velocity tends to decrease quickly over time. Additionally, the observed deformation is regular and incremental regarding time (Figure 11b). The sponge's homogeneous colour and regular geometry help the "object detection" module of our approach to extract the object's region from the points cloud of the scene. It should be noted that the sponge's size decreases gradually due to the properties mentioned above.

This test is a points cloud sequence of a grasping movement with 250 frames. The fingers of the robot hand do not move in the first 25 frames. The following frames show a sudden, sharp hand movement; this can be seen between Frames 26–115. Then, from frame 116 until the end, the hand movement is not considered relevant to this experiment because the deformation can be detected in the previous frames.

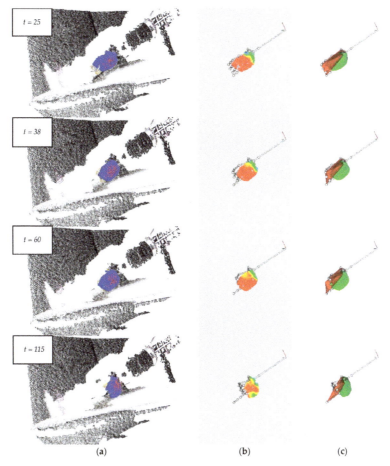

Figure 11. A sequence of video frames showing the grasping of a sponge: (**a**) The points cloud of the scene and the tracking process in the object's detection; (**b**) A curvature map to measure the deformation level of the object; (**c**) The intersection for measuring the visual contact between the robot hand and object.

On the one hand, Figure 11a shows the tracking of the detected object. In particular, it shows how the target model of the sponge evolves dynamically during the deformation, fitting and tracking the object while its size is changing as a result of the contact force. On the other hand, Figure 11b shows the traces of the curvature map in which the regions with low curvature levels slowly disappear, but, simultaneously, the deformation regions close to the fingers indicate that the curvature levels are constant from frame 38 onwards. Additionally, Figure 11c shows the decrease in both the region of the hand and the object estimated as a hull.

7.2. Experiment 2: Brick

In contrast with the sponge's features, the brick's features are not homogeneous. The brick is made of several carton caps and their number, shape and structure determine the level of the brick's elasticity, flexibility and stiffness. The most rigid points of its structure are its edges and the least rigid ones are the centre of the sides. These properties generate irregular deformation, limiting the spread of deformations to the brick's edges (Figure 12b).

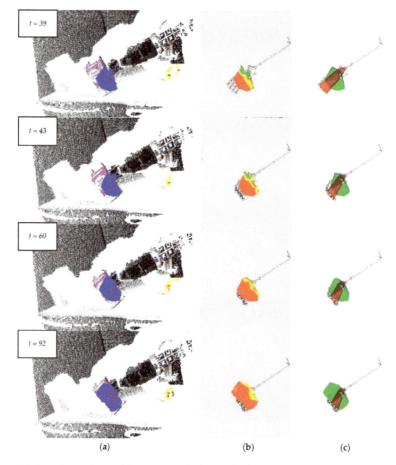

Figure 12. Some frames from a video sequence depicting a brick: (**a**) The points cloud of the scene and the tracking process in the object's detection; (**b**) A curvature map to measure the deformation level of the brick's surface; (**c**) The intersection to measure the visual contact between the robot hand and the object.

We have chosen a brick because it is a very common household object and has been used widely in recognition experiments by other researchers in the field. A brick is also usually painted with serigraphs of different colours. Its colour is not homogeneous and this fact entails more complexity and a new challenge for the object's recognition and the tracking process (Figure 12a).

As above, the test includes 250 frames; however, the robot is motionless in the first 40 frames and it moves between Frames 41 and 60, in which it is possible to determine the brick's deformation level and the interaction between the hand and the brick (Figure 12b,c). The other frames do not provide interesting information for the surveillance of this grasping task.

Figure 12a highlights the small amount of variability in the frames; this fact indicates the robustness of the object tracker. Besides this, Figure 12b shows slight variations of deformation in the first few frames but remains almost constant after the deformation. Additionally, the intersection changes only a little because of the brick's rigidity (Figure 12c).

7.3. Experiment 3: Plastic Glasses

This experiment has been designed to show the behaviour of our approach when using another household object, such as a plastic glass. Its structure and manufacturing materials define new elastic and flexibility properties compared with those of the objects above. In particular, the grasping tasks cause irregular deformations of the glass. Its features of a homogeneous colour and simple geometry expedite the creation of a vision system for suitable recognition. As our aim is not to present new techniques for object recognition, we have always used objects that do not have complex shapes and many colours. In this case, the glass allows the vision system to check the behaviour of the newly implemented methods when the object size is smaller than a sponge or brick, and is made of another material.

The test also has 250 frames but, here, the first 18 frames show the robot hand when it is motionless, frames 19 to 38 show the robot making a little movement and from here until frame 133, the fingers make a grasping movement (Figure 13). As in previous experiments, the rest of the frames do not supply new data of interest for the detection of deformation and the control of grasping movements. In this experiment, little variability in the tracker's evolution is observed in Figure 13a, in contrast with a great variability in the curvature map, shown in Figure 12b.

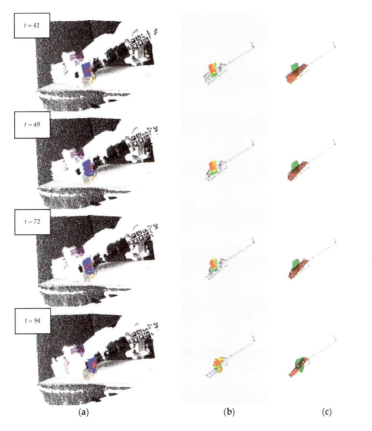

(a)   (b)   (c)

Figure 13. Some frames of a video sequence of a plastic glass: (**a**) The points cloud of the scene and the tracking process in the object's detection; (**b**) A curvature map to measure the deformation level of the glass's surface; (**c**) The intersection to measure the visual contact between the robot hand and the object.

7.4. Experiment 4: Shoe Insole

The last experiment was carried out in a different environment with a different ambient lighting. In contrast to the three previous experiments, here, both robot hand and RGBD sensor are mounted at the end of industrial robot arms (Figure 14a).

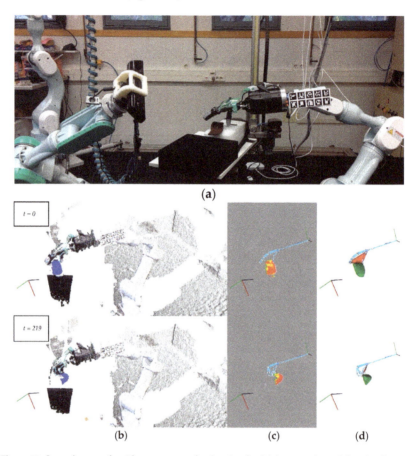

Figure 14. Some frames of a video sequence of a shoe insole: (**a**) An overview of the visual system working in a new workspace; (**b**) The points cloud of the scene and the tracking process in the object's detection; (**c**) A curvature map to measure the deformation level of the glass's surface; (**d**) The intersection to measure the visual contact between the robot hand and the object.

Then, the recognition process of search a marker board is no longer necessary since The fingers pose are calculated from the spatial location of the industrial robot arm which is equipped with de RGBD sensor as:

$$^{C}T_{f_i} = {}^{C}T_R \times {}^{R}T_E \times {}^{E}T_{FA} \times {}^{FA}T_{f_i} \qquad (15)$$

where R is the robot frame located at its base and E is the effector-robot. ${}^{C}T_R \times {}^{R}T_E$ are the transformation of the robot frame with respect to the camera and the transformation of the effector-robot with respect to the robot frame, respectively.

This experiment (Figure 14b) has 250 frames as above experiments. By comparison with previous Experiments 6.1–6.3, the main challenge of this experiment is to prove that our visual system works when a scenario is different. Here, hand-robot is mounted at the end of industrial robot, the ambient lighting has more intensity, the pose of camera has also changed and the object is another.

8. Discussion and Analysis of the Results

The previous experiments, shown in Figures 11–13 were analysed to determine the main capability of how well our visual monitoring strategy works in discovering deformations and determining a good grasping motion. This goal was achieved by the two new methods presented here. First, the proposed method of *"detection deformation"* with *"Δ-modulation"* is able to measure deformations with regards to time, from RGBD data using a curvature distribution. The second method of *"visual monitoring of grasping"*, based on the intersection of regions, is used to determine whether an object is being correctly grasped, with the absence of contact between the robot hand and the object. The results of this analysis are shown in Figure 15.

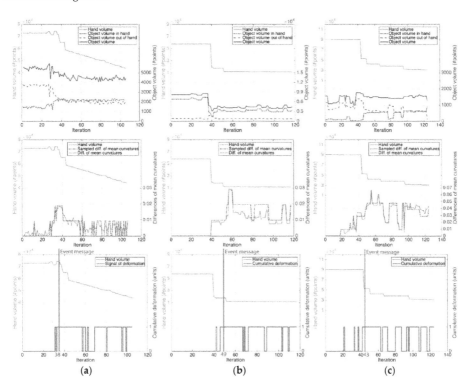

Figure 15. Grasping and deformation results: (**a**) Sponge; (**b**) Brick; (**c**) Plastic glass.

For each experiment, as it is noted above, the results of the visual monitoring approach are shown in the first row of the figure. The charts show the relationship between the $H(t)$ and $O(t)$ volumes, as well as $I(t)$ and $O(t) - I(t)$. It is clear that $H(t)$ tends to decrease due to the hand's movement. This is a typical case of a grasping task in which the robot hand is closed in order to wrap around and grip an object. In contrast, $O(t)$ does not follow that same trend. For example, in the case of the sponge, its volume does not change much. This fact is due because the area of the object located close to the fingers is compressed and so the other parts located further away are expanding. Furthermore, these charts show the grasp-monitoring results. Thereby, if the red dashed lines $I(t)$ tend towards zero, then the hand is losing its grip on the object. Consequently, $O(t) - I(t)$ tending towards $O(t)$ means the same thing. Apart from that result, the second row of the figure shows how the average of the curvature distribution changes over time. These curves show the relationship between the average of the curvature distribution at the initial time and the rest of time t, ensuring the consistency of the results shown in the third row of Figure 15.

Moreover, the results of our method "*deformation detection*" with "Δ-*modulation*" are shown in the third row of Figure 15. The three plots present a view of how the deformation signature evolves in connection with $H(t)$, *i.e.*, how the object is deformed while the hand is working. These results are used to analyse the sensitivity, specificity and accuracy (Table 1) of our approach for detecting deformations by using the hand's movement. Three sequences for each object (nine tests) with 250 frames each one was used to carry out this study.

Table 1. Information retrieved from the charts of Figure 15 for 3 tests of each object were carried out with different hand poses.

Experiment	Sensitivity [1] Average	Specificity [2] Average	Accuracy [3] Average
Tests 1–3: Sponge	0.7750	1	0.8286
Tests 4–6: Brick	0.8333	1	0.8889
Tests 7–9: Plastic glass	0.8837	0.7160	0.7742

[1] Sensitivity measures the proportion of positives that are correctly identified as such; [2] Specificity measures the proportion of negatives that are correctly identified as such; [3] Accuracy is the level of measurement that yields true and consistent results.

The three statistical features give information about the behaviours of our method. The sensitivity measures the proportion of times that the visual system has determined correctly that the object is deformed. The sensitivity is computed as:

$$sensitivity = \frac{TP}{P} = \frac{TP}{TP + FN} \tag{16}$$

where P is the number of times that the system detects a deformation, in contrast, N is the number of times that the system does not detect deformation. Then TP is the number of times that the system succeeds about deformation and FP is the number of times that the system fails detecting deformations. Similarly, TN is the number of times that the visual system succeeds when this determines if the object is not deformed and FN is the times that the visual system confuses an undeformed object with a deformed object. Using this relation, the specificity is defined as:

$$specificity = \frac{TN}{N} = \frac{TN}{FP + TN} \tag{17}$$

In contrast with the sensitivity, specificity measures the proportion of times that the visual system has determined correctly that the object is not deformed. Also, the accuracy is calculated in this work as:

$$accuracy = \frac{TP + TN}{P + N} \tag{18}$$

Ideally, the behaviour of the deformation signal in these experiments should be like a step function with just one rising edge, since the robot hand makes only one grasping movement. This rising edge would have to occur in the moment at which the hand starts to grasp the object and its $H(t)$ starts to change. Hence "1" values before the ideal rising edge are considered to be false positive "0" values; after that, they are considered to be false negative ones.

9. Conclusions

The use of visual data in grasping tasks and intelligent robot manipulation is still an emerging topic. In the past, force and tactile sensors have often been used for these tasks without considering visual information. Although there are some approaches that use some visual data, they were just designed to recognize the object to be grasped or manipulated but never to check or supervise the task during the grasping process in order to detect when deformations occurred, or when there was a loss of contact if it was caused by displacement of the object within the hand. In this work, the experiments focused on 3D sensors, such as RGBD, combined with a multi-fingered robot hand, without considering tactile data from another kind of sensor. Furthermore, there are still some challenges remaining when the grasping tasks are directed to solid objects as much as to elastic ones.

The proposed approach is motivated by the need to develop new strategies to solve problems throughout the grasping tasks and to supply robustness. Thus, this paper presents a novel sensorized approach in order to carry out robot-hand manipulation of an unmarked object whose flexibility properties are unknown. This new algorithm is based on geometric information of the object and the curvature variations on its surface, and it is used to procure suitable grasping motions even when the object is being deformed. Using visual data with the help of robot-hand kinematics, this new approach allows us to check when a deformation is being caused by the multi-fingered robot and whether there is a lack of contact between the hand and object from a visual sensing point of view. The experiments show the behaviour of the methods in several grasping tasks in which the object's deformation can be measured using a visual sensor. In the future, our approach could be combined with a hybrid tactile/visual control system for a reactive adjustment of pressure and the contact of fingers, whenever tactile data are not adequate for the manipulation process.

Acknowledgments: The research leading to these result has received funding from the Spanish Government and European FEDER funds (DPI2015-68087R), the Valencia Regional Government (PROMETEO/2013/085) as well as the pre-doctoral grant BES-2013-062864.

Author Contributions: The contribution presented in this work is the result of the joint work of all authors comprising the research team. Each of the members contributed in varying degrees to each step of this research work, such as analysis, design, development, implementation and the testing of the system. In particular, the research concept, the design of the system's architecture and the design of tests were carried out mostly by Carlos Mateo, Pablo Gil and Fernando Torres. In particular, Carlos Mateo and Pablo Gil devised the implementation of an approach based on the new method of deformation detection and the intersection of regions, and Carlos Mateo tested the system using data and household objects. Furthermore, all of the authors analysed the results and provided insights into the integration of the method in a vision system for grasping and intelligent manipulation. All of the authors reviewed the final version of the manuscript and approved its publication.

Conflicts of Interest: The authors declare no conflict of interest.

References

1. Saut, J.P.; Sidobre, D. Efficient models for grasp planning with a multi-fingered hand. *Robot. Auton. Syst.* **2012**, *60*, 347–357. [CrossRef]
2. Popovic, M.; Kraft, D.; Bodenhagen, L.; Baseski, E.; pugeault, N.; Kragic, D.; Asfour, T.; Krüger, N. A strategy for grasping unknown objects based on co-planarity and colour information. *Robot. Auton. Syst.* **2010**, *58*, 551–565. [CrossRef]
3. Montesano, L.; Lopes, M. Active learning of visual descriptors for grasping using non-parametric smoothed beta distributions. *Robot. Auton. Syst.* **2012**, *60*, 452–462. [CrossRef]
4. Bohg, J.; Morales, A.; Asfour, T.; Kragic, D. Data-driven grasp synthesis—A survey. *IEEE Trans. Robot.* **2014**, *30*, 289–309. [CrossRef]
5. Chitta, S.; Sturm, J.; Piccoli, M.; Burgard, W. Tactile sensing for mobile manipulation. *IEEE Trans. Robot.* **2011**, *27*, 558–568. [CrossRef]
6. Yousef, H.; Boukallel, M.; Althoefer, K. Tactile sensing for dexterous in-hand manipulation in robotics—A review. *Sens. Actuators A Phys.* **2011**, *167*, 171–187. [CrossRef]
7. Kappassov, Z.; Corrales, J.A.; Perdereau, V. Tactile sensing in dexterous robot hands—Review. *Robot. Auton. Syst.* **2015**, *74*, 195–220. [CrossRef]
8. Morales, A.; Prats, M.; Felip, J. Sensors and methods for the evaluation of grasping. *Grasping Robot.* **2013**, *10*, 77–104.

9.   Li, Q.; Elbrechter, C.; Haschke, R.; Ritter, H. Integrating vision, haptics and proprioception into a feedback controller for in-hand manipulation of unknown objects. In Proceedings of the IEEE/RSJ International Conference on Intelligent Robots and Systems (IROS), Tokyo, Japan, 3–7 November 2013; pp. 2466–2471.

10. Kim, H.; Han, I.; You, B.-J.; Park, J.-H. Towards cognitive grasping: Modelling of unknown objects and its corresponding grasp types. *Intell. Serv. Robot.* **2011**, *4*, 159–166. [CrossRef]

11. Bimbo, J.; Seneviratne, L.; Althoefer, K.; Liu, H. Combining touch and vision for the estimation of and object's pose during manipulation. In Proceedings of the IEEE/RSJ International Conference on Intelligent Robots and Systems (IROS), Tokyo, Japan, 3–7 November 2013; pp. 4021–4026.

12. Mkhirtayan, A.; Burschka, D. Vision based haptic multisensory for manipulation of soft, fragile objects. In Proceedings of the IEEE Sensors, Taipei, China, 28–31 October 2012; pp. 1–4.

13. Cappelleri, D.J.; Piazza, G.; Kumar, V. Two dimensional, vision based N force sensor for microrobotics. In Proceedings of the IEEE International Conference on Robotics and Automation (ICRA), Kobe, Japan, 12–17 May 2009; pp. 1016–1021.

14. Luo, Y.; Nelson, B.J. Fusing force and vision feedback for manipulating deformable objects. *J. Robot. Syst.* **2001**, *18*, 103–117. [CrossRef]

15. Khalil, F.F.; Curtis, P.; Payeur, P. Visual monitoring of Surface deformations on objects manipulated with a robotic hand. In Proceedings of the IEEE International Workshop on Robotic and Sensors Environments (ROSE), Phoenix, AZ, USA, 15–16 October 2010; pp. 1–6.

16. Leeper, A.; Hsiao, K.; Chu, E.; Salisbury, J.K. Using near-field stereo vision for robotic grasping in cluttered environments. *Exp. Robot.* **2014**, *79*, 253–267.

17. Boonvisut, P.; Cenk, M.C. Identification and active exploration of deformable object boundary constraints through robotic manipulation. *Int. J. Robot. Res.* **2014**, *33*, 1446–1461. [CrossRef] [PubMed]

18. Calli, B.; Walsman, A.; Singh, A.; Srinavasa, S.; Abbel, P.; Dollar, A.M. Benchmarking in manipulation research. *IEEE Robot. Autom. Mag.* **2015**, *22*, 36–52. [CrossRef]

19. Jia, Y.-B.; Guo, F.; Lin, H. Grasping deformable planar objects: Squeeze, stick/slip analysis, and energy-based optimalities. *Int. J. Robot. Res.* **2014**, *33*, 866–897. [CrossRef]

20. Lin, H.; Guo, F.; Wang, F.; Jia, Y.-B. Picking up soft 3D objects with two fingers. In Proceedings of the IEEE International Conference on Robotics and Automation (ICRA), Hong Kong, China, 31 May–7 June 2014; pp. 3656–3661.

21. Shadow Robot, "Dexterous Hand". Available online: http://www.shadowrobot.com/products/dexterous-hand/ (accessed on 21 January 2012).

22. Yoshikawa, T. Multifingered robot hands: Control for grasping and manipulation. *Ann. Rev. Control* **2010**, *34*, 199–208. [CrossRef]

23. Navarro-Alarcon, D.; Liu, Y.H.; Guadalupe-Romero, J.; Li, P. Model-free Visually servoed deformation control of elastic objects by robot manipulators. *IEEE Trans. Robot.* **2013**, *29*, 1457–1468. [CrossRef]

24. Navarro-Alarcon, D.; Liu, Y.H.; Guadalupe-Romero, J.; Li, P. On the visual deformation servoing of compliant objects: Uncalibrated control methods and experiments. *Int. J. Robot. Res.* **2014**, *33*, 1462–1480. [CrossRef]

25. Berenson, D. Manipulation of deformable objects without modeling and simulating deformation. In Proceedings of the IEEE/RSJ International Conference on Intelligent Robots and Systems, Tokyo, Japan, 3–7 November 2013; pp. 1–8.

26. Alt, N.; Steinbach, E. Navigation and Manipulation Planning using a Visuo-haptic Sensor on a Mobile Platform. *IEEE Trans. Instrum. Meas.* **2014**, *63*, 1–13. [CrossRef]

27. Sun, L.; Aragon-Camarasa, G.; Rogers, S.; Siebert, J.P. Accurate garment surface analysis using and active stereo robot head with application to dual-arm flattening. In Proceedings of the IEEE International Conference on Robotics and Automation (ICRA), Seattle, MA, USA, 26–30 May 2015; pp. 185–192.

28. Khoshelham, K.; Elberink, S.O. Accuracy and Resolution of Depth Data for Indoor Mapping Applications. *Sensors* **2012**, *12*, 1437–1454. [CrossRef] [PubMed]

29. Khalfaoui, D.; Seulin, R.; Fougerolle, Y.; Fofi, D. An efficient method for fully automatic 3D digitization of unknown objects. *Comput. Ind.* **2013**, *64*, 1152–1160. [CrossRef]

30. Delgado, A.; Jara, C.A.; Torres, F.; Mateo, C.M. Control of Robot Fingers with Adaptable Tactile Servoing to Manipulate Deformable Objects. In Proceedings of the Robot 2015: Second Iberian Robotics Conference, Lisboa, Portugal, 19–21 November 2015; pp. 81–92.

31. Kragic, D.; Miller, A.T.; Allen, P.K. Real-time tracking meets online grasp planning. In Proceedings of the on IEEE International Conference on Robotics and Automation, Seoul, Korea, 21–26 May 2001; Volume 3, pp. 2460–2465.
32. Mateo, C.M.; Gil, P.; Torres, F. Visual perception for the 3D recognition of geometric pieces in robotic manipulation. *Int. J. Adv. Manuf. Technol* **2015**, 1–15. [CrossRef]
33. Corrales, J.A.; Torres, F.; Perderau, V. Finger Readjustment Algoritm for Object Manipulation Based on Tactile Information. *Int. J. Adv. Robot. Syst.* **2013**, *10*, 1–9. [CrossRef]
34. Corrales, J.A.; Perderau, V.; Torres, F. Multi-fingered robotic hand planner for object reconfiguration through a rolling contact evolution model. In Proccedings of the IEEE International Conference on Robotics and Automation (ICRA), Karlsruhe, Germany, 6–10 May 2013; pp. 625–630.
35. Mateo, C.M.; Gil, P.; Torres, F. A Performance evaluation of surface normal-based descriptors for recognition of objects using CAD-Models. In Proccedings of the 11th International Conference on Informatics in Control, Automation and Robotics (ICINCO), Vienna, Austria, 2–4 September 2014; pp. 428–435.
36. Aldoma, A.; Marton, Z.C.; Tombari, F.; Wohlkinger, W.; Potthast, C.; Zeisl, B.; Rusu, R.B.; Gedikli, S.; Vincze, M. Tutorial: Point cloud library: Three dimensional object recognition and 6DoF pose estimation. *IEEE Robot. Autom. Mag.* **2012**, *19*, 80–91. [CrossRef]
37. Papon, J.; Abramov, A.; Schoeler, M.; Woergoetter, F. Voxel Cloud Connectivity Segmentation—Supervoxels from PointClouds. In Proceedings of the IEEE Conference on Computer Vision and Pattern Recognition (CVPR), Porland, OR, USA, 23–28 June 2013; pp. 2027–2034.
38. Philps, T.Y.; Rosenfeld, A.; Sher, A.C. O(log *n*) Bimodality Analysis. *Pattern Recognit.* **1989**, *22*, 741–746. [CrossRef]
39. Garrido-Jurado, S.; Muñoz-Salinas, R.; Madrid-Cuevas, F.J.; Marín-Jiménez, M.J. Automatic generation and detection of highly reliable fiducial markers under occlusion. *Pattern Recognit.* **2014**, *47*, 2280–2292. [CrossRef]
40. Fox, D. KLD-Sampling: Adaptive Particle Filters. In Proceedings of the Neural Information Processing Systems Conference (NIPS), Vancouver, BC, Canada, 3–8 December 2001; pp. 713–720.
41. Fox, D. Adapting the sample size in particle filters through KLD-sampling. *Int. J. Robot. Res.* **2003**, *22*, 985–1003. [CrossRef]
42. Mateo, C.M.; Gil, P.; Torres, F. Analysis of Shapes to Measure Surface: An approach for detection of deformations. In Proceedings of the 12th International Conference on Informatics in Control, Automation and Robotics (ICINCO), Colmar, France, 21–23 July 2015; pp. 1–6.

Article

Intelligent Multisensor Prodder for Training Operators in Humanitarian Demining

Roemi Fernández [1,*], Héctor Montes [1,2] and Manuel Armada [1]

[1] Centre for Automation and Robotics (CAR) CSIC-UPM, Ctra. Campo Real, km. 0,200, La Poveda, Arganda del Rey, Madrid 28500, Spain; hector.montes@car.upm-csic.es (H.M.); manuel.armada@csic.es (M.A.)

[2] Facultad de Ingeniería Eléctrica, Universidad Tecnológica de Panamá, Panamá 0819, Panamá

* Correspondence: roemi.fernandez@car.upm-csic.es; Tel.: +34-918-711-900

Academic Editor: Gonzalo Pajares Martinsanz
Received: 23 May 2016; Accepted: 20 June 2016; Published: 24 June 2016

Abstract: Manual prodding is still one of the most utilized procedures for identifying buried landmines during humanitarian demining activities. However, due to the high number of accidents reported during its practice, it is considered an outmoded and risky procedure and there is a general consensus about the need of introducing upgrades for enhancing the safety of human operators. With the aim of contributing to reduce the number of demining accidents, this paper presents an intelligent multisensory system for training operators in the use of prodders. The proposed tool is able to provide to deminers useful information in two critical issues: (a) the amount of force exerted on the target and if it is greater than the safe limit and, (b) to alert them when the angle of insertion of the prodder is approaching or exceeding a certain dangerous limit. Results of preliminary tests show the feasibility and reliability of the proposed design and highlight the potential benefits of the tool.

Keywords: intelligent feedback prodder for training; force exerted; prodder's angle; humanitarian demining

1. Introduction

Unfortunately, after the end of a war, it is quite frequent that the affected populations have to confront the legacy of landmines planted in its territory. Anti-personnel mines, anti-tank mines, cluster munitions, ERW and IEDs can remain active for decades, and hurting or killing indiscriminately any living being that accidentally activates them. The most recent report from the Landmine and Cluster Munition Monitor organization indicates that in 2014 a total of 3678 mine/ERW casualties were recorded, making an average of 10 casualties per day and a 12% increase from 2013. From this total, the 80% were civilians, and 39% were children. Taking into account that in many states and areas, numerous casualties go unrecorded, it is possible to affirm that the true casualty figure is significantly higher [1].

Apart from the human casualties, the presence of landmines also produces negative economic effects, as it denies access to the affected areas and their resources, causing deprivation and social problems among the affected populations. Therefore, elimination of antipersonnel mines is a vital requirement for the recovery of the affected regions [2]. Humanitarian demining, unlike the military demining, where a clearance rate of 80% to 90% is well-accepted [3], requires the complete removal of all mines, with a demining rate of 100%, so that the cleared minefields may be reverted to normal use [4–6]. This turns humanitarian demining into a long and difficult process.

Today, a mine-clearer's work still depends on metal detectors and prodders, with all the danger this entails. Prodders are mainly used as complement to metal detectors, so that once a possible target has been detected, the prodder allows to locate it precisely in the terrain, providing information on the depth, size, shape and orientation of the target before attempting to safely excavate it. Finding mines with a prodder involves pushing tool into the ground and relaying on tactile feedback to identify an obstruction that maybe a mine [7]. An expert operator is even able to characterize explosives and housing materials with it. Nevertheless, prodding of landmines is a major cause of demining accidents, especially in those countries where the soil is hard or rocky.

Conventional prodder has been improved from the original soldier's bayonets to a lightweight, non-magnetic, and wear-resistant instrument consisting of a stainless rod and a stainless or wooden handle. There are several commercial models that vary in the style of the handle, the length and the hand protection. In addition, there have been several attempts to enhance conventional prodders with advanced functionalities. For instance, DEW Engineering and Development Ltd manufactured for a brief period a prodder with an ability to discriminate between plastic, rock and metal. The SmartProbe™ (DEW Engineering and Development Ltd., Ottawa, Canada) was based on the use of an acoustic pulse to characterize the material under contact. However, several shortcomings in the ergonomic design, ruggedness and performance, discouraged the acquisition of the SmartProbe™ for field use [8].

In the scientific and academic community, there are also several relevant prototypes under research. First solutions included the integration of a miniature ultrasonic sensor capable of materials characterization into the hand probe [9,10]. The tip of the prodder was used as a waveguide for transmitting the ultrasonic pulse and receiving reflected energy from the examined object. Recognition among diverse object was performed by using statistical analysis of large number of samples collected from different objects under varying conditions. However, more recent prototypes are based on PZTs and accelerometers, such as the device proposed in [11]. This prodder vibrates a buried target via the PZT actuator. Then, depending on the stiffness of the target, specific accelerations return through the stick and are measured by the accelerometer mounted on the prodder. Thus, the stiffness of the buried target being in touch with the prodder tip can be determined by analyzing the acceleration signals. In [12,13], the authors present a novel smart prodder that is capable of recognizing the material of the suspected object. This prodder consists of: (i) a pair of piezoelectric transducers, used as actuator and sensor; (ii) a force sensor to guarantee a constant application force, and consequently, a good repetitiveness of the piezoelectric response; and (iii) an inclinometer to improve the reliability of the contact. A prodder system based on tactile augmentation is described in [14]. The device is equipped with an accelerometer to sense vibrations and a piezoelectric actuator to amplify the measured acceleration and to generate the tactile sensation to the operator. A different solution is presented in [15,16], where authors state that basic parameters of landmines buried in sandy desert can be accurately estimated according to the contact pressure sensed by a rolling cylinder and by using a PNN. The rolling cylinder is endowed with a with fixable pressure mat, and is pushed to roll over the sand with a constant pressure (less than the activation pressure of any landmine). In consequence, a pressure distribution is generated on the sand surface due to the difference between the Young's modulus of the sand and the landmine. This pressure distribution is measured by the rolling cylinder and then utilized as input for the PNN, which is responsible of the landmine characterization.

The design of innovative prodders may contribute to increase the rate of mines detected but does not definitively increase the safety of the deminers. Training is then one of the most crucial aspects in order to improve the safety and effectiveness of the landmine detection activities performed by human operators [17]. However, in the literature there is not any study that has been devoted to the improvement of training tools utilized during prodding activities. Prodding has the disadvantage that unlike other aspects of training, the use of excessive force cannot be detected simply by a supervisor observing the trainee. Some studies [18] involving field measurements of the force exerted by the operators showed that deminers, and even senior training staff, had no real idea of the force they were using and consistently underestimated the force they were exerting by large amounts. Therefore, to alleviate this situation, this paper presents an intelligent prodder that provides useful information to the deminers in two critical issues: (a) the amount of force exerted on the target and if it is greater than the safe limit and, (b) to alert them when the angle of insertion of the prodder is approaching or exceeding a certain dangerous limit. In this way, the tool will contribute to improve the deminers' competencies and consequently, will help to reduce the number of demining accidents during close-in detection tasks. The work presented here has been carried out within the framework of the TIRAMISU project, funded by the European commission in the 7th Framework Programme.

The rest of the paper is organized as follows. Section 2 describes the design and implementation of the proposed intelligent multisensor prodder for training. Section 3 presents the results obtained from the experimental tests carried out with a prototype of the tool. Section 4 discusses the main results of this work and finally, Section 5 summarizes major conclusions.

2. Materials and Methods

The first step accomplished for the design and development of the proposed tool was the analysis of the present status of training programs held in humanitarian demining for close-in detection tasks, the current role of training tools, the identification of the current problems confronted during the training sessions, and the gathering of the requirements and needs expressed by the end-users. The compiled information was based on both findings from literature and on dedicated interviews and workshops with representatives of fifteen different institutional and private organizations concerned with training for humanitarian demining. International standards (IMAS 06.10), guidelines (CWA 15465), and national standards (NMAS) outline training contents and curricula. Close-in detection and disposal are well covered in the existing standards and regulations for EOD training, though particularly the duration of trainings varies considerably. The equipment and material used during close-in detection training resembles the ones utilized in actual operation as much as possible. Therefore, metal detectors and prodders are the tools most commonly used. However, an analysis of accidents in humanitarian demining shows that 37% of them could have been avoided by better training in close-in detection methods. Thus, after analyzing all the compiled information, the key teaching points that were identified for prodders are the following [19]:

- The prodder shall be inserted into the soil using an angle less than 30°–45° (this limit angle varies depending on local conditions of soil) and every 2–3 cm for each trial, so it can hit the mine laterally. A greater angle could be unsafely, since the detonator of the mine could be achieved by the prodder.
- Probing shall be done softly and gradually so a fuse would not be activated due to an excessive pressure.
- In order to conduct a complete and safe search and marking of the spots where mines and UXO have been found, probing frequency is as follows:

   ○ 4–5 probes on 1 dm^2 when looking for antipersonnel mines
   ○ 15–20 probes on 1 m^2 when looking for antitank mines.

- Prodding needs a lot of experience and requires particular skill in hard, stony ground.

Thus, an intelligent multisensory system that provides information about the amount of force exerted, and alerts deminers when the prodder's angle is approaching or exceeding a certain limit is proposed for the improvement of training tasks and consequently for reducing demining accidents [20]. The proposed tool consists of an instrumented prodder, a USB DAQ module, an electronic module for signals conditioning and a HMI. All basic parts of the instrumented prodder (sensors, the rod with the sharp spike, the handle and the extension) are separable with the ability of replacing different extensions in order to obtain different versions of the prodder, depending on the demining training needs.

For the design of the instrumented prodder, two main types of sensors have been selected and evaluated: a compression load cell and a wireless IMU. Table 1 summarizes the main technical specifications of the selected compression load cell, whereas Figure 1 shows the custom-made installation that has been designed to embed this compression load cell in the prodder. It is possible to see that the force sensing capability has been incorporated in the connection between the handle and the rod.

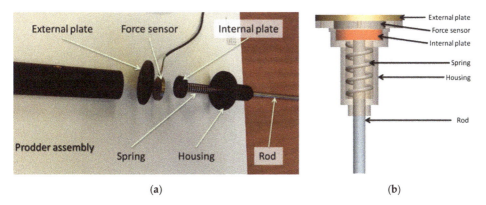

Figure 1. Custom-made installation of the load cell: (**a**) Real disassembled view of the different elements that compose the described design; (**b**) Detailed drawings of the custom-made installation.

Table 1. Main technical specifications of the compression load cell.

Non-linearity	Hysteresis	Thermal Zero Shift	Thermal Sensitivity Shift
±1% FSO	±1% FSO	±2.5 mV/50 °C	±2.5%/50 °C

Deflection at "FS"	Operating Temperature	Thickness	Diameter
<0.013 mm nom.	(−40 to 120) °C	3.81 mm	25.4 mm

As in the first 10% of the full scale of operation, the sensor has a non-linear behaviour, a mechanism that preloads the sensor has been incorporated. This simple mechanism consists of a spring and two plates that enclose the sensor. In this way, the sensor will be functioning always in its linear region of operation. The shape and the dimensions of the selected load cell allows and easy adaptation and installation to the currently used prodders, and provides proper reliability, sensitivity and resolution for the required application.

Table 2 gathers the main technical characteristics of the IMU that was also installed in the prototype. This unit is able to measure the following parameters:

- Pitch, roll and yaw angles.
- Angular and linear velocities in the 3 axes of the Cartesian coordinate system.
- Accelerations in the 3 axes of the Cartesian coordinate system.

Table 2. Main technical specifications of the IMU.

Parameters	Orientation Performance	Parameters	Angular Velocity	Acceleration
Dimensions	pitch, roll, yaw	Dimensions	3 axes	3 axes
Full scale	±180°	Full scale	±120°/s	±1600 m/s²
Angular resolution	0.05°	Linearity	0.1% FS	0.2% FS
Dynamic accuracy	2° RMS	Alignment error	0.1°	0.1°

The electronic module for signals conditioning is responsible of filtering and amplifying the force sensor output in order to meet the requirements of the next stage, in which the DAQ module converts the resulting analogic signal into a digital one for further processing. The USB DAQ module is connected to a PC where the HMI is installed, so that the HMI can be able to process, record and display the trainee data. The sampling frequency was chosen to be 100 Hz for the IMU signals and 500 Hz for the force signal. Figure 2 shows a block diagram of the intelligent feedback prodder with the IMU and the force sensor, the electronic module for signals conditioning and the DAQ module. Figure 3 displays a real photo with the main components of the proposed tool.

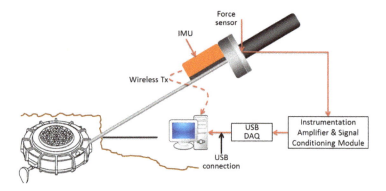

Figure 2. Main components of the proposed tool.

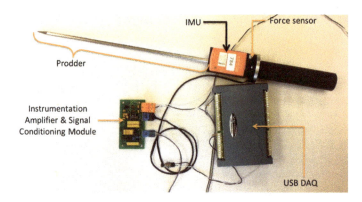

Figure 3. Main components of the proposed tool.

In addition, it is important to mention that the instrumented prodder is in conformity with the technical requirements stipulated in books of rules and regulations for devices and equipment used in Humanitarian Demining [21]. These technical requirements are the following:

- The stick with the sharp spike is made of stainless steel—prochrome (antimagnetic), the extensions are made of aluminum alloy, and the handle made of steel coated with rubber or aluminum alloy.
- The length of the stick with sharp spike is at least 400 mm.
- The thickness of the stick with sharp spike (diameter) is at least 8 mm.
- The length of the handle is at least 90 mm.
- The length of the extension is at least 400 mm.
- The outer diameter of the extension is at least 20 mm.
- The total mass of the feedback prodder is less than 600 g.
- The prodder has a rugged, sealed construction for field-portability.

- The feedback prodder for training is cost-effective, reliable and safe.

On the other hand, the HMI is responsible of collecting the data acquired by the sensors of the instrumented prodder, processing, analyzing and monitoring the measured performance variables, and presenting the essential information required during the training sessions, including the activation of relevant alarms when at least one of the following conditions is fulfilled:

- The operator exceeds a pre-defined maximum force. The value of this maximum force will depend on the soil type, the soil conditions and the kind of target.
- The angle of insertion of the prodder exceeds 45°, since that could activate the mine detonation. Nevertheless, this limit angle can be adjusted depending on the local conditions.

The HMI design process consisted of three differentiated phases:

- Assessment of HMI needs and requirements. Thus, the design process started with a review of the SOP, the functional needs, system requirements and the objectives of the training activities.
- Design of the graphical user interface. Thus, main functionality and key components of the HMI were identified and defined, and a draft version of the HMI was developed.
- Finally, system capabilities were evaluated to ensure that the HMI design would fulfil all the needs and requirements identified for it.

Figure 4 shows a screenshot of the designed HMI. The HMI console is the principal mechanism though which instructor interacts with and controls the performance of trainees. Some important sections of the HMI console are described below [22].

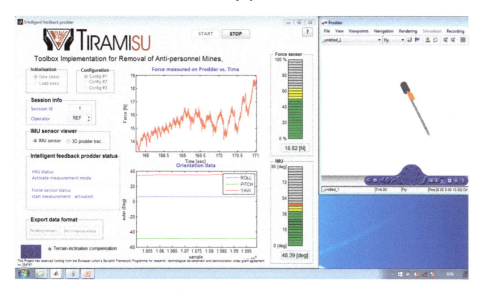

Figure 4. HMI for the intelligent feedback prodder.

2.1. Configuration

This section provides the possibility of loading different configuration files for modifying the objectives of the training session (reference values for the force exerted and the insertion angle of the prodder) according to the soil type and the kind of target to be detected (see Figure 4). Thus, for instance, in highly compact, hard or rocky soil, objectives are set to achieve an approach at a shallow angle. Force reference values are defined taking into consideration the type of mine, its dimensions and its activation load or activation pressure [16,23,24].

2.2. IMU Sensor Viewer

This section contains two radio buttons that are not mutually exclusive (see Figure 4). The first one, named "IMU sensor" opens and displays a window for calibrating the IMU. The second one, named "3D prodder tracking" opens and displays the Prodder Monitoring window, which is described below.

2.3. Prodder Monitoring

This VRML graphic reconstructs in real time the orientation of the prodder carried out by the human operator. The graphic enables the instructor to check if each prodding is being performed with a proper angle of insertion. Figure 5 shows different snapshots from the prodder monitoring.

Figure 5. Different snapshots from the prodder monitoring.

2.4. Force and Orientation Graphics

These two graphics display in real time the force exerted by the human operator while prodding, in N, and the roll, pitch, yaw angles in degrees, describing the orientation of the prodding during the training session (see Figure 6).

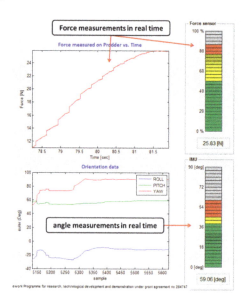

Figure 6. Close-up view of the force and angle data provided by the HMI.

2.5. Force and Angle Data

In this section, data acquired by the intelligent feedback prodder is turned into useful information that will help the instructor to monitor the current situation. Two performance variables are utilized for this purpose: the force exerted by the human operator while prodding in N, and the angle of insertion of the prodder in degrees, which is given by the pitch angle measured independently of the roll and yaw angles. It is also important to mention that this angle is measured with respect to the ground by incorporating a second IMU that is placed on the terrain where prodding is being conducted, and by indicating this option in the designed HMI (see Figure 4). In this way, the second IMU measures the inclination of the terrain, given once again by the pitch angle, and this measurement is then utilized in the HMI for compensating the resulting angle of insertion of the prodder that is provided to the user. Analogic representation of force and angle values, indicating their position relative to normal, abnormal and alarm conditions are displayed on the graphical user interface. The alarms included for each variable will enable the operator to quickly detect values outside the safety range, so he wouldn't have to relay in his memory and mentally compare each value to its corresponding defined range to discover deviations of trainee objectives. In addition, colors are utilized in two bar graphics to indicate if the performance is holding or not within the training objectives: green is used for indicating that all the evaluated variables are within the training objectives, yellow for warning that the values are starting to deviate from the goals and red for values out of the defined safety ranges (see Figure 6).

Therefore, with the proposed easy-to-use interface, the instructor is capable of:

- Monitoring the performance variables of the operator, which are the force exerted and the angle of insertion of the prodder.
- Recording all the acquired information in a database.
- Assessing the performance operation of the trainees.
- Recording long data-runs without data loss.

3. Results

Several experimental tests have been carried out under controlled conditions in order to evaluate the technical performance of the proposed tool. Figure 7 shows the prototype that has been used during the tests. This prototype consists of an instrumented prodder, a box that contains the signal conditioning module and the DAQ module, and a HMI. The HMI is installed on a fully rugged laptop that has autonomy of approximately 3 h working continuously. The box that contains the electronic modules has autonomy of 6 h. In addition, the instrumented prodder has been designed in such a way that its length, thickness and total mass are in conformity with technical requirements stipulated in books of rules and regulations for equipment used in Humanitarian Demining. Therefore, the proposed training tool can be easily deployed on the field (see Video S1).

The first part of the experimentation was devoted to the assessment of the force feedback provided by the intelligent prodder. Figure 8 displays several force measurements obtained with the instrumented prodder. In this test, the prodder was located vertically and was gradually loaded with masses of 1.25 kg, 2.5 kg, 3 kg, 4 kg and 5 kg. The mean values measured by the sensory system were 12.5 N, 25 N, 30 N, 40 N and 50 N, approximately. Figure 9 shows the absolute and the relative errors for the obtained force measurements. Thus, results provided by the intelligent feedback prodder demonstrate that the force sensor has an accurate dynamic response.

For the second part of the experimentation, a robotic arm was utilized to handle the intelligent feedback prodder and to carry out repetitive and exhaustive robotic testing (see Video S2). In this way, it was possible to control the angle in each prodding with high accuracy and to use this information as ground truth data during the comparison with the results measured by the intelligent sensory system. Figure 10 shows the set-up utilized for this experimental phase, which consists of the instrumented prodder installed on the robotic arm and connected to the HMI through the signal conditioning and DAQ module and the wireless LAN network connection implemented between the IMU and the HMI.

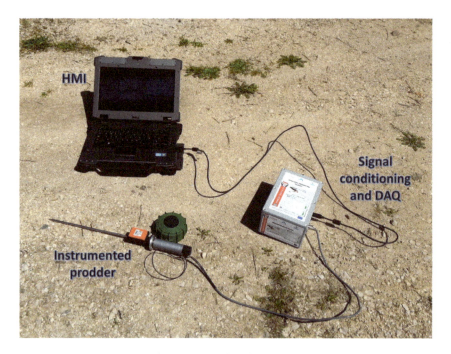

Figure 7. Prototype of the propose tool used during the experimental stage.

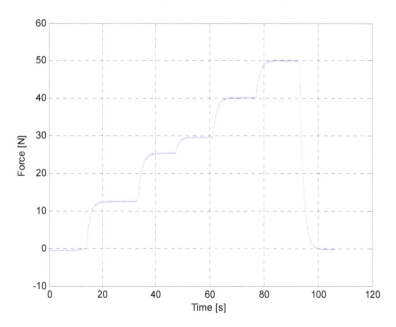

Figure 8. Measurements obtained with the force sensor of the instrumented prodder.

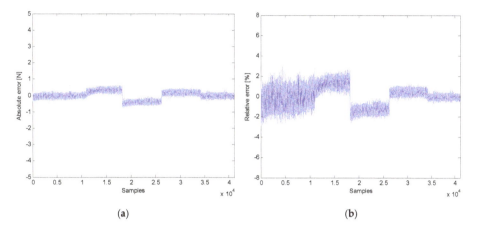

(a)   (b)

Figure 9. Force measurement errors: (**a**) Absolute error; (**b**) Relative error.

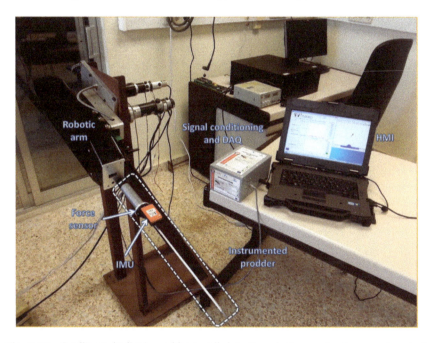

Figure 10. Intelligent feedback prodder installed in the robotic arm for the second part of the experimentation.

Figures 11 and 12 illustrate results of two of the experimental tests that were carried out. Black lines represent the reference angle trajectory programmed for the first degree of freedom of the robotic arm, which corresponds to the movement of the shoulder, whereas the red squares plotted on the red lines represent the measurements acquired with the sensory system of the proposed prodder. On these examples, the robot was programmed to move its shoulder 30° from the horizontal position, and once this position is achieved, come back to the original configuration. Note also that on Figure 11, sampling rate was set to 120 Hz for the measurement of the prodder's angle, while in Figure 12 the sample rate was decreased to 20 Hz. In both cases, the angle tracking provided by the intelligent feedback prodder exhibits a high accuracy.

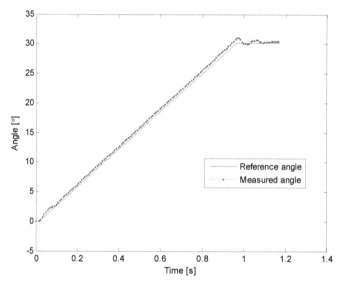

Figure 11. Inclination angle vs. time—first trial.

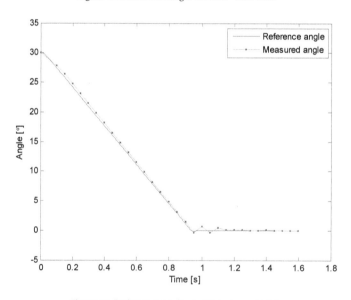

Figure 12. Inclination angle vs. time—second trial.

The last set of trials was devoted to verifying the correct activation of alarms when the operator exceeds the pre-defined maximum force, or when the angle of insertion of the prodder exceeds 45°, which can activate the mine detonation. The result presented corresponds to a test where the prodder is inserted in the terrain until a buried metallic piece is found. Figure 13 shows the scenario of the experiment, whereas Figure 14 presents the obtained results. The blue solid line in Figure 14a represents the acquired inclination angle of the prodder, black solid line in Figure 14b represents the exerted force, and finally, dotted lines in Figure 14a,b represent the behavior of the alarms for the intelligent feedback prodder, in such a way that green color indicates that values are within the training objectives, yellow color warns that values are starting to deviate from the goals and red colors informs that values are out of the defined safety ranges. This feedback is crucial to teach trainees to establish good working habits. For instance, if there is a chance that maximum force is exceeded during training session in highly compact, hard, or rocky soil, trainee should learn to apply some techniques that are commonly used in practice to avoid accidents, such as: (i) softening the ground before prodding, (ii) excavating, or (iii) penetrating soil with low force by slowly pressing the tip of the prodder into the ground and loosen pebbles with a slow twist of the wrist.

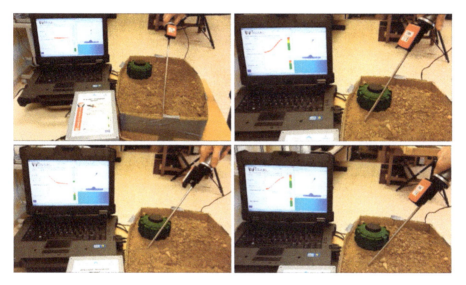

Figure 13. Scenario for the third sequence of experiments.

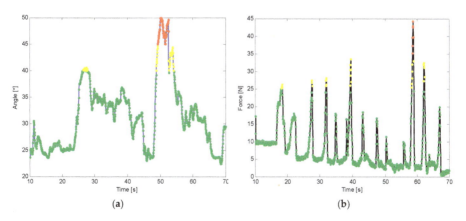

(a) (b)

Figure 14. Activation of alarms: (**a**) Inclination angle measurements; (**b**) Force measurements.

4. Discussion

Gathering together the quantitative results obtained from the experimental tests presented in the previous section, it is possible to highlight the high accuracy of the proposed intelligent sensory system to estimate in real-time the force exerted by the operator as well as the angle of inclination of the prodder. Figure 15 shows the distribution of the absolute and relative errors of the force measurements acquired in the experimental stage. Results demonstrated that the proposed tool is able to provide information about the tip forces in the rod direction with a maximum absolute error of ±0.7 N and a maximum relative error of ±3% in dynamic conditions. Thus, the force feedback provided by the training tool will be very useful to teach trainees what a safe prodding force feels like, in order to establish good working habits.

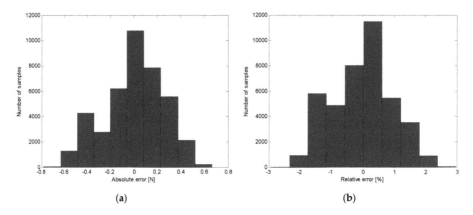

(a) (b)

Figure 15. Distribution of force measurement errors: (**a**) Absolute error; (**b**) Relative error.

Figure 16 shows the distribution of the absolute and relative errors of the prodder's angle measurements acquired during the experimental tests. Results demonstrated that the proposed tool is able to provide information about the angle of insertion of the prodder with a maximum absolute error that ranges from −0.6° to 0.8° and a maximum relative error that goes from −6% to 7.9% in dynamic conditions. These highly accurate measurements of the prodder's angle will be very important to aware deminers when they are approaching or exceeding a certain limit that can drive to the mine detonation.

Lastly, the third phase of the experimentation confirmed that alarms provided by the HMI are activated in 100% of the cases. These alarms were designed for alerting the trainee when he/she is approaching or exceeding a certain limit that can drive to the mine detonation, and thus, play a crucial role during training activities.

During the experimental tests, the HMI also demonstrated its versatility and reliability in gathering, analyzing, presenting and consolidating the information acquired with the instrumented prodder that has been especially conceived and implemented for interacting with this application. The friendly graphic user interface presents the data received in an efficient format, maximizing the instructor's ability for monitoring, processing and evaluating the trainee performance, and consequently, reducing the total cognitive load required.

Despite all the advantages provided by the proposed tool, and despite its design and implementation have been carried out taking into consideration the needs and the requirements of the end-users, its introduction in the current training programs can be hindered by the strong rootedness to the traditional methods. One of the largest barriers to the adoption of instrumented prodders by the demining community is the rigid adherence to existing operating procedures. The demining community perceives the current operating procedures as safe and refuses the introduction of new equipment that does not have pre-existing safety record. Hence the importance of disseminating the obtained results, with the aim of increasing the degree of acceptance of the end-users and facilitate the adoption of the proposed tool. In this way, instructions will be based on scientific knowledge rather than on personal introspections and intuitions.

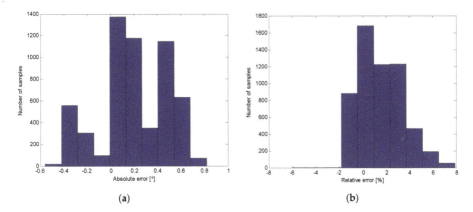

Figure 16. Distribution of measurement errors from prodder's inclination angle: (**a**) Absolute error; (**b**) Relative error.

5. Conclusions

This paper addressed the design, implementation and technical validation of a novel intelligent multisensory system for improving the training activities of humanitarian demining operations carried out with prodders. The main objective of the training tool is to provide force and angle feedback to teach trainees how a safe prodding should be conducted, in order to establish good working habits, and consequently contribute to the reduction of the number of accidents during the practice of this risky and dangerous task. An outline of the main features, functions and components of the system has been described in detail. The proposed tool consists of a prodder instrumented with a custom-made force sensor and an IMU, an electronic module for signals conditioning, a DAQ module, and a HMI. All basic parts of the instrumented prodder are separable and easily interchangeable in order to obtain different versions of the prodder, depending on the demining training needs. Experimental tests demonstrated that with the proposed training tool it is possible to determine in real-time and with high accuracy, the tip forces in the rod direction as well as the angle of inclination of the prodder. Thus, the proposed tool is able to provide useful information about the amount of force exerted and alert deminers when the angle of insertion of the prodder is approaching or exceeding a certain limit while they are prodding. In addition, the designed HMI has the advantage of providing an overview of the entire operation conducted with the prodder and a limited number of well-defined alarms. In this way, the instructor or the trainee will be able to see the entire operation almost at-a-glance. Therefore, the proposed tool will improve the instructor's ability for monitoring, processing and assessing the performance data of the training, reducing the total cognitive load. The intelligent multisensor prodder proposed in this work represents a significant advance in the current state-of-the-art, since it enables the development and implementation of training sessions based on scientific analysis of the problem and the formative and summative assessment of trainees, rather than on personal introspection and intuitions of the training designers.

Future work will be directed to design and implement a RF module that allows us to eliminate the cable that connects the force sensor installed on the instrumented prodder with the box that contains the signal conditioning module, providing enhanced flexibility to the proposed tool. In addition, validation tests will be conducted in real scenarios with demining end-users in order to confirm the benefits of the intelligent feedback prodder for training.

Supplementary Materials: The following are available online at http://www.mdpi.com/1424-8220/16/7/965/s1, Video S1: Intelligent feedback prodder for training, Video S2: Robotic testing.

Acknowledgments: Authors would like to thank Carlota Salinas for her help with the implementation of the HMI. Special thanks are also due to Javier Sarria for his assistance during the manufacturing of the 3D printed elements for the force sensor. Authors acknowledge funding from the European Commission under 7th Framework Programme (TIRAMISU Grant Agreement N° 284747) and partial funding from the CSIC project Robótica y sensors para los retos sociales (ROBSEN—PIE 20165E050) and the Robocity2030-III-CM project (Robótica aplicada a la mejora de la calidad de vida de los ciudadanos. Fase III; S2013/MIT-2748), funded by Programas de Actividades I+D en la Comunidad de Madrid and cofounded by Structural Funds of the EU. Roemi Fernández acknowledges the financial support from Ministry of Economy and Competitiveness under the Ramón y Cajal Programme. Héctor Montes acknowledges support from Universidad Tecnológica de Panamá.

Author Contributions: The work presented here was carried out in collaboration between all authors. Roemi Fernández designed the study and wrote the manuscript. Héctor Montes, Manuel Armada and Roemi Fernández conceived and designed the experiments. Héctor Montes designed and implemented the force sensor for the instrumented prodder. Roemi Fernández designed the sensory part devoted to the measurement of the prodder's angle, as well as the HMI. Héctor Montes and Roemi Fernández performed the experiments for the technical evaluation of the proposed tool. All authors contributed to the review of the manuscript.

Conflicts of Interest: The authors declare no conflict of interest.

Abbreviations

The following abbreviations are used in this manuscript:

ERW	Explosive remnants of war
IED	Improvised explosive device
PZT	Piezo-electric transducer
PNN	Perceptron Neural Network
TIRAMISU	Toolbox implementation for removal of antipersonnel mines, submunitions and UXO
UXO	unexploded ordnance
IMAS	International Mine Action Standards
CWA	CEN workshop agreement
CEN	European Centre for Standardization
NMAS	National mine action standards
EOD	explosive ordnance disposal
FSO	Full scale output
FS	Full scale
USB	Universal serial bus
HMI	Human machine interface
IMU	Inertial measurement unit
VRML	Virtual reality modelling language
DAQ	Data acquisition
LAN	Local area network

References

1. International Campaign to Ban Landmines—Cluster Munition Coalition. Landmine Monitor 2015. Available online: http://www.the-monitor.org/media/2152583/Landmine-Monitor-2015_finalpdf.pdf (accessed on 2 May 2016).
2. Habib, M.K. Mine Clearance Techniques and Technologies for Effective Humanitarian Demining. Available online: http://www.jmu.edu/cisr/journal/6.1/features/habib/habib.htm (accessed on 2 May 2016).
3. Rosengard, U.; Dolan, T.; Miklush, D.; Samiei, M. Humanitarian Demining Nuclear Techniques May Help the Search for Landmines. Available online: https://www.iaea.org/sites/default/files/publications/magazines/bulletin/bull43-2/43205031619.pdf (accessed on 2 May 2016).
4. Habib, M.K. Humanitarian demining mine detection and sensors. In Proceedings of the 2011 IEEE International Symposium on Industrial Electronics, Gdansk, Poland, 27–30 June 2011; pp. 2237–2242.
5. Hussein, E.M.; Waller, E.J. Landmine detection: The problem and the challenge. *Appl. Radiat. Isotopes* **2000**, *53*, 557–563. [CrossRef]

6. Shimoi, N. *Technology for Detecting and Clearing Landmines*; Morikita Shuppan, Co. Ltd: Austin, TX, USA, 2002.

7. Smith, A. Understanding the Use of Prodders in Mine Detection. Available online: http://www.jmu.edu/cisr/journal/18.1/notes/smith.shtml (accessed on 2 May 2016).

8. Alternatives for Landmine Detection. Available online: https://www.rand.org/content/dam/rand/pubs/monograph_reports/MR1608/MR1608.pref.pdf (accessed on 2 May 2016).

9. Antonic, D. Analysis and Interpretation of Ultrasonic Prodder Signal. In Proceedings of the MATEST 2003—Achievements & Challenges, Brijuni, Hrvatska, 28–30 September 2003.

10. Stepanic, J.; Maric, G.; Schauperl, Z. Improving Integration of Ultrasonic Sensor and Hand Probe. In Proceedings of the 9th European Conference on Non Destructive Testing, Berlin, Germany, 25–29 September 2006.

11. Ishikawa, J.; Iino, A. Experimental test and evaluation of an active sensing prodder. In Proceedings of the 8th International Symposium "Humanitarian Demining 2011", Sibenik, Croatia, 26–28 April 2011.

12. Baglio, S.; Cantelli, L.; Giusa, F.; Muscato, G.; Noto, A. The development of an inteligent manual prodder for material recognition. In Proceedings of the 11th International Symposium "Mine Action 2014", Zadar, Yugoslavia, 23–25 April 2014.

13. Baglio, S.; Cantelli, L.; Giusa, F.; Muscato, G. Intelligent prodder: Implementation of measurement methodologies for material recognition and classification with humanitarian demining applications. *IEEE Trans. Instrum. Meas.* **2015**, *64*, 2217–2226. [CrossRef]

14. Iwatani, A.; Shoji, R.; Ishikawa, J. Development of prodder for humanitarian demining based on tactile augmentation. In Proceedings of the 12th International Symposium "Mine Action 2015", Biograd, Croatia, 27–30 April 2015; pp. 43–51.

15. Ali, H.F.M.; Fath El Bab, A.M.R.; Zyada, Z.; Megahed, S.M. Inclination Angle Effect on Landmine Characteristics Estimation in Sandy Desert Using Neural Networks. In Proceedings of the 10th Asian Control Conference (ASCC), Sabah, Malaysia, 31 May–3 June 2015; pp. 1–6.

16. Ali, H.F.M.; Fath El-Bab, A.M.R.; Zyada, Z.; Megahed, S.M. Estimation of landmine characteristics in sandy desert using neural networks. *Neural Comput. Appl.* **2016**. [CrossRef]

17. Fernández, R.; Montes, H.; Salinas, C.; Santos, P.G.D.; Armada, M. Design of a training tool for improving the use of hand-held detectors in humanitarian demining. *Ind. Robot Int. J.* **2012**, *39*, 450–463. [CrossRef]

18. Gasser, R. Technology for Humanitarian Landmine Clearance. Ph.D. Thesis, University of Warwick, Warwick, UK, September 2000.

19. CROMAC. Book of Rules and Regulations on the Training and the Competence Examination of the Employees in Humanitarian Demining. Available online: https://www.hcr.hr/en/pravilnici.asp (accessed on 2 May 2016).

20. Fernández, R.; Montes, H.; Gusano, J.; Sarria, J.; Armada, M. Force and angle feedback prodder. In Proceedings of the 17th International Conference on Climbing and Walking Robots, CLAWAR 2014, Poznan, Poland, 21–23 July 2014.

21. Jungwirth, O. Book of Rules and Regulations on Technical Requirements and Conformity Assessment of Devices and Equipment Used in Humanitarian Demining. Available online: https://www.hcr.hr/en/pravilnici.asp (accessed on 2 May 2016).

22. Fernández, R.; Salinas, C.; Montes, H.; Sarria, J.; Armada, M. Design of a human machine interface for training activities with prodders. In Proceedins of the 12th International Symposium Mine Action, Biograd, Croatia, 27–30 April 2015; pp. 161–164.

23. Jaeger Platoon Website. Available online: http://www.jaegerplatoon.net/landmines2.htm (accessed on 2 May 2016).

24. Absolute Astronomy Website. Available online: http://www.absoluteastronomy.com/topics/List_of_landmines (accessed on 2 May 2016).

 *sensors*

Article

Functional Analysis in Long-Term Operation of High Power UV-LEDs in Continuous Fluoro-Sensing Systems for Hydrocarbon Pollution

Francisco Jose Arques-Orobon [1,*], Neftali Nuñez [1,2], Manuel Vazquez [1,2] and Vicente Gonzalez-Posadas [1]

[1]   ETSIST—Technical University of Madrid, Madrid 28031, Spain; neftali.nunez@upm.es (N.N.); manuel.vazquez@upm.es (M.V.); vgonzalz@diac.upm.es (V.G.-P.)
[2]   Instituto de Energía Solar—Technical University of Madrid, Madrid 28040, Spain
*   Correspondence: jose.arques@upm.es; Tel.: +34-913-365-500; Fax: +34-913-367-784

Academic Editor: Gonzalo Pajares Martinsanz
Received: 20 January 2016; Accepted: 22 February 2016; Published: 26 February 2016

Abstract: This work analyzes the long-term functionality of HP (High-power) UV-LEDs (Ultraviolet Light Emitting Diodes) as the exciting light source in non-contact, continuous 24/7 real-time fluoro-sensing pollutant identification in inland water. Fluorescence is an effective alternative in the detection and identification of hydrocarbons. The HP UV-LEDs are more advantageous than classical light sources (xenon and mercury lamps) and helps in the development of a low cost, non-contact, and compact system for continuous real-time fieldwork. This work analyzes the wavelength, output optical power, and the effects of viscosity, temperature of the water pollutants, and the functional consistency for long-term HP UV-LED working operation. To accomplish the latter, an analysis of the influence of two types 365 nm HP UV-LEDs degradation under two continuous real-system working mode conditions was done, by temperature Accelerated Life Tests (ALTs). These tests estimate the mean life under continuous working conditions of 6200 h and for cycled working conditions (30 s ON & 30 s OFF) of 66,000 h, over 7 years of 24/7 operating life of hydrocarbon pollution monitoring. In addition, the durability in the face of the internal and external parameter system variations is evaluated.

Keywords: continuous real-time sensing; pollution identification; UV-LED sensing; fluorescence spectroscopy; water pollution

---

1. Introduction

Water is one of the most important resources necessary for sustaining life on earth. Population growth and the consequent creation of infrastructures, sewage treatment plants, crops, industries, intensive agriculture, *etc.*, and the associated traffic generated implies a relevant/critical hazard for inland water. In the case of inland water pollution by hydrocarbons and other pollutants which emit fluorescence under high energy light, new developments and analyses are being conducted [1–12]. The conventional methods used in recognizing a hydrocarbon pollutant involves transporting a sample from the area of water pollution and identifying the pollutant in the laboratory, most often using the fluorescence spectroscopic method [13,14].While this is an accurate method, it involves too much time for a rapid, and in some cases, programmable automatic action, to mitigate the pollution effect. Therefore, new flexible non-contact and continuous real-time systems need to be developed to identify the hydrocarbons and avoid or mitigate dangerous pollution episodes.

Hydrocarbons are volatile compounds which respond clearly with fluorescence when they are excited with a light source of a specific wavelength range, usually the ultraviolet range, due to its high energy of radiation. This characteristic has been utilized over the decades to identify the hydrocarbon type and its compounds, initially in health applications [15,16], and later in petroleum industrial applications [17]. The basic method involved in fluorescence spectroscopy is to apply an excitation source (light) to the sample and read and analyze the spectra emitted in response [14]. The classical equipment consists of optical, electrical, and mechanical elements, with several alternatives for each, as described in [13]. Light sources include [13]: xenon lamps, high pressure mercury vapor lamps, xenon-mercury arc lamps, lasers, and LEDs (Light Emitting Diodes).

The HP (High Power) UV-LED (Ultraviolet Light Emitting Diode) has significantly increased its optical output power over the last few years, being adopted as an alternative light source for the spectroscopic sensing systems, as it shows some advantages over the classical exciting lights, *viz.*, small size, less cost, quick output optical power stabilization and long life. The UV-LEDs emits in a narrow band of wavelength [18–20]. To obtain the fluorescence signature at different wavelengths sources the use of several LEDs are needed [2–4], or the combination with other light sources [21]. The classical spectrometer laboratory equipment [14], has been used for decades in the identification of pollutants by fluorescence in a wide wavelength range. However, the employment of LED or UV-LED has enjoyed only a short field experience in the fluoro-sensing applications in real-time [2–4,7,8,10]. This work proposes to move a step further from the occasional real-time measurements to continuous real-time detection and pollutant identification.

This work studies the functional long-term working of fluoro-sensing using HP UV-LED as the source light, for non-contact pollution detection and identification in inland water, in real-time 24/7 (24 h, 7 days), focusing on hydrocarbon analysis. The future short-term objectives are recognition of the signature pattern and integration of the whole experimental system in automatic water quality station of the Jarama-Tagus Rivers (Spain) [22,23], working 24/7 (Figure 1). In these automatic stations, the authors test different disciplines related to the development of new capabilities, such as the present work, which analyzes the HP UV-LED in the long-term, non-contact, continuous, real-time fluoro-sensing identification of the pollutants [20,24,25], as well as others, like the development of compact, low-power consumption high security warning communications systems with redundant dual-band based mobile phone systems (380 MHz and 960 MHz) [26,27].

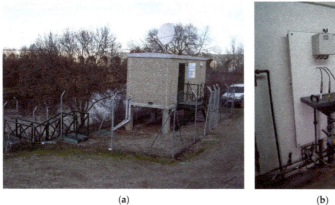

(a)   (b)

Figure 1. (**a**) Automatic station for recording water parameter measurements; (**b**) The analysis-sink in the automatic station for the multi-parametric meters (pH, conductivity, temperature, and dissolved oxygen) on the Jarama-Tagus Rivers, Spain.

In these automatic stations (Figure 1a), the water collected via a pump is conducted to an analysis-sink where the main parameters *viz.*, temperature, conductivity, dissolved oxygen, and pH are measured (Figure 1b). We proposed that the non-contact hydrocarbon identification sensor system would share this installation, and all the parameters measured of the automatic station [12], could be used to improve the identification signature of the hydrocarbon by our fluoro-sensing system by increasing its individual potentialities [28]. At present, the quality parameter measurements can indirectly identify the presence of water pollution, but not the pollutant type. From our experiments, a fluoro-sensing real-time 24/7 pollutant detection and identification system will provide a significant improvement to the existing analyzers in the automatic stations, as it does not involve significant additional cost or maintenance, but it does increase the real-time warning and safety in environmental monitoring.

Furthermore, the spectroscopy workbench instrumentation developed for this HP UV-LED work analysis is in accordance with the new generation of compact, real-time, low cost [7,8], low and easy power, fixed [9] or portable [2,3,12], autonomous systems [4], that connect to a PC or single-board computer. The experiments over the last few years dealt with the detection of substances such as chlorophyll and phytoplankton [2–4,7–9], besides pollutant detection [6,20].

In this work, we analyzed the suitability of using HP UV-LEDs for long-term continuous detection of the hydrocarbons in real-time utilizing fluoro-sensing. To achieve this, we have described several activities, *viz.*, the method and laboratory workbench, analysis of absorbance and pollutants fluoro-sensing signatures, influence of viscosity and temperature, and processing spectra readout. The final analysis is related with the functional operation and durability for long-term of two 3W 365 nm HP UV-LED technologies, based on the temperature accelerated life tests (ALT) under two different working conditions: continuous and cycled with 30 s ON & 30 s OFF power.

The results and conclusions of this functional analysis are oriented to continuous real-time hydrocarbon detection in automatic river stations; however, they can be essentially extended to other pollutants in water, and other alternative systems either portable or submersible systems.

2. Materials and Method

2.1. Materials

The hydrocarbons analyzed are the ones commonly used in Europe in gas stations and heating boilers. The samples that were drawn in Spain include two types of gasoline with different octane ratings, two Diesel A types for automotive use (containing different additives, with the ability to cause changes in the spectral response) present in the gas stations, Diesel B for agricultural vehicles and industrial uses, and Diesel C which has other uses including heating boilers because of its better calorific value. The complete list is given below:

- Gasoline 95 and 98 octanes.
- Diesel types A and A Plus, B, and C.

The samples were collected several times during two years from different gas stations of the same multinational European brand, and stored in glass laboratory bottles, inside chamber to keep its temperature at of 15 °C to prevent vaporization due to their high volatility. No differences were observed between the signatures of the samples during these years. The samples were renewed every few months, by to the modification of the signatures due to evaporation in the container, especially in gasolines. Several micro-spectrometer aperture times were tested, and 10 ms with five average measurements was selected as a compromise between number of signature counts and measurement time.

In order to arrive at a standardized response during the experiments, all the measurements were recorded in a laboratory at 24 °C using a Pyrex beaker filled with water and 1 mm uniform thickness of each different type of hydrocarbon (Figure 2a).

The 1 mm thickness election has been used in a previous work [20] and selected as low limit of identification. Larger thicknesses of hydrocarbon imply higher absolute fluorescence signature values. We suppose that a sample taken in a short distance to source pollution will have a thickness significantly higher than 1 mm. To our knowledge, there is not references related with hydrocarbon equilibrium thickness at long distances in inland water, but in seawater the evaluated hydrocarbon equilibrium thickness at t∞ is between 2 mm and 12.5 mm [29]. Therefore, the selection of 1 mm thickness is a conservative value to identify real hydrocarbon pollution.

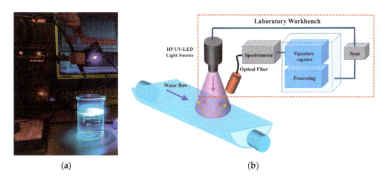

(a)   (b)

Figure 2. (**a**) Photograph laboratory workbench: optical fiber (left of UV-LED) HP UV-LED and Pyrex beaker containing the fluorescent hydrocarbon; (**b**) Diagrammatic representation of the instrumentation elements of the laboratory workbench.

2.2. Laboratory Workbench Instrumentation

The choice of a non-contact system instead of the peristaltic system employed in the other applications to identify diluted materials [3,4], was because of the future integration of the present workbench instrumentation in the analysis-sink with open fluent water, as shown in Figure 1b. This method is advantageous because it is clean and the LED light analyzes fluent water; however, the drawback is that it requires a higher optical output power because the HP UV-LED needs to be separated from the water in order to avoid the fluent water staining the LED lens.

In Figure 2a a photograph of laboratory workbench instrumentation measurement exciting fluorescent hydrocarbon is shown. In Figure 2b the block diagram representing the laboratory workbench is displayed.

The main elements shown in the laboratory workbench instrumentation block diagram are:

- UV light source, HP UV-LED. A 3W 365 nm LED was chosen for its simplicity and better beam light concentration than a circuit with several LEDs. The HP UV-LED is placed perpendicular at water surface at a distance of 8 cm, the view angle of the UV-LEDs is $2\theta_{1/2} = 70°$. The optical fiber has an angle of 30° with respect UV-LED perpendicular line, and a 6 cm distance to water surface.
- Spectrometer and optical fiber. Fluorescence detection systems based on compact CCD spectrometer are emerging as a tool in research field applications [2–4,20]. In this work two compact spectrometers (Avantes® Ava-Spec 2048-USB2-UA, and Mightex® HRS-BD1-025) have been used, the first one for workbench instrumentation in signature readout, and the second one for analysis of UV-LED temperature ALT [24,25]. The spectra readout is obtained via a specific optical fiber of 0.6 mm Ø, solid angle $\Omega = 0.15$ sr, numerical aperture NA = 0.22 (Figure 2a). Considering the perpendicular distance to the water is 6 cm, the measurement area of fluorescence is 6.73 cm². The wavelengths range of Avantes® Ava-Spec 2048-USB2-UA is 200–1100 nm, and the Mightex® HRS-BD1-025 with wavelengths range 300–1050 nm, proprietary measurement software, power supply, and PC connection by USB cable. This element is the only significant system cost, around 2500$.
- Processor block. This element is implemented in a Windows laptop computer under the software proprietary of Avantes® integrated in MATLAB® (MathWorks®, Natick, MA, USA). The micro-spectrometer readout is loaded in a MATLAB® application, which processes the signal

for pollutant identification. The objective of this work is to make a qualitative identifications of hydrocarbons in an early warning system, therefore the measurements of the time-dependent decay of fluorescence have not been developed.

The temperature ALT instrumentation is explained in [24] where the instrumentation and preliminary results of ALT are described. The HP UV-LEDs used their characteristics, and electrical and optical operational working conditions are exhaustively described in [25].

3. Results

3.1. Wavelength Analysis for Hydrocarbon Identification

The introduction of LED as light source for spectrometric instrumentation has increased over the last years, because it is much simpler, it minimizes the instrumentation costs, it has lower power consumption and shows higher reliability. Currently, there are High Power 1–3 W (electrical) UV-LEDs up to 365 nm from different manufacturers (Nichia, Led Engin, LG, Lumileds). Today, this is the working wavelength limit for non-contact high area fluent water fluorescent analysis. Commercial LEDs with wavelength lower than 365 nm LED (Nichia, Crystal, Roithner Lasertechnik, Qphotonics, and LG for 285 nm) involve higher costs and significantly less electrical and optical power, in the order of mW, with low external quantum efficiency, are used for certain scientific applications [30].

As the wavelength limitation of HP UV-LEDs is known, we can analyze by absorbance the wavelength range in which the hydrocarbons could emit fluorescence. Absorption spectroscopy is an analytical chemistry tool mainly used to determine the presence of a particular substance [31], but also it contributes to see if the substance could be fluorescent. To produce the fluorescence effect, the substance to be detected must emit at a wavelength greater than the light source used, and in order to emit photons, it must first absorb a part of the light, which usually works closer to ultraviolet [13]. Figure 3 shows the absorbance response for the pollutant substances performed according to criteria ISO/IEC 17025. (General requirements for the competence of testing and calibration laboratories, International Organization for Standardization) using a PerkinElmer's Lambda 25, obtaining the absorbance response shown in Figure 3. This implies that it is possible to determine whether the energy provided by the photons is absorbed by the compound; furthermore, a portion of this energy is emitted as light of a higher wavelength (fluorescence) and another portion of energy as heat in the compound.

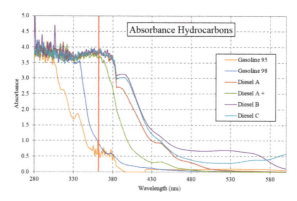

Figure 3. Absorbance of different hydrocarbons compounds used in Europe. Vertical red line at 365 nm.

In Figure 4, the absorbance curve (dotted line) with HP UV-LED emitted light (3W UV-LED with a narrow peak at 365 nm) and the fluorescence (continuous line) response of the hydrocarbon are shown. The absorbance and fluorescence response are normalized at the maximum count number of the spectrometer readout. From Figure 4, it can be observed that the diesels have higher fluorescence than the gasolines, and the signatures of each hydrocarbon is unequivocal.

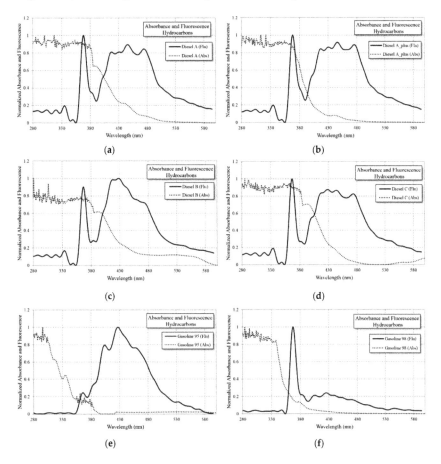

Figure 4. Hydrocarbons normalized fluorescence (continuous line) and absorbance (dotted line) with LED@365nm as the light excitation source. The peaks at 365 nm coinciding with the emission wavelength are due to scattering and reflection. (**a**) Diesel A; (**b**) Diesel A Plus; (**c**) Diesel B; (**d**) Diesel C; (**e**) Gasoline 95; (**f**) Gasoline 98.

These results seem to confirm the relation between absorbance and fluorescence. All the compounds except Gasoline 98 have very high signature levels, as shown later, although in this case, the 365 nm HP UV-LED is sufficient to identify the hydrocarbon; however, a UV-LED source below 365 nm will be much more effective in identifying this hydrocarbon. The limitation of output optical power of UV-LEDs under 365 nm will show significant improvement over the next few years due to the extensive use of UV-LEDs technology, which is continuously and intensively being developed [19].

3.2. Viscosity and Temperature Influence in Hydrocarbon Identification

Several factors are found to affect hydrocarbon fluorescence, among which temperature and viscosity are the main variables that can affect the hydrocarbons diluted in inland water. Therefore, it is necessary to quantify the impact of these factors on the spectrometer readout results, in order to consider this in the signature identification.

Temperature can decrease the fluorescence power between 1% and 5% per degree Celsius increase. As the temperature increases, the viscosity decreases and the frequency of collisions increases, increasing the probability of deactivation as non-radiant energy. In the case of hydrocarbons, it is necessary to measure the change in viscosity with temperature, although it is expected that because these compounds are highly volatile, they naturally have very low viscosity.

In Figure 5a, the viscosity measurements are shown for three types of fuel, observing no great difference among them in the whole temperature range. Measurements are performed according to ASTM D-445 (Standard Test Method for Kinematic Viscosity of Transparent and Opaque Liquids (and Calculation of Dynamic Viscosity), ASTM International). Details regarding the temperature range at which they can occur in water are shown in Figure 5b in which the viscosity variations are between 3 and 5 cSt.

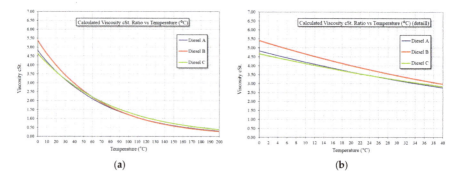

(a)   (b)

Figure 5. Viscosity measures of Diesel types in two temperature ranges. (**a**) Wide temperature range; (**b**) Detail of temperature range in inland waters.

To corroborate the results presented, some fluorescence measurements taken using UV-LED as the light source, recorded under the same conditions at different temperatures, clearly highlight the influence of temperature on fluorescence. The temperatures selected are 5, 21, and 35 °C, values that can be observed in different field conditions.

In Figure 6, four representative hydrocarbons are shown as examples. As evident from the results, at lower temperatures a greater fluorescence response is obtained. However, from Figure 5, two different cases can be observed:

- In the case of Diesel A, the response at 35 °C is different from the others at lower temperatures because the fluorescence peak at 425 nm is lower at 35 °C. This difference cannot be corrected with a normalization of the signal because it depends on the wavelength, in this case a different master signature for low, medium, and high temperatures are required, and furthermore, a real-time measured temperature is needed for a correct identification of the signature [12,28].
- In the case of Diesel A Plus and Gasoline 98 octane, the difference in the optical power between the high and low temperatures is much greater than for the other hydrocarbons.

The main conclusion drawn is that if, during a year, extreme temperature ranges occur in the river waters, a complementary temperature measurement is needed in order to identify a direct relationship between the fluorescence response and the pollutant.

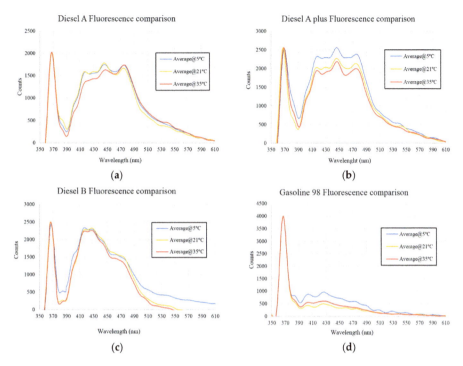

Figure 6. Fluorescence of the hydrocarbons at different temperatures with LED@365nm as the light source. (**a**) Diesel A; (**b**) Diesel A Plus; (**c**) Diesel B; (**d**) Gasoline 98 octane.

3.3. Processing of Spectra Readout

The readout of the CCD micro-spectrometer has two components; reflected UV-LED signature and significant fluorescence signature. The UV-LED spectrum is represented in Figure 7 of [25], observing a very low optical power of UV-LED at wavelengths higher than 400 nm, for that, we considered the significant fluorescence signature in the range between 400 nm and 650 nm. Over 650 nm, the fluorescence signature CCD counts are near to zero.

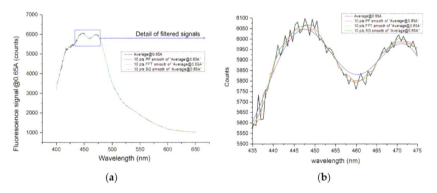

Figure 7. (**a**) Different filtering methods tested for significance of fluorescence; (**b**) The detail of the signatures are shown in black line readout, red line median smooth, blue line Fast Fourier time filter, and green line Savitzky-Golay smooth.

The second step involves reducing the noise and improving the signature, and several filter types have been tested, between themselves: median smooth, Fast Fourier Time, and Savitzky-Golay (S-G) smooth. Figure 7 shows the results for Diesel A in the UV-LED normal working condition.

The Avantes® micro-spectrometer as shown in Figures 7a and 8c set the minimum count value in 1000 counts; after filtering the signal it is normalized to the maximum value between 400 and 650 nm. The final results are shown in Figure 8d. This process, reduces a great part of the changes in the readout due to the UV-LED different working conditions; ambient temperature, LED temperature, situation of the LED optical output power degradation in the long-term, *etc.*, producing a significant improvement for signature identification.

As evident in Figure 7b, a detail of the S-G smooth filter, widely used in the chemometrics spectrometer analysis, reveals the best results [32–34], it does not reduce the peaks and increase valleys of the fluorescence (Fast Fourier time filter) and does not modify the position of the wavelength peaks; furthermore, it shows the best smooth as the median smooth. In conclusion, an S-G filter with 10 samples (one sample each 0.589 nm) and polynomial of 2 degrees is elected for processing all the readout signatures, post normalization

3.4. Influence of HP UV-LED Parameters in the Spectrometer Readout

The HP UV-LEDs are more advantageous than the lamps, but is a less mature technology and currently undergoing development of materials and processes [18]. HP UV-LEDs are designed for working over a range of current supply, the voltage level is defined by the defined current supply by means of the I-V curve, and the output optical power of LED, an increase in the current (power supply) of UV-LEDs supposes an increase in the optical output power. However, an increase in the current supply implies a higher current stress and temperature which affect the LED efficiency and reliability. In Figure 8, the fluorescence response is represented for the different currents, which implies different outputs of optical power. Furthermore, the high current stress of the HP UV-LED (tested at 3W UV-LEDs over a current density of 60 A/cm^2) makes it necessary to have a large sink to avoid the high semiconductor junction temperature. Both HP UV-LEDs types have high thermal conductivity ceramic encapsulation and with one unique chip. The description of the complete package and thermal state of the HP UV-LEDs under test are given in [25].

During long-term operations, the LEDs suffer degradation in the semiconductor and encapsulation materials which affect the LEDs efficiency. As it is not a mature technology, UV-LEDs of the same type have similar characteristics but show differences for the same binning in optical power of 5%–10% among them. Therefore, it is necessary to record the fluorescence master signature prior to normalization in the normal working conditions for a typical HP UV-LED emission of one type, as it cannot be the same for the other UV-LEDs, these variations will be considered in the analysis [35]. Thus, it is necessary to consider these optical output power variations, the long-term output optical power degradation, and other negative effects as due to the cloudiness of the LED lens. As these parameters influence and affect the optical output power in a similar way to the current, and the influence of currents on fluorescence readout are analyzed. To achieve this, a new 3W HP UV-LED was supplied with power of different current levels, emulating the different states of degradation during life.

The normal working condition defined for the two technologies (A and B) of 3W HP UV-LEDs are 600 mA [25], that is 85% of the 700 mA defined as the nominal and maximum value for the continuous working condition according to the manufacturers in the datasheet, other parameters of two LEDs types appear in Table 1. Voltage is defined for the characteristic I-V curve of the HP UV-LED. The optical output power of both the UV-LEDs types under test are similar, with a difference of over 15%.

Table 1. Main electrical and optical parameter of HP UV-LEDs at nominal and at working condition.

HP UV-LED Type	Nominal Electrical/ Optical Power (W)	Electrical/Optical Power at 600 mA (W)	Irradiance at 8 cm & 600 mA (mW/cm^2)
A	≈3/0.8	≈2.4/0.65	9.11
B	≈2.8/1	≈2.25/0.75	10.40

For a conservative election, we measure the signature with the UV-LED type of minor output optical power. The relation between current applied and output optical power measured at 8 cm for 600 mA, with Juno-USB equipment and 3A optical sensor, both of Ophir®, is 9.11 mW/cm^2, for other supply conditions used in fluorescence characterization the values are: 14.45 mW/cm^2 at 0.99 A, 11.92 mW/cm^2 at 800 mA, 9.86 mW/cm^2 at 650 mA, 8.26 mW/cm^2 at 540 mA, 6.68 mW/cm^2 at 440 mA, 4.97 mW/cm^2 at 330 mA, and 3.24 mW/cm^2 at 220 mA. As it can be seen, the light output is not proportional to current, the best relation is over the 0.6A, and it is worse at higher currents with elevated temperatures, this also is reflected in the fluorescence measured in hydrocarbons Figure 8b.

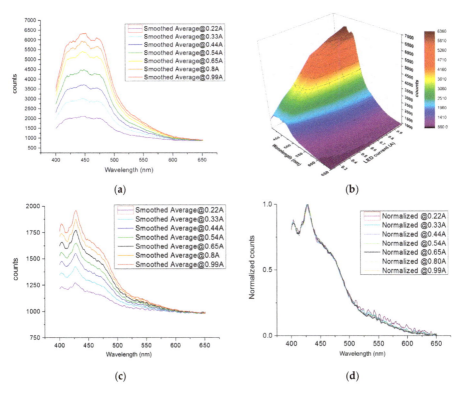

Figure 8. (**a**) Fluorescence signatures after the signal was processed without normalization of Diesel A Plus at different current supplies; (**b**) 3D fluorescence graphic for Diesel A Plus; (**c**) Fluorescence signatures after the signal was processed without normalization of Gasoline 98; (**d**) Fluorescence signatures processed with normalization for Gasoline 98.

In Figure 8, it is evident that the signatures of the two hydrocarbons are different at different HP UV-LEDs power supplies (current injections). The hydrocarbons selected are: Diesel A Plus that has a good fluorescence spectrum and shows the highest values, whereas Gasoline 98 with a relatively poor fluorescence level shows the lowest values. In the case of gasoline, the relation between signal and noise is minor, and recognition is more difficult at low currents of operation, which emulates a higher level of degradation; for example, the readout at 440 mA is equivalent at a maintained optical power output of 73% compared with the initial nominal condition (6.68*100/9.11 = 73%), at 330 mA it is 54.44% and at 220 mA it is 25.6%.

In order to define a level of optical output power degradation that enables the functionality of the HP UV-LED for identification of fluorescence pollution several issues have been considered:

- The variations in the initial optical outputs of power of the new LEDs identified earlier.
- The optical output power degradation during the life of the LED, and the additional optical output power losses during real operation due to a dirty lens or dust.
- A 30% optical output power degradation with respect to the initial optical output power value is the standard definition of degradation failure in white illumination LEDs [36]. The reliability datasheets of the LEDs that include the standard report IES-LM-80-08 (Approved Method for Measuring Lumen Maintenance of LED Light Sources) [37] also select this failure limit evaluating the $MTTF_{(L70)}$ (Mean Time to Failure assuming failure when optical output power decays below 70% of the initial optical output power). The other, less frequent failure modes of the LED are the catastrophic failures due to short or open circuit. During temperature ALT, no catastrophic failures have been observed, all detected failures are due to light output degradation over 30%.
- In Figure 8d, the normalized signatures in the worst fluorescence signature case (Gasoline 98), for different currents have been represented. The signature at 220 mA (violet) has been observed to have higher noise than the fluorescence curve at nominal current (650 mA black line), the 330 mA (cyan) fluorescence signature (with a 54.4% of the optical output power with respect to the nominal current) has significantly less noise and therefore, we will consider that 220 mA is not a functional power at which to identify Gasoline 98. Therefore, although 50% or 60% of the initial optical output power would suffice, considering the other possible factors of optical output power that have been explained in the prior paragraph, we conservatively selected 70% of the initial optical output power, which is the standard value [36] and it is contemplated in the reliability datasheet of the commercial LEDs.

3.5. Long Life UV-LED Test for Hydrocarbons Detection

ALT (Accelerated Life Test) is an alternative method to assess the life of a device in a short period of time. Temperature ALT that offers an estimation of the total life in some thousands of hours has been developed in order to assess the life of the HP UV-LEDs in order to identify the hydrocarbons.

This acceleration of time working is based in that the catastrophic or degradation failure mechanisms of LED technology is accelerated by temperature, and this time acceleration is modeled physically and mathematically by the Arrhenius model [38]. This model is the general application used in LEDs by manufacturers [39,40], with standardization of degradation life prediction IES-TM-21-11 (Projecting Long Term Lumen Maintenance of LED Light Sources) [41], and research reliability of the III-V optical semiconductors devices [42–46]. By this model, the failure distribution of the HP UV-LED population under test at a nominal temperature of 25 °C, is reproduced in accelerated time by the Arrhenius model at higher temperatures. To achieve this, the process involved estimating the complete life evolution of a population of LEDs in nominal working conditions, involving the inverse process, to develop accelerated tests at higher ambient temperatures, in this case 60 °C, 75 °C, and 90 °C at the same nominal working conditions of the HP UV-LEDs and characterizing periodically the electrical, thermal, and output optical power parameters during the tests [25], for final analysis (Figure 9).

In this case, we estimate the life of two types of 3W 365nm HP UV-LEDs. For temperature ALTs three ovens with three (Ta) ambient temperatures (60 °C, 75 °C, 90 °C) were used during 4000 h for each test. The LEDs have been tested under two different working modes at 600 mA; continuous and cycled (30 s ON & 30 s OFF) completely analyzed in [25].

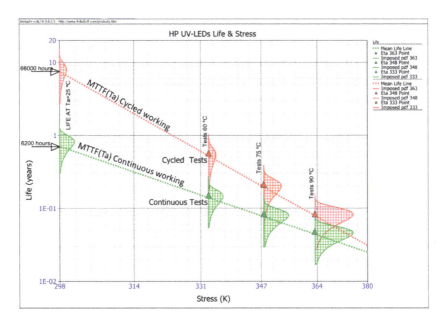

Figure 9. Evolution of the HP UV-LEDs life *vs.* ambient temperature (Ta) stress. The continuous line represents failure density probability function, the dotted line MTTF$_{(L70)}$(Ta) shows continuous (green) working and cycled (red) working (30 s ON & 30 s OFF).

During ALT, we measured the instant time of degradation failure (70% of the initial power) for the LEDs with high degradation, and the extrapolation of this time for those LEDs that did not get degraded below 70% initial power during the temperature ALT. With the failure times of the LEDs in all the experiments (three temperatures and two working modes) the results were analyzed, applying the Arrhenius model. To accomplish this, we employed the specific software for the temperature ALT reliability experiments, Weibull++/ALTA of Reliasoft® (ALTA—Accelerated Life Test Analysis application) [47].

The complete results by Weibul++/ALTA obtained are shown in Figure 9, in the continuous working mode (green color) and cycled working mode (red color) for the HP UV-LED type with the higher MTTF$_{(L70)}$ (25 °C), Mean Time To Failure at Ta = 25 °C and 70% limit of maintaining light output. This Figure shows the failure distribution of the population under test (Figure 9, red and green vertical draw functions), at 60 °C, 75 °C, 90 °C Ta, and the extrapolation by the Arrhenius model at the nominal Ta = 25 °C. In Figure 9, representations were also done by the red and green continuous lines of MTTF$_{(L70)}$(Ta) at each temperature. For the other type of HP UV-LED under test, the results for continuous working are very similar, and sensibly minor for the cycled condition, but in any case, several times better than for continuous working. Other relevant aspects during the tests related with the failure analysis are the degradation of the silicone encapsulation of the UV-LEDs, a slow progressive degradation of the optical output power, and a slight peak shift to the higher wavelengths (1.5 nm), that are described in [25].

The MTTF$_{(L70)}$ obtained for the HP UV-LED type with the longest life at 25 °C ambient temperature, is 66,000 h (seven-and-a-half years of working life) for cycled working conditions 30 s ON & 30 s OFF [25] and 6200 h for the continuous mode (eight-and-a-half months of working life). For the other technology, HP UV-LED, in cycled working mode the MTTF$_{(L70)}$ was over 2,9000 h, and 5700 h for the continuous working mode.

The results of ALT enable the following conclusions:

- For the continuous working mode HP UV-LEDs MTTF life of eight-and-a-half months, is better than or similar to continuously working spectrometric lamps.
- For the cycled mode, the MTTF life of more than seven years for one type of the HP-UV-LEDs and four years for the other type, are considerably better than that of the conventional spectrometric lamps in the continuous operation mode. It must be considered that the conventional UV-lamps cannot operate in the cycled (1 min) mode, due to the great stabilization times.
- The continuous working mode has a short life compared with the cycled working condition, where for 30 s of each minute the water pollutants are not sensed. A partial solution will be to switch the power supply every 30 s to two different LEDs to excite the fluorescence, and although the time for stabilizing the optical output power is over 7 s and the cost of each 3W HP UV-LEDs is roughly $35.

We have also compared, after ALT, the LEDs that have been degraded with the brand new LEDs working at different currents based on power supply. The results are consistent with those recorded in Section 3.4, observing the good quality of fluorescence pollutant detection when the optical output power drops below 30%. To accomplish this, we proposed as the failure degradation limit a conservative value of 70% of the initial optical output power.

The process involved in obtaining these results has been extended over time: the authors studied the HP UV-LEDs with respect to the hydrocarbons comparing them with other classical light sources in [20], and later developed a series of experiments with two types of new commercial technologies 3W 365 nm HP UV-LEDs, characterizing the encapsulating, thermal, and electrical response in two working modes: continuous and cycled (30 s ON & 30 s OFF) [25], indicating the method and the preliminary results of the ALT [24,25] with these two types of UV-LEDs and the two working conditions in [24].

4. Conclusions

The main objective of this work is to validate the continuous real-time detection and identification of the hydrocarbon pollutant, during the long-term operation of the HP UV-LEDs as the excitation light of the fluoro-sensing systems. Extrapolation of these results for any pollutant detected using a similar real-time technology is relevant, because any improvement in the reaction time in a water pollution accident could be critical.

The main conclusions drawn from this work are given below:

- The materials, process, and laboratory equipment workbench used have been discussed. The detection system proposed has been performed without any contact with the fluent water, which offers several advantages, as it avoids soiling of the intermediate elements (glass, lens) of UV-LED while it analyzes a large area. The sensitivity and detection area could be improved by reducing optical losses and increasing the solid angle for fluorescence detection, using a lens or lens system.
- The absorbance and fluorescence signature of the different hydrocarbons have been analyzed using the selected HP UV-LED, and the influence of temperature and viscosity in the signature have been studied.
- After testing several types of signal processing, it has been proposed that a Savitzky-Golay filter and signal normalization in the range of 400 to 650 were the best method to obtain a repetitive signature under different conditions of the parameters, including the possible optical output power degradation of the HP UV-LED.
- The influence of HP UV-LED long-term output optical power degradation by current supply emulation, in fluorescence signature detection error was analyzed and the degradation limit of 30% was noted. Then maintaining 70% of the initial optical output power of the HP UV-LED was selected as the conservative limit for the repetitive identification of all the hydrocarbon signatures.

Degraded LEDs in the ALT with optical output power degradations below 30% value show an optimum fluorescence signature.

- In order to validate the long-term life of the HP UV-LEDs, temperature ALTs have been developed. The complete setup of the parameters in this test are explained in [24,25], and it has been assessed and extrapolation of failure distribution is shown (Figure 9). From these data, it can be concluded that the $MTTF_{(L70)}$ for the cycled working mode (30 s ON & 30 s OFF) is several times better than that obtained for the continuous working mode.

- For long term operation, the HP UV-LEDS with continuous working mode has an estimated $MTTF_{(L70)} \approx 6200$ h working, and for cycled working it depends on the type technology of LED, $MTTF_{(L70)} \approx 66,000$ h for HP UV-LED with long life type. In all cases, the results of $MTTF_{(L70)}$ are very promising, mainly for the cycled 30s ON & 30s OFF working condition.

As the technology of HP UV-LED is very innovative, new generations of UV-LEDs will significantly improve. In this work, we have validated HP UV-LEDs as excitation light sources of non-contact, continuous real-time fluoro-sensing pollutants for long-term applications, thus, these results open up new possibilities in fluoro-sensing.

At the current state, we have probed the viability of detecting saturated hydrocarbons in a laboratory environment, in order to transfer this technology to real conditions, it will be necessary to analyze the hydrocarbons signatures influenced by real samples of river water, with different aqueous matrices (e.g., various organics, biomolecules, nutrients, *etc.* which are in water miscible/soluble). Other future work is related with the identification of mixtures of hydrocarbons in different proportions, emphasizing the hydrocarbons mixed with car oil, since it is a standard type of pollution.

Acknowledgments: The Water Authority of the Tagus River (Spain) has assisted in this study by providing their laboratory for testing measures of absorbance, as well as the Automatic Station. We would also express our sincere gratitude to Fuels and Petrochemical Laboratory and Gomez Pardo Foundation (Spain) for their collaboration making viscosity measures and the tremendous support they extended throughout the process.

Author Contributions: F.J.A.-O. and V.G.-P. conceived and designed the experiments and the setup; F.J. A.-O. and N.N. performance the fluorescence signature characterization and ALT; F.J.A.-O and V.G.-P analyzed the fluorescence results; N.N. and M.V analyzed the ALT results. The manuscript was mainly drafted by F.J.A.-O., N.N. and M.V.. The manuscript was revised and corrected by all co-authors. All authors have read and approved the final manuscript.

Conflicts of Interest: The authors declare no conflict of interest.

References

1. O'Toole, M.; Diamond, D. Absorbance Based Light Emitting Diode Optical Sensors and Sensing Devices. *Sensors* **2008**, *8*, 2453–2479. [CrossRef]
2. Puiu, A.; Fiorani, L.; Menicucci, I.; Pistilli, M.; Lai, A. Submersible Spectrofluorometer for Real-Time Sensing of Water Quality. *Sensors* **2015**, *15*, 14415–14434. [CrossRef] [PubMed]
3. Ng, C.-L.; Teo, W.-K.; Cai, H.-T.; Hemond, H.F. Characterization and Field Test of an in Situ Multi-Platform Optical Sensor. *Limnol. Oceanogr. Methods* **2014**, *12*, 484–497. [CrossRef]
4. Ng, C.-L.; Senft-Grupp, S.; Hemond, H.F. A Multi-Platform Optical Sensor for in Situ Sensing of Water Chemistry. *Limnol. Oceanogr. Methods* **2012**, *10*, 978–990. [CrossRef]
5. Maraver Abad, P.; Vassal'lo Sanz, J.; Gutiérrez Ríos, J.; Esteban Orobio, A.; Soto Macía, I.; Vassal'lo Sanz, J.; Gallego García, E.; Medrano Gil, A. Detector a distancia de sustancias fluorescentes. Utility model. No ES1075980 U, 19-Jan-2012. Available online: http://digital.csic.es/handle/10261/53851 (accessed on 18 January 2016).
6. Ibáñez, G.A.; Escandar, G.M. Luminescence Sensors Applied to Water Analysis of Organic Pollutants—An Update. *Sensors* **2011**, *11*, 11081–11102. [CrossRef] [PubMed]
7. Leeuw, T.; Boss, E.; Wright, D. In Situ Measurements of Phytoplankton Fluorescence Using Low Cost Electronics. *Sensors* **2013**, *13*, 7872–7883. [CrossRef] [PubMed]
8. Kissinger, J.; Wilson, D. Portable Fluorescence Lifetime Detection for Chlorophyll Analysis in Marine Environments. *IEEE Sens. J.* **2011**, *11*, 288–295. [CrossRef]

9. Ji, J.W.; Xu, M.H.; Li, Z.M. Research and application on chlorophyll fluorescence on-line monitoring technology. *Adv. Mater. Res.* **2010**, *139*, 2550–2555. [CrossRef]

10. Fernandez-Jaramillo, A.A.; Duarte-Galvan, C.; Contreras-Medina, L.M.; Torres-Pacheco, I.; Romero-Troncoso, R.J.; Guevara-Gonzalez, R.G.; Millan-Almaraz, J.R. Instrumentation in Developing Chlorophyll Fluorescence Biosensing: A Review. *Sensors* **2012**, *12*, 11853–11869. [CrossRef] [PubMed]

11. Sighicelli, M.; Iocola, I.; Pittalis, D.; Luglié, A.; Padedda, B.M.; Pulina, S.; Iannetta, M.; Menicucci, I.; Fiorani, L.; Palucci, A. An Innovative and High-Speed Technology for Seawater Monitoring of Asinara Gulf (Sardinia-Italy). *Open J. Mar. Sci.* **2014**, *4*, 31–41. [CrossRef]

12. Hur, J.; Hwang, S.-J.; Shin, J.-K. Using Synchronous Fluorescence Technique as a Water Quality Monitoring Tool for an Urban River. *Water Air Soil Pollut.* **2008**, *191*, 231–243. [CrossRef]

13. Lakowicz, J.R. *Principles of Fluorescence Spectroscopy*, 3rd ed.; Springer-Verlag: New York, NY, USA, 2006; pp. 623–673.

14. Measurement of Environmental Contaminant Hydrocarbons by Fluorescence Spectroscopy. Princenton Instrumets. Available online: http://www.princetoninstruments.com/Uploads/Princeton/Documents/Library/UpdatedLibrary/EEM_for_Environmental_contaminant_hydrocarbons.pdf (accessed on 18 January 2016).

15. Jones, R.N. The Spectrographic Analysis of Carcinogenic Hydrocarbons and Metabolites. III. Distribution of 1,2,5,6-Dibenzanthracene in Rats Following Subcutaneous Injection in Olive Oil. *Cancer Res.* **1942**, *2*, 252–255.

16. Van Duuren, B.L. Identification of some polynuclear aromatic hydrocarbons in cigarette-smoke condensate. *J. Natl. Cancer Inst.* **1958**, *21*, 1–16. [PubMed]

17. Northrop, D.C.; Simpson, O. Electronic Properties of aromatic hydrocarbons. II. Fluorescence transfer in solid solutions. *Proc. R. Soc. London A Math. Phys. Eng. Sci.* **1956**, *234*, 136–149. [CrossRef]

18. Muramoto, Y.; Kimura, M.; Nouda, S. Development and Future of Ultraviolet Light-Emitting Diodes: UV-LED Will Replace the UV Lamp. *Semicond. Sci. Technol.* **2014**, *29*, 84004–84011. [CrossRef]

19. Venugopalan, H. UVC LEDs enable cost-effective spectroscopic instruments. *Laser Focus World* **2015**, *51*, 81–85.

20. Arques Orobon, F.J.; Gonzalez Posadas, V.; Jimenez Martin, J.L.; Gutierrez Rios, J.; Esteban Orobio, A. Fluoro-sensing applied to detection and identification of hydrocarbons in inland waters. In Proceedings of the 10th IEEE International Conference on Networking, Sensing and Control (ICNSC), Evry, France, 10–12 April 2013; pp. 193–198.

21. Montes-Hugo, M.; Fiorani, L.; Marullo, S.; Roy, S.; Gagné, J.-P.; Borelli, R.; Demers, S.; Palucci, A. A Comparison between Local and Global Spaceborne Chlorophyll Indices in the St. *Lawrence Estuary. Remote Sens.* **2012**, *4*, 3666–3688. [CrossRef]

22. Tagus Hydrographic Confederation, Spanish Ministry of Agriculture, Food and Environment Minister. Available online: http://hispagua.cedex.es/en/instituciones/confederaciones/tajo (accessed on 18 January 2016).

23. Tagus Hydrographic Confederation, (Confederación hidrográfica del Tajo). Red SAICA (Sistema Automático de Información de Calidad de las Aguas—Automatic Information System Water Quality). Available online: http://www.chtajo.es/Informacion%20Ciudadano/Calidad/AguasSup/Paginas/RedSAICA.aspx (accessed on 18 January 2016). (In Spanish)

24. Arques-Orobon, F.J.; Nuñez, N.; Vazquez, M.; González-Posadas, V. UV LEDs Reliability Tests for Fluoro-Sensing Sensor Application. *Microelectron. Reliabil.* **2014**, *54*, 2154–2158. [CrossRef]

25. Arques-Orobon, F.J.; Nuñez, N.; Vazquez, M.; Segura-Antunez, C.; González-Posadas, V. High-Power UV-LED Degradation: Continuous and Cycled Working Condition Influence. *Solid-State Electron.* **2015**, *111*, 111–117. [CrossRef]

26. Jimenez Martin, J.L.; Gonzalez-Posadas, V.; Gonzalez-Garcia, J.E.; Arques-Orobon, F.J.; Garcia Munoz, L.E.; Segovia-Vargas, D. Dual Band High Efficiency Class Ce Power Amplifier Based on Crlh Diplexer. *Prog. Electromag. Res.* **2009**, *97*, 217–240. [CrossRef]

27. Jiménez-Martín, V.; Gonzalez-Posadas, F.J.; Arques-Orobon, L.E.; Garcia-Muñoz, D. Dual Band High Efficiency Power Amplifier Based on CRLH Lines. *Radioengineering* **2009**, *18*, 567–578.

28. Pajares, G. Sensors in collaboration increase individual potentialities. *Sensors* **2012**, *12*, 4892–4896. [CrossRef] [PubMed]

29. Boniewicz-Szmyt, K.; Pogorzelski, S.; Mazurek, A. Hydrocarbons on sea water: Steady-state spreading signatures determined by an optical method. *Oceanologia* **2007**, *49*, 413–437.

30. Hirayama, H. Building brighter and cheaper UV LEDs. *Semicond. Compd.* **2013**, *19*, 43–48.

31. Melendez-Pastor, I.; Navarro-Pedreño, J.; Gómez, I.; Koch, M. Identifying Optimal Spectral Bands to Assess Soil Properties with VNIR Radiometry in Semi-Arid Soils. *Geoderma* **2008**, *147*, 126–132. [CrossRef]

32. Savitzky, A.; Golay, M.J.E. Smoothing and Differentiation of Data by Simplified Least Squares Procedures. *Anal. Chem.* **1964**, *36*, 1627–1639. [CrossRef]

33. Schafer, R.W. What is a Savitzky-Golay filter? [lecture notes]. *IEEE Signal Process. Mag.* **2011**, *28*, 111–117. [CrossRef]

34. Melendez-Pastor, I.; Almendro-Candel, M.; Navarro-Pedreño, J.; Gómez, I.; Lillo, M.; Hernández, E. Monitoring Urban Wastewaters' Characteristics by Visible and Short Wave near-Infrared Spectroscopy. *Water* **2013**, *5*, 2026–2036. [CrossRef]

35. Slade, W.H.; Boss, E. Spectral Attenuation and Backscattering as Indicators of Average Particle Size. *Appl. Opt.* **2015**, *54*, 7264. [CrossRef] [PubMed]

36. Taylor, J. Industry Alliance Proposes Standard Definition for LED Life. *LED's Mag.* **2005**, *2*, 9–11.

37. *Approved Method for Lumen Maintenance Testing of LED Light Source*; Illuminating Enginering Society: New York, NY, USA, 2008.

38. Arrhenius, S. Über die Reaktionsgeschwindigkeit bei der Inversion von Rohrzucker durch Säuren. *Z. Physikal. Chem.* **1889**, *4*, 226–248.

39. Philips Lumileds. Luxeon reliability, Reliability Datasheet RD25, Section: Estimating Failure Rates over Temperature 2006. Available online: https://www.olino.org/wp-content/uploads/2008/articles/thousand_lumen_led_Luxeon_RD25.pdf (accessed on 18 January 2016).

40. Packard, H. Reliability of Precision Optical Performance AlInGaP LED Lamps in Traffic Signals and Variable Message Signs. Application Brief I-004. 1999. Available online: www.semiconductor.agilent.com (accessed on 18 January 2016).

41. IES-TM-21-11. In *Standard. Projecting Long Term Lumen Maintenance of LED Light Sources*; Illuminating Engineering Society: New York, NY, USA, 2011.

42. Glaab, J.; Ploch, C.; Kelz, R.; Stölmacker, C.; Lapeyrade, M.; Ploch, N.L.; Rass, J.; Kolbe, T.; Einfeldt, S.; Mehnke, F.; *et al.* Degradation of (InAlGa)n-Based UV-B Light Emitting Diodes Stressed by Current and Temperature. *J. Appl. Phys.* **2015**, *118*, 094504. [CrossRef]

43. Vázquez, M.; Núñez, N.; Nogueira, E.; Borreguero, A. Degradation of AlInGaP Red LEDs under Drive Current and Temperature Accelerated Life Tests. *Microelectron. Reliabil.* **2010**, *50*, 1559–1562. [CrossRef]

44. Chang, M.-H.; Das, D.; Varde, P.V.; Pecht, M. Light Emitting Diodes Reliability Review. *Microelectron. Reliabil.* **2012**, *52*, 762–782. [CrossRef]

45. Mukai, T.; Morita, D.; Yamamoto, M.; Akaishi, K.; Matoba, K.; Yasutomo, K.; Kasai, Y.; Sano, M.; Nagahama, S. Investigation of Optical-Output-Power Degradation in 365-Nm UV-LEDs. *Phys. Status Solidi* **2006**, *3*, 2211–2214. [CrossRef]

46. Núñez, N.; González, J.R.; Vázquez, M.; Algora, C.; Espinet, P. Evaluation of the Reliability of High Concentrator GaAs Solar Cells by Means of Temperature Accelerated Aging Tests. *Prog. Photovol. Res. Appl.* **2013**, *21*, 1104–1113. [CrossRef]

47. ReliaSoft Corporation, Weibull++ and ALTA PRO, Software Packages, Tucson, AZ. Available online: www.ReliaSoft.com (accessed on 18 January 2016).

Article

Laser Spot Detection Based on Reaction Diffusion

Alejandro Vázquez-Otero [1,2,*], Danila Khikhlukha [2], J. M. Solano-Altamirano [2,3], Raquel Dormido [1] and Natividad Duro [1]

1   Department of Computer Sciences and Automatic Control, UNED, C/ Juan del Rosal, 16, 28040 Madrid, Spain; raquel@dia.uned.es (R.D.); nduro@dia.uned.es (N.D.)

2   ELI Beamlines, Institute of Physics ASCR, Na Slovance 2, Prague 8, 18221, Czech Republic; Danila.Khikhlukha@eli-beams.eu (D.K.); jmanuel.solano@correo.buap.mx (J.M.S.-A.)

3   Facultad de Ciencias Químicas, Benemérita Universidad Autónoma de Puebla, 14 Sur y Av. San Claudio, Col. San Manuel, 72530 Puebla, Mexico

*   Correspondence: alejandro.vazquez-otero@eli-beams.eu; Tel.: +420-266-051-312

Academic Editor: Gonzalo Pajares Martinsanz

Received: 31 December 2015; Accepted: 23 February 2016; Published: 1 March 2016

Abstract: Center-location of a laser spot is a problem of interest when the laser is used for processing and performing measurements. Measurement quality depends on correctly determining the location of the laser spot. Hence, improving and proposing algorithms for the correct location of the spots are fundamental issues in laser-based measurements. In this paper we introduce a Reaction Diffusion (RD) system as the main computational framework for robustly finding laser spot centers. The method presented is compared with a conventional approach for locating laser spots, and the experimental results indicate that RD-based computation generates reliable and precise solutions. These results confirm the flexibility of the new computational paradigm based on RD systems for addressing problems that can be reduced to a set of geometric operations.

Keywords: reaction diffusion models; Turing patterns; reaction diffusion computers; reaction diffusion computation; Fitzhugh-Nagumo model; laser spot position; laser spot detection; laser beam detection

1. Introduction

Nowadays, self-organization is a prolific area of research that throughout the last century has struggled with developing models to reproduce the complexity found in nature. Reaction Diffusion (RD) system models belong to this class, and they aim to describe chemical reactions, that under some circumstances [1], may exhibit stable pattern formation, or even more rich dynamics [2,3]. All of these interesting properties have already been explored in the context of problem-solving by means of diverse experimental setups [4–10]. Unlike its experimental counterparts, computational implementations do not rely on energy dissipation, for there is no consumption of chemical species as is required in the *thermodynamics of out-of-equilibrium*. Therefore, it results in the appealing possibility of taking advantage of all the properties of RD models' dynamics, such as their inherent robustness against noisy data and automatic dismissal of the typical drawbacks of experimental setups. Some of these properties include natural parallelism of the model propagation; for instance, the frontwaves that separate stable states into a bistable configuration propagate with constant velocity, and all possible solutions may be explored in parallel [11,12]. Also, due to its strong nonlinear character, the interactions between these wavefronts produce counterintuitive results. It has been shown that such interactions can be used to encapsulate the underlying logic of different types of geometric operations, and indeed, this has led to the development of a novel computational framework capable of addressing robotics-related problems [13–15].

In this paper, the problem of finding the center of a laser spot with high accuracy by means of the aforementioned RD-based computational framework is addressed, extending its range of applicability beyond robotic approaches. Traditionally, this problem is carried out in several steps. The algorithms are diverse, depending on the specific applications (e.g., gaming, laser measurement, laser triangulation method, cutting edge laser physics experiments such as laser plasma, X-ray production, sub-atomic particle creation, and even military applications such as target designation or ammunition guidance—See [16] and references therein). However, two main steps can be clearly stated: first, a spot detection, commonly based on recognizing circular- or ellipsoidal-like shapes combined with color selectivity, followed by the actual computation of the spot center. Here, only the computation of the center is considered, since we are interested in applications where high accuracy is required, such as high-tech laser experiments, wherein the laser spot is known to occupy the majority of the detection system's output (such as an image provided by a CCD sensor). Contrastingly, in the latter we expect notable aberrations caused by complex production, alignment, and focusing systems. In a research facility such as ELI Beamlines [17], a laser beam is expected to encounter tens of optical devices, such as mirrors (static and deformable), telescopes, lenses, *etc.*, from its production and during its transport towards experimental targets. Therefore, a highly accurate method for performing calibration of the beams is needed. In addition, this method may also be used for post processing of experimental data, wherein accuracy, above speediness, is required for fine-tuning and high-quality experimental characterization of the produced beams.

This paper is organized as follows. The next section provides a brief introduction to the RD systems and a short explanation about how the external (*i.e.*, sensory) information can be introduced into the model dynamics. Section 3 introduces a standard algorithm for the location of a laser spot center, intended to provide a suitable comparison method for the RD-based approach, along with the description of the RD algorithm. The experimental results, as well as a brief discussion of our findings, are presented in Section 4. This paper concludes in Section 5 with a summary of contributions and conclusions, as well as our future directions.

2. The Reaction Diffusion System

A detailed description of the relevant information about RD systems and their dynamics, required for the comprehension of this text, can be found in our previous work [15]; nevertheless, for the sake of clarity and self-containment, some fundamental concepts will be reproduced in this and the following sections. A RD model can be written in the simplest way as a system of two equations:

$$
\begin{aligned}
\dot{u} &= f(u,v) + D_u \triangle u \\
\dot{v} &= g(u,v) + D_v \triangle v
\end{aligned}
\tag{1}
$$

Computationally, this system describes the spatio-temporal evolution of both state variables $u = u(\vec{x}, t)$ and $v = v(\vec{x}, t)$ over an integration grid. Originally, the functions $f(u,v)$ and $g(u,v)$ represent biochemical aspects of the system, describing how the substances (u,v) are converted into each other over time. Whilst the so-called diffusive coupling between cells (denoted by the laplacian terms $\triangle u$ and $\triangle v$) describes how the substances are spread out in space. The two-variable FitzHugh–Nagumo (FHN) model [18,19] has remained as one of the simplest available RD cell models. Its basic formulation is shaped by the following set of coupled equations, where the particular parameters of the model are denoted as α, β, ϵ, and ϕ:

$$
\begin{aligned}
\dot{u} &= \epsilon(u - u^3 - v + \phi) + D_u \triangle u \\
\dot{v} &= (u - \alpha v + \beta) + D_v \triangle v
\end{aligned}
\tag{2}
$$

The first step to analyze its dynamics involves plotting the so-called nullclines, defined as the geometric shape for which $\dot{u} = 0$ and $\dot{v} = 0$ in the absence of diffusion (*i.e.*, $D_u = D_v = 0$). The nullclines provide information about the asymptotic behaviour of each isolated cell, generating a phase space with information about trajectories and fixed points that may be stable or unstable, depending on the model configuration. Upon that, the dynamics of each cell are governed by a combination of the asymptotic behaviour, defined by its nullcline configuration plus the spatial interactions with other cells due to the diffusive coupling. In addition, an interesting feature of these systems is the possibility of codifying external information in the model dynamics. On the one hand, this can be done through the so-called external forcing, which allows the introduction of binary or even gradient-like information in the computational grid permanently. For example, in [14] this method was used for introducing the environmental information coming from sensors in a RD-based robotics exploration algorithm. On the other hand, the external information can be introduced simply as the initial values for the state variables u, v. In such a case the information is volatile, as it will be overwritten by the spatio-temporal evolution of the model. This last option is the one used throughout this work.

3. Laser Spot Centering

3.1. Conventional Algorithm for Laser Spot Centroid Calculations

Conventional spot center-finding algorithms are based on matching basic geometrical shapes such as circles or ellipses. The spot is expected to be quite similar to one of these patterns (only small deviations are considered) and the spot shape is fitted as best as possible to the reference shape. For the fitting part of the method, Least-squares, Hough transforms [20], Fourier transforms, Zernike moments [21], Gaussian functions [16], *etc.*, are commonly used. Many of these methods are also based on interpolation schemes for achieving sub-pixel accuracy. However, all of these algorithms implicitly assume that the laser spot shape is close to a circle. Further refinements of center-location algorithms consider embedding of the spot in noisy environments, fast-acquisition techniques, and multi-spot scenarios. However, to the best of our knowledge, there is no algorithm for processing spots which considerably deviate from circular/ellipsoidal shapes.

In this paper we use a conventional image moment method (IMM) for benchmarking the proposed RD-based algorithm. Although the above mentioned methods [16,20,21] may provide better accuracy for a centroid location, the advantages of IMM are its simplicity and performance. This method can be easy implemented for almost all platforms, including field-programmable gate arrays (FPGA). Moreover, in order to produce a high-performance algorithm, it is feasible to use an iterative process to locate the centroid of the laser spot. Using heuristic optimization techniques (e.g., Hill Climbing [22]), one can recalculate the centroid location after each step of the optimized algorithm. In this case there is no need for such a high accuracy.

The implementation of the image moment method is described below.

- An image acquired from a CCD camera is transformed into a single channel 8-bit gray image
- A simple thresholding is applied to produce a binary image. Each pixel with a value greater than I_{th} is set to $2^7 - 1$
- Using the binary data we can find the outer contour of the spot. The topology information, such as nested contours, is skipped for the sake of simplicity
- Now we can calculate the center of mass of the area surrounded by this contour using the definition of the centroid:

$$\bar{x} = \frac{\sum\limits_{i,j} x_i I_{ij}}{\sum\limits_{i,j} I_{i;j}} \quad \bar{y} = \frac{\sum\limits_{i,j} y_j I_{ij}}{\sum\limits_{i,j} I_{i;j}} \tag{3}$$

where $I_{i,j}$ is an image matrix

Figure 1 illustrates different stages of the above algorithm.

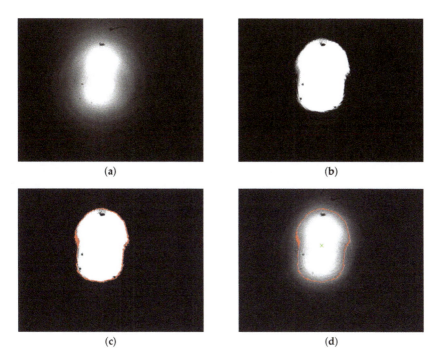

Figure 1. Steps of the image moment method applied to the original image *sample19*: (**a**) 8-bit single channel gray scale image; (**b**) binary image, obtained using the threshold value $I_{th} = 2^7$; (**c**) contour detection using the binary data; (**d**) centroid calculation.

3.2. RD-FHN Laser Spot Centering

3.2.1. Exemplification of the FHN-Based Computation

The technique for calculating the coordinates of the laser spot by means of the FHN model uses the switch-of-phase mechanism introduced in [14,15] that analyses the nullclines in a bistable asymmetric configuration (see Figure 2). For a bistable nullcline configuration, it is possible to locally modify the relative stability of both stable points in order make one of them more stable than the other. In a broad description, the relative stability can be related to the area under the curve as shown in Figure 2a,b. The notation $(SS_{\{+,-\}})_{\{a,b\}}$ denotes the most and least stable points respectively, for both configurations, a and b. Based on this, the following description aims to showcase the mechanism by which the RD model is used for computation:

- First, a two-dimensional integration grid, where each cell has a bistable nullcline configuration similar to the one depicted in Figure 2a is prepared. As initial conditions, the concentration values of u, v for all the grid points are set in the less stable state SS_-. As a result of this configuration, the system remains static in those levels of concentration for all cells in the integration grid.
- Then, a geometric figure is introduced in the model using u, v. This means that for a few cell points corresponding to the shape of the desired pattern, the values of u, v are set in SS_+. Consequently, a frontwave will be triggered, as can be seen in the series of Figure 3a–c, where the model is evolving from SS_- towards SS_+. The diffusive term in Equation (2) explains this movement. Although the asymptotic behaviour is defined by the nullcline configuration, when a cell point changes its concentration levels it transfers this change to its neighbor's by means of the diffusive link between them. Hence, the global result is a wavefront that moves the whole system towards SS_+.

- Depending of the particular algorithm used, the *END* condition for the current *phase* will be different. For instance, it can be stopped after a specific number of iterations, or wait until the model stops its evolution after reaching a static situation. Thus, in the previous step the spatio-temporal evolution was arbitrarily stopped after 1000 iterations.
- Finally, switching the nullcline configuration to the one depicted in Figure 2b produces a shift in the relative stability of the stable states. What in the previous step was SS_+ now becomes SS_-, and *vice versa*. This is possible due to the fact that the stable points in both configurations $(SS)_{\{a,b\}}$ are very close to each other, as depicted in Figure 2c, where both configurations were superimposed. Therefore, after a short period of adaptation, $(SS_-)_a$ becomes $(SS_+)_b$, and $(SS_+)_a$ becomes $(SS_-)_b$. As can be expected, the system starts evolving all its grid points towards the concentration values of u, v, corresponding to the new SS_+. In short, the original wavefront changes its direction of propagation. This process can be seen in the series of Figure 3c–e, where the model evolution was again stopped after 1000 iterations.

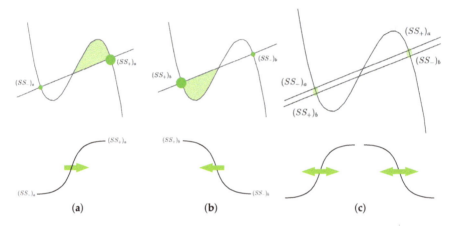

Figure 2. Nullcline representations for a bistable system. (**a**) bistable asymmetric configuration in which the second stable point is the most stable—therefore, in a system endowed with such a configuration, the wavefronts will always travel towards $(SS_+)_a$; (**b**) on the contrary, in this case, the first stable point is the most stable; (**c**) both asymmetric configurations are now superimposed, thus exposing the proximity of their stable points.

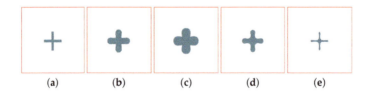

Figure 3. Model spatio-temporal evolution in a two-dimensional grid. (**a**) original pattern introduced as the initial condition by means of u; (**b**,**c**) the model is left to freely evolve for a limited number of iterations; (**d**,**e**) a switch of the nullclines configuration reverses the wavefront's direction of propagation, and the original pattern contracts over itself.

Needless to say, reversing the direction of propagation of the wavefronts will not have any other result than modifying the size of the original pattern that was introduced as initial conditions, as it has been shown in the whole series of pictures depicted in Figure 3. Therefore, in the current approach, endowing the system with computing capabilities means to encapsulate geometric operations in the wavefront propagation. In other words, it is necessary to find configurations for the FHN-model whose wavefront interactions can be related with specific geometric operations, instead of simply producing a change in the direction of propagation of the wavefronts. This fact turns out to be the cornerstone of the RD-based computation, at least in the approach that is exemplified above.

Herein, four different phases in correlation with four different model configurations are used. Based on the above description of the algorithm, each of these phases can be considered as a filter that performs a geometric operation over the original image. The succession of phases transforms a raw image coming from the camera into a small spot corresponding to the expected location of the laser spot center. Additionally, the spot size is increased in the last phase of the algorithm, to reach a desired size for the section of the laser beam.

3.2.2. Thresholding Camera Data

The raw data coming from sensors has to be *adapted* somehow before being used in any algorithm. In the present case, the cameras provide a gray scale pictures with 8 bits. Besides, for the location of the laser center spot, only the information that represents laser light is relevant, because only such information (with respect to the current approach) will trigger the wavefronts used in the RD computation. Therefore, all information that does not represent laser light has to be filtered based on the particular value of a grayscale threshold.

The histograms for the 19 images that will be used in Section 4 are shown in Figure 4.

It can be seen that the majority of the information corresponding to the background is below 30 in the abscissa. This is approximately the starting point from which the plateau of the histogram starts for all the images. Values above this threshold represent laser light. The histogram also acts as a calibration method for both cameras and the RD algorithm.

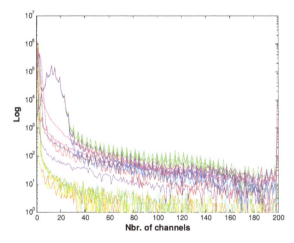

Figure 4. Overlaid histograms corresponding to the 22 samples. Notice that the Y-axis representing the number of pixels is in logarithmic scale.

3.2.3. Algorithm Description

As described in Section 2, the FHN model accepts external information by simply using the initial concentration values of the state variables u, v. This will introduce a "pattern" in the system, as discussed in Section 3.2.1. Then, the system will evolve and clearly will overwrite such a pattern by the new u, v values. The evolution will be different in regards to the particular model configuration. In the present case, the "pattern" to be introduced corresponds the profile of the laser beam that will be transferred at the first stage of the algorithm using the u variable. For that purpose, the matrix containing the camera data has to be mapped from the original [30 200]. After that, by means of the previously described *switch-of-phase* mechanism, the algorithm will perform smooth transitions from one *phase* to the next. Also, when a specific *phase* of the algorithm relies on a nullcline configuration similar to the one depicted in Figure 2a, it will be called *expansion-phase*, whilst the nullclines depicted in Figure 2b will correspond to a *contraction-phase*. This terminology was first introduced in [23], and it has been maintained in subsequent publications for the sake of clarity. The different stages of the FHN-based laser spot centering algorithm are described below:

- Firstly, a contraction phase where the result of the model evolution will be removal of the background noise. Besides, the edges of the beam section become smoother. An exemplification of this process is depicted in the series of pictures in Figure 5a–d. This phase runs for only 1500 iterations, a value that was determined experimentally.
- The second phase, an expansion in this case, removes all the remaining patches of wavefronts that appear as isolated regions and also further smooths the contour of the laser beam. This process can be seen in the series of pictures depicted in Figure 5e–h. In this case, 5000 iterations is the appropriate value.
- The third phase is again a contraction, where the wavefront representing the laser beam contour retracts over itself, decreasing its size until the supposed laser spot center. The result of this phase is a small spot that coincides with the centering of the laser beam. The process is shown in the series of pictures depicted in Figure 5i–o. The end condition of this phase is determined by the system reaching a static situation in the model evolution. The main reason underlying this behaviour is the constant speed of the wavefront propagation; therefore, the velocity of contraction is the same in all directions.
- A final expansion phase results in the growth of the small spot that was achieved in the previous step until it reaches the size of the supposed original laser beam contour. Although this phase is non-essential for obtaining the coordinates of the laser spot, it shows how the FHN-based computational approach is able to process geometric information. The process is represented in the last series of pictures shown in Figure 5p–t. In this last case, the number of iterations varies according to the desired size of the beam section, from which it is also trivial to extract the beam contour.

It must be noticed that all images in Figure 5 have been cut out around the area of interest to include a larger number of them in the figure. Also the chosen representation aims to give a 3D glimpse of what is actually a 2D grid, which makes it easier to see the RD evolution, but also slightly deforms the final perspective.

The set of FHN model values for each phase is represented in Table 1.

Table 1. Set of values for the FitzHugh–Nagumo (FHN) model, in order to reproduce the specific behaviors in each of the four stages of the Reaction Diffusion (RD)-based algorithm.

		FHN Parameters				
		α	β	ϵ	D_u	D_v
Phase	contraction	5	0.1	10	0.1	1.5
	expansion	4	−0.5	40	0.45	2
	contraction	5	0.1	10	0.1	1.5
	expansion	5	−0.2	10	0.1	1.5

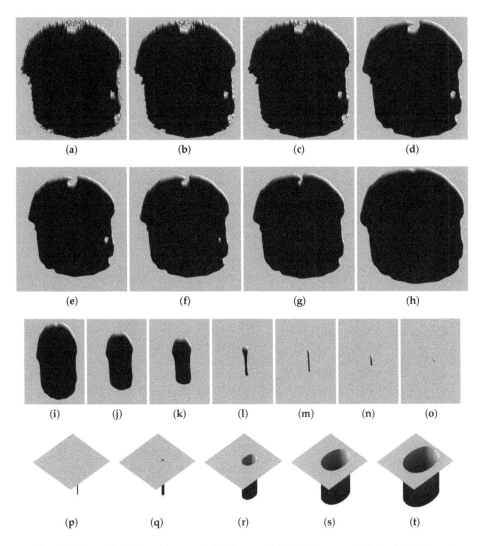

Figure 5. Steps of the RD-based laser spot algorithm applied to the image *sample19*: (**a–d**) first stage of the algorithm, where a contraction phase smoothes all of the noisy contour; (**e–h**) second stage, where an expansion phase removes undesirable isolated spots that are interpreted as noise; (**i–o**) third stage where an cotraction phase reduces its size until the supposed laser spot center; (**p–t**) final and extra stage that reconstructs the expected circular beam.

4. Results and Discussion

In our experience, selection of the right metrics is a cumbersome task when trying to compare the outcome of a RD-based algorithm with the results of a standard one. However, in this work, the offset vector between centroids calculated by IMM and FHN-based method has been chosen. Since the same images were used as an input for both calculations, such vector being normalized to the number of pixels in each direction provides a meaningful metrics. The results are summarized in Table 2 which shows a good agreement between the standard image moment method and its FHN-based counterpart. Figure 6 depicts the results in terms of distances between centroids calculated by both methods.

Table 2. The comparison between the image moment method (IMM) and the RD method. Coordinates are in pixels. The offset value is normalized to the number of pixels in each direction.

		Centroid Coordinates					
		IMM x	IMM y	RD x	RD y	Offset x (10^{-3})	Offset y (10^{-3})
	1	342.330	584.317	342.452	584.337	−0.095	−0.021
	2	353.200	590.375	353.304	590.349	−0.081	0.027
	3	342.330	584.317	342.452	584.337	−0.095	−0.021
	4	366.411	585.350	366.306	585.524	0.082	−0.181
	5	373.349	594.326	373.339	594.661	0.008	−0.349
	6	568.699	552.488	562.710	562.493	4.679	−10.422
	7	649.642	536.352	648.890	547.549	0.588	−11.664
	8	646.670	599.226	644.034	615.806	2.059	−17.271
	9	131.899	460.435	132.180	460.296	−0.220	0.145
Image	10	662.283	453.218	661.256	458.574	0.802	−5.579
	11	662.263	453.169	661.517	458.714	0.583	−5.776
	12	662.642	453.075	661.465	458.493	0.920	−5.644
	13	213.383	477.399	216.049	480.340	−2.083	−3.064
	14	655.300	510.910	653.975	520.126	1.035	−9.600
	15	187.473	472.137	188.648	473.333	−0.918	−1.246
	16	655.563	511.081	654.045	520.999	1.186	−10.331
	17	625.730	463.174	624.991	469.963	0.577	−7.072
	18	578.307	531.479	586.126	523.829	−6.109	7.969
	19	574.611	380.773	577.164	380.872	-1.995	-0.103

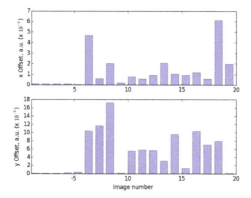

Figure 6. Illustration of the results depicted in Table 2. The height of each bar corresponds to the distance between centroids calculated by both methods.

It is remarkable that many of the traditional center-location algorithms tacitly assume that the laser spot outer contour can be approximated to a circle or an ellipse. In some cases, the beam is intentionally defocused in order to fit a Gaussian profile (whose contour maps are also circles—See for instance [16]), which may introduce large errors in the center position of non-symmetrical spots. On another hand, while using ellipses to approximate the shape of the spot contour may improve the calculation, one would expect the algorithm to produce non-neglectable errors for non-symmetrical spots as well.

Regarding the RD-based method, it is noteworthy that no fine tuning at all was necessary in any phase of the algorithm, in any case. As long as the threshold value obtained from a collection of samples is well defined, the algorithm works seamlessly for finding the center in all cases. Indeed, three samples were removed from the collection used in the comparison, as the IMM method was not able to properly handle those samples. However, this fact is not really relevant, because the IMM algorithm can be improved as much as necessary in regards to the particular optical set up (beam path, optical elements, *etc.*) that produces the misalignments, or can also be combined with other algorithms to increase its accuracy. It is, however, worth mentioning the robustness of the FHN-based solution. This quality makes the algorithm especially interesting for being integrated in other systems, for instance in an automatic alignment setup, where the robustness of the method (understood as the fact that a solution will be always provided) is preferable over accuracy.

5. Conclusions and Future Work

The proposed algorithm can accurately find the center of a spot, regardless of its shape or whether the spot image is highly noisy. As shown in the previous section, the algorithm behaves very well when the images are noisy, and when the spot presents strong arbitrary aberrations. In this context, the RD-based method may be useful during the calibration of beams in research facilities, or in data post-processing, wherein robustness of the method is preferred.

From a broader perspective, the RD-based computation has been in the spotlight of many works regarding non-linear science for a long time, due to the numerous and interesting properties exhibited by the spatio-temporal evolution of the models. The main problem faced by RD computation was the lack of stable states as the outcome of the computation. This stems from the fact that using excitable nullcline configurations has been the norm so far for implementing RD-based computation. However, such configurations only generate transient states that have to be checked iteratively, a fact that undermines the usefulness of the algorithm. The herein-presented approach based on the switch-of-phase mechanism in a bistable asymmetric nullcline configuration overcomes that problem, generating final stable states as the output of the computation.

Furthermore, the presented results are an extension of our previous studies, where different behaviors have been successively included in a RD-based computational core. The addition of the laser spot detection algorithm adds new capabilities to that novel approach, and represents another step in the direction of consolidating a computational framework completely based in RD computation. For that reason, it will be interesting to continue pursuing applications for the developed framework, to extend the range of applicability, and also with the aim of demonstrating the feasibility of using RD models to develop general purpose computation. This will also make interesting a port to specialized hardware.

Acknowledgments: This work has been funded by the National Plan Projects DPI2011-27818-C02-02, DPI2012-31303, and DPI2014-55932-C2-2-R, under the supervision of the Spanish Ministry of Science and Innovation and the FEDER funds. Supported by the project ELI: Extreme Light Infrastructure Beamlines (CZ.02.1.01/0.0/0.0/15_008/0000162) from European Regional Development Fund.

Author Contributions: Alejandro Vázquez-Otero, Danila Khikhlukha and J.M. Solano-Altamirano conceived and designed the experiments, performed the experiments and analyzed the data. Alejandro Vázquez-Otero, Danila Khikhlukha and J.M. Solano-Altamirano wrote the paper. Natividad Duro and Raquel Dormido contributed analyzing the data and writing the paper.

Conflicts of Interest: The authors declare no conflict of interest.

References

1. Turing, A.M. The Chemical Basis of Morphogenesis. *Philos. Trans. R. Soc. Lond. Ser. B Biol. Sci.* **1952**, *237*, 37–72.
2. Vanag, V.K.; Epstein, I.R. Pattern formation mechanisms in reaction-diffusion systems. *Int. J. Dev. Biol.* **2009**, *53*, 673–683.
3. Epstein, I.R.; Berenstein, I.B.; Dolnik, M.; Vanag, V.K.; Yang, L.; Zhabotinsky, A.M. Coupled and forced patterns in reaction-diffusion systems. *Philos. Trans. R. Soc. A: Math. Phys. Eng. Sci.* **2008**, *366*, 397–408.
4. Krinsky, V.I.; Biktashev, V.N.; Efimov, I.R. Autowave principles for parallel image processing. *Phys. D* **1991**, *49*, 247–253.
5. Kuhnert, L.; Aglazde, K.I.; Krinsky, V.I. Image-Processing Using Light-Sensitive Chemical Waves. *Nature* **1989**, *337*, 244–247.

6. Steinbock, O.; Kettunen, P.; Showalter, K. Engineering of dynamical systems for pattern recognition and information processing. *J. Phys. Chem.* **1996**, *100*, 18970–18975.

7. Perez-Munuzuri, V.; Perez-Villar, V.; Chua, L. Autowaves for image processing on a two-dimensional CNN array of excitable nonlinear circuits: Flat and wrinkled labyrinths. *IEEE Trans. Circuits Syst. I Fundam. Theory Appl.* **1993**, *40*, 174–181.

8. Steinbock, O.; Showalter, K.; Kettunen, P. Navigating complex labyrinths: Optimal paths from chemical waves. *Science* **1994**, *100*, 868–871.

9. Adamatzky, A.; Holland, O. Voronoi-like Nondeterministic Partition of a Lattice by Collectives of Finite Automata. *Math. Comput. Model.* **1998**, *28*, 73–93.

10. De Lacy Costello, B.; Ratcliffe, N.; Adamatzky, A.; Zanin, A.L.; Liehr, A.W.; Purwins, H.G. The Formation of Voronoi Diagrams in Chemical and Physical Systems: Experimental findings and theoretical models. *Int. J. Bifurc. Chaos* **2004**, *14*, 2187–2210.

11. Muñuzuri, A.P.; Vázquez-Otero, A. The CNN solution to the shortest-path-finder problem. In Proceedings of the 11th International Workshop On Cellular Neural Networks and Their Applications, Santiago de Compostela, Spain, 14–16 July 2008; pp. 248–251.

12. Adamatzky, A.; De Lacy Costello, B.; Asai, T. *Reaction-Diffusion Computers*; Elsevier Science: Philadelphia, PA, USA, 2005.

13. Vázquez-Otero, A.; Faigl, J.; Munuzuri, A.P. Path planning based on reaction-diffusion process. In Proceedings of the 2012 IEEE/RSJ International Conference on Intelligent Robots and Systems (IROS), Vilamoura-Algarve, Portugal, 7–11 October 2012; pp. 896–901.

14. Vázquez-Otero, A.; Faigl, J.; Duro, N.; Dormido, R. Reaction-Diffusion based Computational Model for Autonomous Mobile Robot Exploration of Unknown Environments. *Int. J. Unconv. Comput.* **2014**, *4*, 295–316.

15. Vázquez-Otero, A.; Faigl, J.; Dormido, R.; Duro, N. Reaction Diffusion Voronoi Diagrams: From Sensors Data to Computing. *Sensors* **2015**, *15*, 12736–12764.

16. Dong, H.; Wang, L. Non-Iterative Spot Center Location Algorithm Based on Gaussian for Fish-Eye Imaging Laser Warning System. *Optik* **2012**, *123*, 2148–2153.

17. Mourou, G.A.; Korn, G.; Sandner, W.; Collier, J.L. *ELI White Book*; Andreas Thoss: Meudon Cedex, France, 2011.

18. Fitzhugh, R. Impulses and Physiological States in Theoretical Models of Nerve Membrane. *Biophys. J.* **1961**, *1*, 445–466.

19. Nagumo, J.; Arimoto, S.; Yoshizawa, S. An Active Pulse Transmission Line Simulating Nerve Axon. *Proc. IRE* **1962**, *50*, 2061–2070.

20. Ruifang, Y.; Chun, L.; Zhenheng, L.; Xizhao, L.; Xiaolan, C.; Yuanquing, H. New Algorithm of Sub-pixel Locating Laser Spot Center. In Proceedings of the 4th International Symposium on Advanced Optical Manufacturing and Testing Technologies: Advanced Optical Manufacturing Technologies, Chengdu, China, 20 May 2009.

21. Cui, J.W.; Tan, J.B.; Ao, L.; Kang, W.J. Optimized Algorithm of Laser Spot Center Location in Strong Noise. *J. Phys. Conf. Ser.* **2005**, *13*, 312–315.

22. Russell, S.J.; Norvig, P. *Artificial Intelligence: A Modern Approach*; Prentice-Hall, Inc.: Upper Saddle River, NJ, USA, 1995.

23. Vázquez-Otero, A.; Muñuzuri, A.P. Navigation algorithm for autonomous devices based on biological waves. In Proceedings of the 2010 12th International Workshop on the Cellular Nanoscale Networks and Their Applications (CNNA), Berkeley, CA, USA, 3–5 February 2010; pp. 1–5.

Review

Robot Guidance Using Machine Vision Techniques in Industrial Environments: A Comparative Review

Luis Pérez [1,*], Íñigo Rodríguez [1,†], Nuria Rodríguez [1,†], Rubén Usamentiaga [2] and Daniel F. García [2]

[1] Fundación PRODINTEC, Avda. Jardín Botánico 1345, 33203 Gijón (Asturias), Spain; irf@prodintec.com (I.R.); nrl@prodintec.com (N.R.)

[2] Department of Computer Science and Engineering, Universidad de Oviedo, Campus de Viesques, 33203 Gijón (Asturias), Spain; rusamentiaga@uniovi.es (R.U.); dfgarcia@uniovi.es (D.F.G.)

* Correspondence: lcp@prodintec.com; Tel.: +34-984-390-060 (ext. 1403)

† These authors contributed equally to this work.

Academic Editor: Gonzalo Pajares Martinsanz

Received: 13 January 2016; Accepted: 26 February 2016; Published: 5 March 2016

Abstract: In the factory of the future, most of the operations will be done by autonomous robots that need visual feedback to move around the working space avoiding obstacles, to work collaboratively with humans, to identify and locate the working parts, to complete the information provided by other sensors to improve their positioning accuracy, *etc*. Different vision techniques, such as photogrammetry, stereo vision, structured light, time of flight and laser triangulation, among others, are widely used for inspection and quality control processes in the industry and now for robot guidance. Choosing which type of vision system to use is highly dependent on the parts that need to be located or measured. Thus, in this paper a comparative review of different machine vision techniques for robot guidance is presented. This work analyzes accuracy, range and weight of the sensors, safety, processing time and environmental influences. Researchers and developers can take it as a background information for their future works.

Keywords: machine vision; 3D sensors; perception for manipulation; robot guidance; robot pose; part localization

---

1. Introduction

Since the end of the 18th century with the first Industrial Revolution through the introduction of mechanical production facilities powered by water and steam, factories have experimented big changes in their production systems [1]. The second Industrial Revolution, in the start of the 20th Century, introduced mass production based on the division of labor powered by electrical energy [2]. The third Industrial Revolution of the start of 1970s introduced the use of electronics and information technologies for a further automatization of production [3]. Nowadays, we are involved in the fourth Industrial Revolution, commonly called "Industry 4.0", based on cyber-physical production systems (CPS) and embracing automation, data exchange and manufacturing technologies. These cyber-physical systems monitor the physical processes, make decentralized decisions and trigger actions, communicating and cooperating with each other and with humans in real time. This facilitates fundamental improvements to the industrial processes involved in manufacturing, engineering, material usage and supply chain and life cycle management [4].

Inside each revolution, several milestones have been achieved incrementing the innovation level. For instance, inside Industry 4.0, robot-based automatization has experimented its own revolutions. The first robotic revolution was the industrial automatization, the second one was the introduction of sensitive robots for safe automatization, the third one was the mobility with mobile manipulators, and the fourth and the last one is based on intelligent and perceptive robot systems [5].

The European Commission has set as objective for the Horizon 2020 Work Programme to achieve the leadership in industrial technologies (*i.e.*, Public-Private Partnership Factories of the Future or PPP FoF [6]). For this purpose, process automation and decreased accident rates are both important. Productivity and safety were limited by manual processes in the traditional industry; automatization and intelligent robots drive modern industry towards efficiency, resulting in a rapid increase in productivity, major material and energy savings and safer working conditions. Manufacturing demonstrates a huge potential to generate wealth and to create high-quality and highly skilled jobs [7]. In fact, several technology enablers have been identified within PPP FoF, such as advanced manufacturing processes and technologies, mechatronics for advanced manufacturing systems, including robotics, information and communication technologies (ICT), manufacturing strategies, knowledge-workers and modelling, simulation and forecasting methods and tools. Moreover, the European Commission has identified robotics as a key issue due to its importance for European economy: Europe is one of the world leading regions in industrial robotics with a share of more than 25% of supply and use. It is expected that robotics growth reaches 32 B$ by 2016 and a direct impact on job creation as forecasts determine that each industrial robot needs at least four people to run, maintain and service it [8]. Robotics directly improves society and better living conditions as it addresses global concerns such as climate change, sustainable transport, affordable renewable energy, food safety and security, and coping with an ageing population. Some examples of European projects in this area are TAPAS [9,10], VALERI [11,12] and SYMBIO-TIC [13].

Vision systems are widely used in industry, mainly for inspection and quality control processes [14]. Their use has been increased in applications related to improving the safety of workers in the industrial environment and for robot guidance [15]. Robots need machine vision to move around the working space avoiding obstacles, to work collaboratively with humans, to identify and locate the working parts, to improve their positioning accuracy, *etc.* Depending on the objective, the vision system can be scene-related or object-related [16]. In scene-related tasks the camera is usually mounted on a mobile robot and applied for mapping, localization and obstacle detection. In object-related tasks, the camera is usually attached to the end-effector of the robot manipulator (eye-in-hand configuration), so that new images can be acquired by changing the point of view of the camera.

Industrial robots are able to move to a position repeatedly with a small error of 0.1 mm, although their absolute accuracy can be several mm due to tolerances, eccentricities, elasticities, play, wear-out, load, temperature and insufficient knowledge of model parameters for the transformation from poses into robot axis angles [17–19]. In the automotive industry the accuracy requirement for operations such as spot welding will be of the order of 1 mm. The aerospace industry provides a challenging environment to apply robotics as its accuracy requirements are at least a factor of ten- to twenty-fold higher [20]. Conventional robots are not capable of achieving this accuracy. To improve the accuracy, optical calibration methods, such as laser tracker systems, photogrammetry or vision systems with multiple high resolution cameras, are used to detect the spatial position of the tool tip and to correct the robot motion. The combination of a measurement system with a robot is an optimal solution as it makes use of the robot ability for precise movement and overcomes the accuracy deficiencies. Moreover, working parts can be positioned slightly different from what the robot is expecting. In order to successfully navigate a mobile robot, obtaining detailed information on its immediate environment is the main concern. If the collected data is correct and sufficiently detailed, the creation of a 3D appearance model is possible and developers are given the opportunity to create accordingly sophisticated navigation software.

At this point, machine vision techniques and robotics become the main actors in the industrial scenarios. Thus, this work is a comparative review of different machine vision techniques and their applications for robot guidance, studying and analyzing the state of the art and the different approaches. The suitability of each vision technique depends on its final application, as requirements differ in terms of accuracy, range and weight of the sensors, safety for human workers, acquisition and processing time, environmental conditions, integration with other systems (mainly the robot), and budget. Main challenges are found in textureless surfaces for the correspondence problem, lighting conditions that may cause brightness, occlusions due to the camera point of view, undetermined moving objects, *etc.* Other comparatives can be found in the literature although they are oriented to commercial purposes, centered in one vision technique, or focused on software and algorithms. They are also analyzed in this work.

The rest of the paper is organized as follows: Section 2 resumes the fundamentals of vision-based 3D reconstruction, Section 3 reviews a wide range of robot guidance applications using machine vision, Section 4 performs a comparative analysis according to the application requirements and discusses advantages and drawbacks, and finally main conclusions are found in Section 5.

2. Fundamentals of 3D Reconstruction

Three-dimensional perception is one of the key technologies for robots. A 3D view of the surroundings of the robot is crucial for accomplishing navigation and manipulation tasks in a fully autonomous way in incompletely known environments. Moreover, tele-operation of robots requires a visualization of the environment in a human-readable way, which is important for an intuitive user interface. Thus, vision systems for robot guidance generally need to obtain 3D information.

Given a point in the scene, its corresponding point in the image can be obtained by mathematical models [21]. This is the direct problem. As it determines a set of parameters that describe the mapping between 3D points in the world coordinate system and the 2D image coordinates, this process is also known as camera calibration. The perspective projection of the world coordinates onto the image is generally modeled using the pinhole camera model [22]. Figure 1 shows a graphical representation. Using this model, the image of a 3D point, P, is formed by an optical ray passing through the optical center and intersecting the image plane. The result is the point P' in the image plane, which is located at a distance f (focal length) behind the optical center.

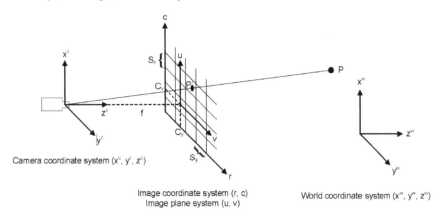

Figure 1. Pinhole camera model [23].

The first step to mathematically describe the projection of 3D points on the 2D image plane is the transformation from the world coordinate system W to the camera coordinate system C. This transformation is given by Equation (1). Using this equation, the camera coordinates of a point $P_c = (x_c, y_c, z_c)^T$ are calculated from its world coordinate $P_w = (x_w, y_w, z_w)^T$ using the rigid transformation $H_{w \to c}$:

$$\begin{pmatrix} P^c \\ 1 \end{pmatrix} = H_{w \to c} \begin{pmatrix} P^w \\ 1 \end{pmatrix} \tag{1}$$

The homogeneous transformation matrix $H_{w \to c}$ includes three translations (t_x, t_y, tz) and three rotations (α, β, γ). These six parameters are called the extrinsic camera parameters and describe the rotation ($R_{w\to c}$) and translation ($t_{w\to c}$) from W to C. Thus, Equation (1) can also be expressed as Equation (2):

$$
\begin{pmatrix} x^c \\ y^c \\ z^c \\ 1 \end{pmatrix} = \begin{pmatrix} R_{w\to c} & t_{w\to c} \\ 0\ \ 0\ \ 0 & 1 \end{pmatrix} \begin{pmatrix} x^w \\ y^w \\ z^w \\ 1 \end{pmatrix}
\tag{2}
$$

Based on the pinhole model, the projection of the point in the camera coordinate system C onto the image coordinate system is calculated using Equation (3):

$$
\begin{pmatrix} u \\ v \end{pmatrix} = \frac{f}{z^c} \begin{pmatrix} x^c \\ y^c \end{pmatrix}
\tag{3}
$$

The pinhole model is only an ideal approximation of the real camera projection. Imaging devices introduce a certain amount of nonlinear distortion [24]. Thus, when high accuracy is required, lens distortion must be taken into account [25]. The final step is the transformation from the image plane coordinate system (u, v) to the image coordinate system (r, c), which is the pixel coordinate system. This transformation is achieved using Equation (4), where S_x and S_y are scaling factors that represent the horizontal and vertical distances between the sensor elements on the CCD chip of the camera and the point $(C_x, C_y)^T$, which is the perpendicular projection of the optical center onto the image plane. Equation (4) reflects the calibration matrix:

$$
\begin{pmatrix} r \\ c \end{pmatrix} = \begin{pmatrix} \dfrac{v}{S_y} + C_y \\ \dfrac{u}{S_x} + C_x \end{pmatrix}
\tag{4}
$$

The projection of a point in the scene on the image can be mathematically calculated as shown in Figure 2a. However, given a point of the image it is not possible to obtain directly its original point in the space, as it is not a one-to-one relationship, but one-to-several, thus the inverse problem is ill-defined [26]. In algebraic terms, the projection of a 3D point on the image is not an injective application. Different points can be projected on the same pixel. What is really obtained by solving the inverse problem is a straight line formed by all points that are represented on the same pixel of the image. This is the projection line shown in Figure 2b.

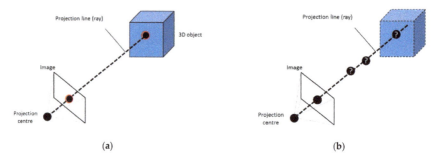

Figure 2. From 3D to 2D. (**a**) Direct problem; (**b**) Inverse problem.

Passive techniques, such as stereo vision or photogrammetry, which only require ambient lighting, solve the problem by looking for the same point in multiple images and computing the intersection of the projection lines. Others project a visible or infrared pattern onto the scene and estimate the depth information from the returning time (time of flight), the deformation of the pattern (light coding) or trigonometric calculations (laser triangulation and structured light). They are active vision. This difference in illumination method is important since the less well-defined features an object may have, the less accurate the system will be when passive vision is used. This is not the case with active vision systems, since a known pattern is used to illuminate the object. Nevertheless, using active vision can result in measurement inaccuracies, especially in the edges or in objects with varying surface finishes [27]. The main disadvantage of non-contact measurement systems is their high sensitivity to various external factors inherent to the measurement process or the optical characteristics of the object [28]. Passive vision techniques need multiple cameras for 3D reconstruction and active ones only use a single camera. However, as it will be shown, some techniques can be passive or active vision or use one or several cameras depending on the application. In addition, some of them are in fact evolutions or improvements of others. Thus, depending on the authors different classifications can be found over the literature. Table 1 shows one possible classification and points out that some technique can be included in more than one group.

Table 1. Vision techniques classification.

	Single Camera	**Multiple Cameras**
Passive vision	2D	Stereo vision Photogrammetry
Active vision	Time of flight Structured light Light coding Laser triangulation	Structured light Projected texture stereo vision

Numerous sensor technologies are available, and each of them provide unique advantages for its use in specific environments or situations. In the following subsections, different 3D vision techniques are presented.

2.1. Stereo Vision and Photogrammetry

Literally photogrammetry consists of measuring real dimensions from a photo of an object. It is a 3D reconstruction technique based on conventional 2D images commonly used in architecture [29–32], topography [33,34], geology [35], archaeology [36], engineering [37,38], and manufacturing [39].

In stereo vision and photogrammetric techniques, the same point has to be found in other image to calculate the intersection of the projection lines and to obtain the 3D position (Figure 3). However, it is recommended that every point could be found at least in three images in order to ensure the detection and to improve the accuracy. The selected points must be the homologous and not any other ones in order to get the right 3D position.

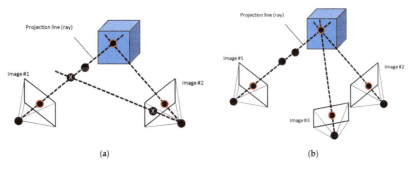

(a)   (b)

Figure 3. From 2D to 3D. (**a**) Homologous points; (**b**) Intersection of the projection lines.

Physical marks, such as stickers or laser points (Figure 4), are necessary over and around the object (the more, the better) and they must be of high contrast in order to ensure the detection. In Figure 5 two images of the same marked part have been taken and processed. The detected marks are printed in red. If the spatial position of the cameras and their calibration parameters are known, the marks can be paired using epipolar geometry and their projection lines and their intersections can be calculated to find the 3D position [40]. Notice that only those marked points are detected and used for the model (marker-based stereo vision).

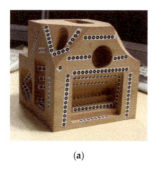
(a)

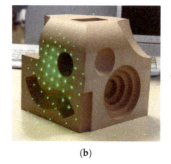
(b)

Figure 4. Physical marks used in marker-based stereo vision. (**a**) Stickers; (**b**) Laser points.

Figure 5. Detection of marks in several images.

In some cases, it is desirable to avoid these physical marks in order to save time and to automate the process. Feature tracking algorithms find, extract and match intrinsic characteristics of the objects between similar or consecutive images- avoiding physical marks, as can be seen in Figure 6 (markerless stereo vision).

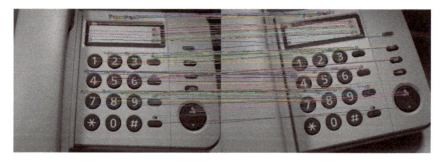

Figure 6. Feature tracking algorithms.

The extracted features depend on the problem or the type of application. In fact, a characteristic feature is a region of interest in the image that provides important information, such as an edge, a corner, bright or dark isolated points, *etc.* These detected points are useful for a subsequent search of correspondences between similar images. Some of the most popular algorithms for features detection, extraction and tracking are Canny [41], Harris [42], KLT [43], SIFT [44], SURF [45], and MSER [46].

Markerless stereo camera systems are widely used in many real applications including indoor and outdoor robotics. They provide accurate depth estimates on well-textured scenes, but often fail when the surface of the object is low-textured or textureless. In this case, it is necessary to project a known static high contrast light on it highlighting points, features, non-visible structures, *etc.* and creating an artificial texture. Then, the reflected light is captured using a stereo camera system and a matching algorithm associates the homologous points to obtain the 3D information [47]. *Ensenso* has developed several series of compact sensors based on this technique [48].

The projected texture is usually pulsed infrared light which is not affected by external light sources. It can take many forms including crosses, circles, squares, dot-matrices, multiple lines and random dot matrices. Finding the optimal texture, that is, the one which provides the best correspondence between features of the images, is a complicated problem, influenced by characteristics of the projector, the pattern, and the stereo cameras [49,50].

2.2. Time of Flight

Active vision techniques obtain the 3D information projecting a visible or infrared pattern on the object as shown in Figure 7. A time of flight (ToF) camera is a range camera that uses light pulses. The illumination is switched on for a very short time. The resulting light pulse is projected on the scene illuminating it and being reflected by the objects. The camera lens captures the reflected light onto the sensor plane. Depending on the distance, the incoming light experiences a delay which can be calculated as shown in Equation (5), where t_D is the delay, D is the distance to the object and c is the speed of light. The pulse width of the illumination determines the maximum range the camera can handle, thus the illumination unit is a critical part of the system. Only with some special LEDs or lasers it is possible to generate such short pulses.

$$t_D = 2 \cdot \frac{D}{c} \tag{5}$$

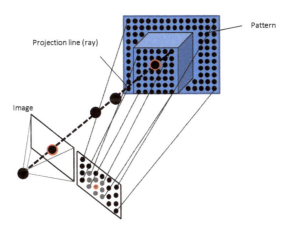

Figure 7. Projecting a pattern on the object.

One of the most common sensors has been developed by *MESA Imaging* [51]. Although ToF sensors open new possibilities and applications as they are quite compact and lightweight, and allow real-time distance acquisition, the quality of raw data is quite noisy and prone to several types of disturbances. They involve major specific challenges: (1) Low resolution compared with other techniques; (2) Systematic distance error; (3) Intensity-related distance error, as the distance is influenced by the incident light; (4) Depth inhomogeneity mainly at object boundaries; (5) Motion artifacts leading to erroneous distance values; (6) Multiple reflections; and (7) Other general aspects of active systems [52].

2.3. Structured Light

Structured light equipment is composed of a light source (the light projector) and one or two information receptors (the cameras). Among all the structured light techniques, there are two main groups [53]: time-multiplexing techniques, which are based on projecting a sequence of binary or grey-scaled patterns, and one-shot techniques, which project a unique pattern. The advantage of time-multiplexing techniques is that, as the number of patterns is not restricted, a large resolution, *i.e.*, number of correspondences, can be achieved. However, their main constraint is that the object, the projector and the camera must all remain static during the projection of the patterns. In one-shot techniques a moving camera or projector can be considered. In order to concentrate the codification scheme in a unique pattern, each encoded point or line is uniquely identified by a local neighbourhood around it [54].

Generally, the projected light is white light which is easily generated and is not dangerous for people unlike laser. This light is modified by grids to create lines or bands with lights and shadows like a zebra (Figure 8) which are recorded by the camera. Depth is obtained from the deviations using a technique similar to triangulation which consists on calculating the intersection between planes and lines [55].

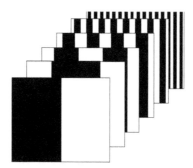

Figure 8. Structured light typical patterns.

Non-contact 3D digitizing scanners derived from structured light projection are increasingly more accurate, fastest and affordable [56]. Recently, new scanners have been launched based on blue light instead of white. As Figure 9 shows, they use a structured blue LED light module and a stereo scanner to generate the 3D point cloud. The LED module produces high contrast patterns allowing a high resolution scanning of the scenario. It is not affected by external light sources and it is safety for people. Sensors as the *HDI 109* and *HDI 120* from *LMI Technologies* can achieve an accuracy of 34 μm and 60 μm, respectively [57].

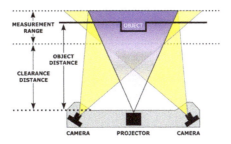

Figure 9. Blue LED sensor components.

2.4. Light Coding

Recently a whole new type of sensor technology called light coding has become available for purchase at only a small fraction of the price of other 3D range finders. Light coding uses an entirely different approach where the light source is constantly turned on, greatly reducing the need for precision timing of the measurements. It can be considered an evolution of structured light.

A laser source emits invisible light (approximately at the infrared wavelength) which passes through a filter and is scattered into a semi-random but constant pattern of small dots which is shown in Figure 10. The reflected pattern is then detected by an infrared camera and analyzed. From knowledge on the emitted light pattern, lens distortion, and distance between emitter and receiver, the distance to each dot can be estimated measuring the deformations in shape and size of the projected points [58]. The technique has been developed by the company *PrimeSense* and its most commonly extended sensor is *Microsoft Kinect*.

Figure 10. Projected pattern in Light coding.

Light coding offers depth data at a significantly low cost, which is a great innovation not only for robotics. However, it has some limitations as these cameras do not provide a dense depth map. The delivered depth images contain holes corresponding to the zones where the sensor has problems, whether due to the material of the objects (reflection, transparency, light absorption, *etc.*) or their position (out of range, with occlusions, *etc.*). The depth map is only valid for objects that are in the range of 1–3 m in order to reduce the effect of noise and low resolution [59]. In addition to this, as it is based on an IR projector with an IR camera, and as the sun emits in the IR spectrum, sunlight negatively affects it.

2.5. Laser Triangulation

In laser triangulation, the point, the camera and the laser emitter form a triangle (Figure 11). The distance between the camera and the laser emitter is known, and because of the angle of the laser emitter corner is also known, the angle of the camera corner can be determined by looking at the location of the laser dot in the camera's field of view. These three pieces of information fully determine the shape and the size of the triangle and give the location of the laser dot corner of the triangle, which is in fact the 3D point. The accuracy depends on the resolution of the CCD sensor, the quality of the lenses, the point size (spot), the laser beam quality, the surface state of the piece and other optical factors [60].

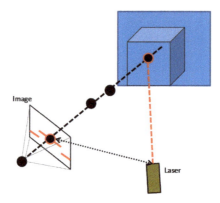

Figure 11. Laser triangulation.

Laser non-contact techniques have changed considerably in recent years. Although initially sensors gave isolated points, the technology has quickly spread to 2D sensors, measuring across a dotted line and collecting multiple points in a single frame. If the piece is moving under the sensor (or *vice versa*), the 3D model of the surface can be generated. This means that, unlike the previously presented techniques where a 3D point cloud can be obtained with a single capture, in this case it is necessary to move the piece or the sensor. In other words, it is necessary to scan the piece. As a drawback, depending on the laser power, laser sensors can be dangerous for people. They are not eye-safe.

3. Robot Guidance in Industrial Environments

Robot guidance using machine vision techniques is a challenging problem as it consists on providing eyes to a machine which is able to move with high repeatability but low accuracy [18] in complex industrial environments with other moving objects or even human workers. This modern technology opens up wholly new possibilities although it also creates new and fairly complex challenges in safety design [61]. Textureless surfaces, lighting conditions, occlusions, undetermined or moving objects, among others, are critical issues which the vision system has to deal with. As shown, 3D point cloud acquisition for robot pose estimation, robot guidance or any other purpose can be achieved by applying many different sensors and techniques. Which one is best suited for each particular application really depends on the needs and requirements [57]. The spatial coordinates of a large number of points can be obtained almost instantaneously or in a few seconds, but they require further treatment. Point clouds must be processed using specific algorithms to identify the geometric elements of the parts to be detected, measured or positioned. Subsequently filtering operations, structuring or interactive segmentation of the point clouds must be carried out [62]. The quality and the robustness of the final application are determined by both processes: the point cloud acquisition and the subsequent treatment.

In this section several approaches of robot guidance using different machine vision techniques are reviewed. As it will be shown, applications are oriented to scene-related tasks for environment reconstruction, including people detection, or to object-related tasks for robot pose estimation and object reconstruction for manipulation or inspection. Figure 12 provides the most common terms of robots that will be mentioned throughout the text.

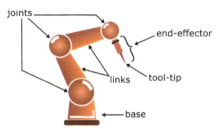

Figure 12. Robot terms.

3.1. Stereo Vision and Photogrammetry

Industrial photogrammetry covers different practical challenges in terms of specified accuracy, measurement speed, automation, process integration, cost-performance ratio, sensor integration and analysis. Main solutions are object-related in the fields of the measurement of discrete points, deformations and motions, 6 DOF (degrees of freedom) parameters, contours and surfaces [63]. Off-line photogrammetry systems can be found at automotive manufacturing (for car body deformation measurement, control of supplier parts, adjustment of tooling and rigs, *etc.*), the aerospace industry (for measurement and adjustment of mounting rigs, alignment between parts, *etc.*), wind energy systems (for deformation measurements and production control), and engineering and construction (for measurement of water dams, tanks, plant facilities, *etc.*). They offer high precise and accurate measurements (the absolute accuracy of length measurements is generally about 0.05 mm for a 2 m object [64]). On the other hand, on-line systems provide 3D information to control a connected process. Some examples include tactile probing (where a hand-held probing device is tracked in 3D space in order to provide the coordinates of the probing tip), robot calibration (where the robot tool center point is observed in order to determine its spatial trajectory), and sensor navigation (where a measurement device, such as a laser profile sensor, is tracked in 6 DOF in order to reconstruct the captured profiles). The accuracy of on-line systems is in the order of 0.2–0.5 mm over a range of 2 m [65], usually less than that of off-line systems. Nowadays, industrial photogrammetric systems are mostly used for off-line measurement of static 3D points in space. Moving from off-line to on-line systems is mainly a matter of speeding up image processing. Image acquisition, transfer, target identification and measurement usually take by far the largest part of the processing time.

There are several approaches in the literature to determine the position and orientation of a robot's end-effector with high accuracy during arbitrary robot motions based on combined and pure photogrammetric solutions. Laser tracking systems combine laser interferometry and photogrammetry [66,67]. As Figure 13 shows, the end-effector of the robot (the probe) is equipped with a number of LED reference targets suitable for camera imaging, as well as a retro-reflector suitable for laser tracking, all with calibrated local coordinates. The 6-DOF pose of the probe is measured by space resection through a camera that is integrated into the laser tracker. Distance information is provided by interferometric laser ranging while the camera measures angular information of the probe [68]. Laser trackers are also used to identify the geometric and dynamic parameters of the robot in order to improve the accuracy of the model, increasing thus the accuracy of the robot [69]. Qu [70] presents a laser tracker application to reduce the relative pose error of the robot of an aircraft assembly drilling process to less than 0.2 mm.

Figure 13. Laser tracker.

In pure photogrammetric solutions, there are three different approaches [18]: (1) Forward intersection, where two or more fixed cameras that are observing target points which are mounted on the end-effector (moving targets); (2) Resection, where alternatively one or more cameras can be mounted on the end-effector observing fixed targets; and (3) Bundle adjustment, which is a combination of both. At a first sight, resection arrangement may seem to be inferior compared to the forward intersection method, because a pose measurement for all possible robot poses requires targets to be placed around the entire workspace. However, most handling tasks require a robot to be accurate only in certain locations, thus the accuracy is not needed in the entire workspace.

Main documented applications are for 6-DOF measurements, robot calibration, object tracking, and robot navigation. Several simultaneously operating solid-state cameras and photogrammetric hardware and software for robot guidance tasks are used in [71]. In this work two different applications were chosen to demonstrate the accuracy, flexibility and speed of the photogrammetric system: 3D object positioning is utilized in the measurement of car body points for accurate seam-sealing robot operation, and a robotized propeller grinding cell uses profile and thickness measuring data to control the grinding process. Hefele [18] shows an off-line photogrammetric robot calibration using a high resolution camera and targets mounted to the robot end-effector. The positioning accuracy of the robot is better than 3 mm in the experiments. First results towards an on-line photogrammetric robot tracking system are also presented in [18], reducing image processing and using intelligent cameras. The bundle adjustment indicates target coordinate RMS values of 0.06 mm in depth. Some improvements were added later in [72].

When accuracy requirements are moderate, [19] presents a digital photogrammetric system for industrial robot calibration tasks. Standard deviations of 0.05–0.25 mm in the three coordinate directions could be obtained over a robot work range of $1.7 \times 1.5 \times 1.0 \text{ m}^3$. In this sense, [20] describes the development of a photogrammetric 6-DOF measurement system mounted on the end-effector of an industrial robot. The functionality of the system has been demonstrated for drilling and assembly operations showing that the scheme is feasible and assesses the capability of such a system to operate within realistic tolerances. The photogrammetric system shown in Figure 14 is proposed in [73] to improve the pose accuracy of industrial robots. Experimental results show that the position error of the robot manipulator is less than 1 mm after being calibrated by the photogrammetric system. Amdal [74] gives a description of an on-line system designed for close range applications. The system has the ability to perform 3D point measurements with one single camera in combination with a measurement tool equipped with photogrammetric targets, which are precalibrated in the local tool coordinate system. Accuracy results obtained from simulation studies and measurement tests are reported. For a camera-to-object distance of 2.5 m, the accuracy was found to be 0.01 mm.

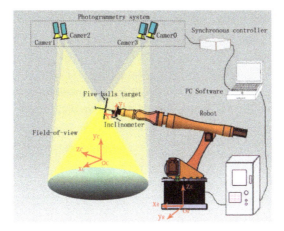

Figure 14. Multiple-sensor combination measuring system [73].

Consequently, photogrammetry is certainly able to determine robot pose accurately [75] although these measurements require special and expensive equipment and processing huge amounts of image data. Because of these reasons, nowadays photogrammetry is mainly limited for calibration, which is performed only once during the setup of the robot at the factory, instead of a continuous tracking of the pose. Recent and future developments are concentrated on higher dynamic applications, integration of systems into production chains, multi-sensor solutions and lower costs.

Projected texture stereo vision technique is mainly used in bin picking applications [76,77]. The task of picking random and unsorted objects from a container or a storage bin presents a number of different challenges. A fast and reliable identification of one or several objects is required in terms of shape, size, position and alignment (Figure 15). This information must be obtained and passed on to the robot controller, which is essential to the *ad hoc* generation of collision-free robot paths. This is the starting point for the use of robots in handling processes.

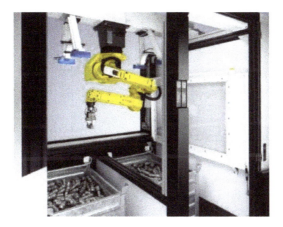

Figure 15. Bin picking [77].

Sturm [78] presents an approach for detecting, tracking, and learning 3D articulation models for doors and drawers using projected texture stereo vision system. The robot can use the generative models learned for the articulated objects to estimate their mechanism type, their current configuration, and to predict their opening trajectory.

3.2. Time of Flight

ToF sensors have several advantages for the development of robotic applications as they are quite compact and lightweight, and allow real-time 3D acquisition with high frame rate. They are used in scene-related tasks, generally involving mobile robots and large displacements, and in object-related tasks, involving instead robotic arms or humanoid-like robots and small depths. However, they involve some challenges as the quality of raw data is quite noisy. To overcome this limitation, some authors apply calibration methods to rectify the depth images in order to obtain better results. Others complement ToF camera information with color cameras to create a 3D point cloud with real features, with grayscale cameras for redundant stereo or with laser scanners.

In the field of scene-related task, ToF camera capabilities in terms of basic obstacle avoidance and local path-planning are evaluated in [79] and compared to the performance of a standard laser scanner. May [80] presents a new approach for on-line adaptation of different camera parameters to environment dynamics. These adaptations enable the usage reliably in real world (changing) environments and for different robotic specific tasks. In [81,82] it is proposed the use of surface normals to improve 3D maps for badly conditioned plane detection. Others, such as [83,84], cope with ToF noisy point clouds using the Iterative Closest Point algorithm to find the relation between two point clouds. Arbeiter [85] performed an environment reconstruction for a mobile robot combining a ToF camera with two color camera, which is the input for a modified fast-SLAM algorithm. This algorithm is capable of rendering environment maps. Kuhnert and Netramai [86,87] combined a ToF sensor and a stereo system for environment reconstruction.

For object-related tasks, ToF cameras have also been successfully used for object and surface reconstruction, where the range of distances is smaller. Depending on the field of view of the camera, multiple 3D point clouds need to be acquired and combined. In fact, the most common setup usually includes a ToF camera mounted on the end-effector of a robotic arm to do the captures. Point cloud registration is more critical in object modeling than in scene modeling. Even if the hand-eye system is precisely calibrated, the displacement given by the robot is usually not enough and the transformation between different point clouds has to be calculated. Some examples of object modeling and object reconstruction can be found in [88–91]. For object manipulation, unknown and unsorted objects have to be identified or categorized in order to be grasped. Generally, it is not necessary to completely reconstruct the object. Some examples are described in [92–94].

Finally, for human-machine interaction, ToF does not require any special background and it is a non-invasive technique, contrary to the widely extended use of special gloves, artificial marks, special attached devices, *etc.* Thus, it is commonly used in human activity recognition than other vision techniques as [95] points out. This work reviews the state of the art in the field of ToF cameras, their advantages, their limitations, and their main applications for scene-related tasks, object-related tasks, and tasks involving humans.

3.3. Structured Light

Positioning a robot with stereo vision depends on features visible from several points of view. Structured light provides artificial visual features independent of the scene, easing considerably the correspondence problem. The main inconvenience for robot guidance is the size of the sensors as they include a projector, which makes them difficult to be attached to the end-effector of a robot. Figure 16 presents a solution where only one camera is attached to the end-effector (eye-in-hand) and a static projector is installed illuminating the working pieces [96].

Figure 16. Robot positioning using structured light [96].

The achieved accuracy is 3 mm, which is enough for this concrete application. This setup also solves a problem related to the breaking of the hot lamp filament of the project due to vibrations if it is moved around. On the other side, an eye-in-hand setup avoids occlusions while it can perceive more details during robot approaching to the scene. The selection of the adequate pattern is also the main focus of most authors. Pagès [54] proposed a coded structured light approach as a reliable and fast way to solve the correspondence problem in another eye-in-hand solution with the projector aside the robot manipulator. In this case, a coded light pattern is projected providing robust visual features independently of the object appearance for robot positioning. Experiments have demonstrated that positioning the robot with respect to planar object provides good results even in presence of occlusions. Results when using non-planar objects show that the camera motion is noisier, slower and less monotonic.

Le Moigne [97] describes some of the important operational considerations and image processing tasks required to utilize a non-scanning structured-light range sensor in path planning for robot mobility. Particular emphasis is placed on issues related to operating in ambient lighting, smoothing of range texture, grid pattern selection, albedo normalization, grid extraction, and coarse registration of images to the projected grid. The created range map can be converted to a topography map for planning short-range paths through the environment while avoiding obstacles.

The new approach of structured blue light is not very extended yet in industrial robotic applications, but it is already being used for part identification and localization in [11], where a mobile and collaborative robot has been developed for aerospace manufacturing industries. Once the robot has reached the working station, it takes the camera, acquires images to get a point cloud (Figure 17) and compares it with a CAD databank, in order to identify the part and its pose. The robot also corrects its own pose to start with the assigned task (apply sealant or do a quality inspection). In this case, the point cloud is highly accurate. However, as the sensor working distance is too short, the robot has to be very close to the part and the scanned area is small. For small parts this is not a problem, but for long ones it requires an accurate CAD matching algorithm in order to avoid deviations in the robot trajectory.

(a)

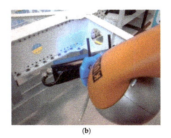
(b)

Figure 17. Mobile robot using a blue light sensor for part localization. (**a**) Mobile robot; (**b**) Sensor operating.

3.4. Light Coding

Nowadays this technique is commonly used in videogames for people tracking. Besides, it is been introduced in more and more industrial applications including robotics although an extra effort is necessary to achieve accuracy for robot pose estimation, and safety requirements for workspace monitoring in human-robot collaboration. It offers visual and depth data at a significantly low cost and, while it is a great innovation for robotics, it has some limitations. Algorithms use depth information for object recognition, but as the sensor produces a noisy point cloud, it is required to improve such information. One possible option is the combination of light coding sensors with HD cameras to obtain high resolution dense point cloud which can be used for robot guidance or pose correction. Experimental results reported in [98] show that this approach can significantly enhance the resolution of the point cloud on both indoor and outdoor scenes. The point cloud is at least ten times denser than the initial one only with the light coding sensor.

To significantly improve the robustness of people detection on mobile robots, light coding cameras have been combined with thermal sensors and mounted on the top of a mobile platform in [99], since humans have a distinctive thermal profile compared to non-living objects. This approach also combines classifiers created using supervised learning. Experimental results of this work have shown that the false positive rate (exclusively achieved using only the light coding sensor) is drastically reduced. As a drawback, some phantom detections near heat sources, such as industrial machines or radiators, may appear. Light coding sensors are also combined with safety certificated 3D zone monitoring cameras in [61]. Wang [100] combined virtual 3D models of robots with the information from the sensor (images of operators) for monitoring and collision detection (Figure 18). 3D models, which are used to represent a structured shop-floor environment and linked to real motion sensors, are driven to mimic the behavior of the real environment. Light coding sensors add unstructured foreign objects, including mobile operators, which are not present in the 3D models.

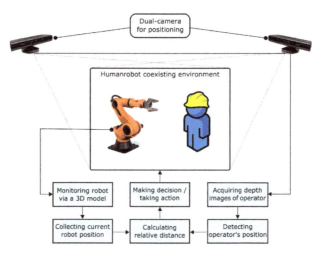

Figure 18. Combining 3D models of robots with information from sensors [100].

One common problem of light coding sensors is that they do not provide a dense depth map. The delivered depth images contain holes corresponding to the zones where the sensor has problems, whether due to the material of the objects (reflection, transparency, light absorption) or their position (out of range, with occlusions).

3.5. Laser Triangulation

This technique has been used in bin picking applications to pick up pieces, where it is necessary to recognize the piece and its pose. As it is necessary to scan the piece for 3D reconstruction, pieces are over a conveyor belt in [101] with a static camera and a static laser or camera and laser are integrated in robot tool for reconstruction and measurement as in [102]. Experimental evaluations of different line extraction algorithms on laser scans for mobile robots in indoor environments are presented in [103,104]. The comparison is carried out in terms of complexity, correctness and precision.

An implementation of a flexible, sensory-controlled robotic welding system is presented in [105]. Conventional, non-adaptive, robot welding systems can only be used when the workpieces are highly repeatable and well fixtured. A steerable cone of laser light and machine vision are used for sensing of the weld joint location and determining the detailed 3D weld joint surface geometry ahead of the welding torch. Robust vision-processing schemes for the detection and recognition of laser stripe features in noisy images are developed and implemented using a pipelined processing architecture. Approaches are proposed and implemented to incorporate the visually determined offsets in robot path planning and to control the welding process parameters. Another example of weld tracking is presented in [106]. This work offers a low-cost system that guarantees satisfactory tracking results even when the welding gap geometry varies strongly.

In [107] a sensor is created by coupling a camera and a laser stripe-. Positioning robotics tasks can be performed with good results, robustness and stability. Nevertheless, there are some constraints such as some restrictions in the laser stripe projection on to the scene. It is necessary to choose the most favorable location of the laser stripe during the calibration to achieve a robotics task under conditions of optimum stability. Pears [108] described a wide field of view range sensor for short range mobile robots maneuvers with an accuracy of 0.15% at l m, 1.3% at 2 m, and 3% at 2.5 m, and an average projected power of 0.9 mW, which is eye-safe. Generally, in robotics applications, it is necessary to take into account safety issues as human workers may be present. Depending on the laser power the sensor can result inappropriate for humans because of safety reasons.

Other works about robot navigation using laser scanners are described in [109–111]. In [109] the problems of self-localization of a mobile robot and map building in an *a priori* unknown indoor environment are addressed. In [110] a method for tracking the pose of a mobile robot using a laser scanner is presented. A new scheme for map building is proposed in [111]. This work describes localization techniques for a mobile robot equipped with a laser rangefinder using line segments as the basic element to build the map. According to the results, line segments provide considerable geometric information about the scene and can be used for accurate and fast localization.

4. Discussion

Depending on the final goal of the application and the type of robot, different considerations and factors need to be taken into account in order to select the most adequate vision technique:

- *Accuracy* of point cloud alignment and *resolution*. They are mainly determined by hardware (sensor) and software (extraction, registration, segmentation, comparison, *etc.*), and in consistence to the size of the object and the application purpose.
- *Range of the sensor*. The working distance will be determined by the accessibility of robot, size of sensor and environment configurations
- *Light weight*. If the sensor is onboard or mounted in the end-effector, the robot has limited max load weight to ensure its full dynamics.
- *Safety issues*. The robot might work closely with human workers, thus sensors should avoid dangerous high-power laser to minimize any risk of accidents.
- *Processing time*. Processing time might be crucial to determine if a system is suitable for a certain application, especially regarding moving robots with safety constraints, *i.e.*, availability to detect

and avoid collisions with humans or obstacles. Some techniques require that object and camera remain static for the capture, thus they are not applicable for moving scenarios.

- *Scanning environment*. Lighting conditions, vibrations, camera movements, *etc.* can disturb the quality of the 3D point cloud in some techniques. It is necessary to avoid these interferences.
- *Hardware and software integration* with other systems. The camera will be automatically controlled by the own robot central control unit or by an external source. *Ad hoc* developments are oriented towards the integration and, nowadays, most of current commercial vision systems are also prepared to be connected to a robot and controlled by external software using SDKs or libraries.
- *Budget*. Outside of technical issues, for a real implementation, budget should also be considered. A trade-off between cost and performance is necessary as most of the previous characteristics can be achieved or improved incrementing the invested amount of money.

A comparison of vision techniques is presented in Table 2 in terms of accuracy, range, weight, safety, processing time, and scanning environmental influences (Env. influences). Some quantitative information is provided according to the referenced specific application. Stereo vision, structured light, and laser triangulation can provide acceptable accuracy under certain conditions for most applications. Except structured white light, active vision techniques need to be closer to the object as they have a short working distance. Nowadays, there are light weight commercial sensors available to be mounted on a robot. Structured white light sensors are in general the biggest. All techniques are not dangerous for people, with the exception of some high power lasers that are not eye-safe. In terms of processing time, photogrammetry requires processing a large amount of images, and structured light techniques require that object and camera remain static during the acquisition process. Time of flight and structured blue LED light are not influenced by environmental lighting conditions.

Table 2. Comparison of vision techniques in terms of accuracy, range, weight, safety, processing time, and environmental influences.

	Accuracy	Range	Weight	Safety	Processing Time	Environmental Influences
Stereo Vision and Photogrammetry	✓ (50 μm [64])	✓	✓	✓	✗ (image processing)	✗ (brightness)
Projected Texture Stereo Vision	✓ (0.1 mm [48])	✗ (0.25–3 m [48])	✓	✓	✗ (image processing)	✗ (brightness)
Time of Flight	✗ (10 mm [51])	✗ (0.8–8 m [16])	✓	✓	✓	✓
Structured White Light	✓ (0.127 mm [112])	✓	✗ (projector [96])	✓	✗ (remain static)	✗ (light, brightness)
Structured Blue LED Light	✓ (34 μm [57])	✗ (157–480 mm [57])	✓	✓	✗ (remain static)	✓
Light Coding	✗ (10 mm [58])	✗ (1–3 m [59])	✓	✓	✓	✗ (sun)
Laser Triangulation	✓	✓	✓	✗ (laser power)	✓	✗ (brightness)

Table 3 compares advantages and disadvantages of the reviewed techniques for robot guidance. Photogrammetry is mainly used in static applications because of its accuracy, but physical marks, such as stickers or laser points, are necessary (marker-based) and it is highly influenced by brightness and lights in industrial environments. Marks can be avoided using feature trackers (markerless), but the density of the point cloud would be low if surfaces are textureless. In fact, low-textured or textureless surfaces are also an inconvenience for stereo vision techniques with conventional 2D cameras and it is necessary to project a high contrast light creating an artificial texture to highlight points, features, *etc.* (projected texture stereo vision). Other 3D active vision techniques, such as light coding and time of flight, have low theoretical accuracy and are not valid for certain applications where the point cloud is compared with a CAD model for accurate part localization, because flat surfaces are represented with rather significant curvature. They can be used for part identification or for people and object tracking considering these accuracy issues. Laser techniques are commonly used in scanning applications where the capture is not a single snapshot, but can be dangerous for people as some laser classes (high power) are not eye-safe. Finally, structured light provides accuracy, although sometimes is influenced by ambient light and has problems to create the 3D model for surfaces of certain colors. Its main disadvantage is that most commercial sensors are quite big to be carried by a robot. Actually, research and development efforts are concentrated on miniaturizing sensors or on separating the projector and the sensor so that only one is onboard the robot. In this sense, the new evolution of this technique called structured blue LED light provides accuracy with a small sensor.

Table 3. Advantages and disadvantages.

	Advantages	Disadvantages
Stereo Vision and Photogrammetry	Commonly used. Accuracy.	Influenced by environment. Physical marks are necessary. The density of the point cloud can be low. Object and camera must be static for the capture.
Projected Texture Stereo Vision	Physical marks are not required.	Influenced by environment. Object and camera must be static for the capture.
Time of Flight	Independent of ambient light. Not necessary that object and camera remain static for the capture.	Low theoretical accuracy.
Structured White Light	Accuracy.	Sometimes influenced by ambient light. Problems to create the 3D model for surfaces of certain colors. Expensive. Sensors can be quite large. Object and camera must be static for the capture.
Structured Blue LED Light	Accuracy. Small sensor.	Short working distance. Object and camera must be static for the capture. Expensive.
Light Coding	Inexpensive. Not necessary that object and camera remain static for the capture.	Low accuracy. Uncertified at industrial level.
Laser Triangulation	Commonly used. Inexpensive (depending on the laser, the accuracy).	Dangerous for people depending on laser power. Usually short working distance. Line scanner.

There are thousands of industrial applications but most of them are confidential for companies and are not documented or widely described in scientific papers, thus they could not be included in this survey. Table 4 summarizes the references of vision techniques for robotics reviewed in this work grouped by scene-related and object-related tasks. According to the survey, main stereo vision applications in robotics are in the field of object-related tasks for robot pose estimation and robot calibration as they may require marks, camera and object must remain static and an important amount of information needs to be processed. Time of flight, which does not require that object and camera remain static, is mainly used for environment and object reconstruction, navigation, obstacle avoidance and indeed people detection. In fact, the main application of light coding is people detection, although it is not certificated for industrial environments. Laser-based sensors are widely used especially for navigation, but also for object-related tasks.

Some other comparative reviews of machine vision techniques for robotic applications can be found in the literature. Wilson [27] reviews 3D vision techniques and solutions in a journal article with commercial purposes providing a list of applications and manufacturers. Sets of questions and advises are presented in [57] in order to help the reader to choose the right 3D vision technique for his/her project. Some of these questions are relative to the size and the surface of the target object, to the accuracy requirements, to the final destination of the obtained 3D data, or to the budget. This classification is quite similar to the one proposed by the authors of this work in Table 2.

Table 4. Common applications in robotics.

	Scene-Related			Object-Related	
	People detection	Environment reconstruction / navigation	Object reconstruction / inspection	Bin picking / object manipulation	Robot pose / calibration
Stereo Vision and Photogrammetry			[74]		[18–20,67–73]
Projected Texture Stereo Vision				[76–78]	
Time of Flight	[95]	[79–87]	[88–91]	[92–94]	
Structured Light					[11,54,96]
Light Coding	[61,98–100]				
Laser Triangulation		[103–111]	[102]	[101]	[107]

Foix [95] focuses on ToF cameras, describing advantages, limitations and applications. It includes 68 references grouped in scene-related, object-related and human-related. Visual human-machine interaction is deeply studied as ToF cameras can be used for gesture recognition. Focusing on software, [113] reviews vision-based control algorithms for robot manipulators and [114] analyzes stereo matching algorithms used in robotics in terms of speed, accuracy, coverage, time consumption, and disparity range. Implementations of stereo matching algorithms in hardware for real-time applications are also discussed.

Robots need flexibility and accuracy to carry out more complex and diverse tasks, such as collision avoidance with static and moving objects during navigation, collaborative work with humans, fine positioning, inspection, *etc.* In all the reviewed applications, each vision system has its single purpose. There has not been found one single vision system able to perform several tasks. Multiple vision systems are used (one for each individual purpose) instead of one single multi-tasking vision system. This is because requirements of each task are quite different and each technique has its scope and is more adequate than others.

5. Conclusions

In this survey, 3D vision solutions used in robotics have been organized, classified and reviewed. The purpose was to provide a compilation of the state of the art and the existing techniques so that future researchers and developers have a background information. Vision techniques for robot guidance have been analysed in terms of accuracy, range, weight, safety, processing time, and scanning environmental influences. Main advantages and main drawbacks have been also presented for each of them. Choosing which type of 3D vision system to use is highly dependent on the parts that need to be located or measured. While laser range finders using time of flight methods can be used to locate distant objects, stereo imaging systems may be better suited to imaging high-contrast objects. Where such objects are highly specular or textureless, it may be more useful to employ projected texture techniques. In addition to this, robot and industrial environments conditions also need to be considered. Each application and each type of robot need a specific vision solution. There is no universal vision technique to perform several tasks. Future woks may focus on multi-tasking or multi-purpose vision systems and their integration with other sensor types and systems.

Robots have become a core element of Industry 4.0 and flexibility can be incorporated to them by vision systems and other sensor technologies in order to achieve the requirements and functionalities of the new applications. New tasks are becoming more or more complex and it is necessary to improve the accuracy and to work collaborative with humans, which means making decisions in real-time and triggering actions. For these goals, visual feedback is the key issue, and this is in fact what vision systems provide to robots. Thus, 3D machine vision is the future for robotics. The idea of considering robot technology as an integral part of production is not new but nevertheless it is a challenge. Whether robots will be able to or should perform all the production steps in future is perhaps less a question of time than of money.

Acknowledgments: The work was supported by European Union H2020 Programme under grant agreement No. 637107, SYMBIO-TIC.

Author Contributions: All authors contributed extensively to the work presented in this paper and wrote the manuscript. Luis Pérez reviewed the state-of-the-art and wrote the initial version of the paper analyzing the vision techniques and their applications for robot guidance. Íñigo Rodríguez and Nuria Rodríguez participated with Luis in the analysis of the vision techniques. Rubén Usamentiaga and Daniel F. García provided suggestions and corrections during the preparation of the submitted paper and the review process.

Conflicts of Interest: The authors declare no conflict of interest.

Abbreviations

The following abbreviations are used in this manuscript:

3D	Three-Dimensional
CPS	Cyber-physical Production Systems
PPP	Public-Private Partnership
FoF	Factories of the Future
ICT	Information and Communication Technologies
CCD	Charge-Coupled Device
2D	Two-Dimensional
CMM	Coordinates Measuring Machine
KLT	Kanade-Lucas-Tomasi
SIFT	Scale-Invariant Feature Transform
SURF	Speeded Up Robust Features
MSER	Maximally Stable Extremal Regions
ToF	Time of Flight
LED	Light Emitting Diode
IR	Infra-Red
DOF	Degrees Of Freedom
RMS	Root Mean Square
SLAM	Simultaneous Localization and Mapping
CAD	Computer-Aided Design
HD	High Definition

References

1. Deane, P.M. *The First Industrial Revolution*; Cambridge University Press: Cambridge, UK, 1979.
2. Kanji, G.K. Total quality management: the second industrial revolution. *Total Qual. Manag. Bus. Excell.* **1990**, *1*, 3–12. [CrossRef]
3. Rifkin, J. The third industrial revolution. *Eng. Technol.* **2008**, *3*, 26–27. [CrossRef]

4. Kagermann, H.; Wahlster, W.; Helbig, J. *Recommendations for Implementing the Strategic Initiative Industrie 4.0: Final Report of the Industrie 4.0 Working Group*; Forschungsunion: Berlin, Germany, 2013.

5. Koeppe, R. New industrial robotics: human and robot collaboration for the factory. In Proceedings of the 2014 European Conference on Leading Enabling Technologies for Societal Challenges (LET'S 2014), Bologna, Italy, 29 September–1 October 2014.

6. European Commission. Factories of the Future in H2020. Available online: http://ec.europa.eu/research/industrial_technologies/factories-of-the-future_en.html (accessed on 5 October 2015).

7. European Factories of the Future Research Association. *Factories of the Future: Multi-Annual Roadmap for the Contractual PPP under Horizon 2020*; Publications office of the European Union: Brussels, Belgium, 2013.

8. European Commission and Robotics. Available online: http://ec.europa.eu/programmes/horizon2020/en/h2020-section/robotics (accessed on 11 October 2015).

9. TAPAS Project. Available online: http://www.tapas-project.eu (accessed on 11 October 2015).

10. Bogh, S.; Schou, C.; Rühr, T.; Kogan, Y.; Dömel, A.; Brucke, M.; Eberst, C.; Tornese, R.; Sprunk, C.; Tipaldi, G.D.; Hennessy, T. Integration and assessment of multiple mobile manipulators in a real-world industrial production facility. In Proceedings of the 45th International Symposium on Robotics (ISR 2014), Munich, Germany, 2–4 June 2014; pp. 1–8.

11. VALERI Project. Available online: http://www.valeri-project.eu (accessed on 11 October 2015).

12. Zhou, K.; Ebenhofer, G.; Eitzinger, C.; Zimmermann, U.; Walter, C.; Saenz, J.; Pérez, L.; Fernández, M.A.; Navarro, J. Mobile manipulator is coming to aerospace manufacturing industry. In Proceedings of the 2014 IEEE International Symposium on Robotic and Sensors Environments (ROSE 2014), Timisoara, Romania, 16–18 October 2014; pp. 94–99.

13. SYMBIO-TIC Project. Available online: http://www.symbio-tic.eu (accessed on 11 October 2015).

14. Labudzki, R.; Legutko, S. Applications of Machine Vision. *Manuf. Ind. Eng.* **2011**, *2*, 27–29.

15. Wöhler, C. *3D Computer Vision: Efficient Methods and Applications*; Springer: Dortmund, Germany, 2009.

16. Alenyà, G.; Foix, S.; Torras, C. ToF cameras for active vision in robotics. *Sens. Actuators A Phys.* **2014**, *218*, 10–22. [CrossRef]

17. Zhang, J.Y.; Zhao, C.; Zhang, D.W. Pose accuracy analysis of robot manipulators based on kinematics. *Adv. Mater. Res.* **2011**, *201*, 1867–1872. [CrossRef]

18. Hefele, J.; Brenner, C. Robot pose correction using photogrammetric tracking. In Proceedings of Intelligent Systems and Smart Manufacturing; International Society for Optics and Photonics, Boston, MA, USA, 12 February 2001; pp. 170–178.

19. Maas, H.G. Dynamic photogrammetric calibration of industrial robots. In Proceedings of Camera and System Calibration, San Diego, CA, USA, 27 July 1997.

20. Clarke, T.; Wang, X. The control of a robot end-effector using photogrammetry. *Int. Arch. Photogramm. Remote Sens.* **2000**, *33*, 137–142.

21. Salvi, J.; Armangué, X.; Batlle, J. A comparative review of camera calibrating methods with accuracy evaluation. *Pattern Recognit.* **2002**, *35*, 1617–1635. [CrossRef]

22. Faugeras, O. *Three-dimensional Computer Vision: A Geometric Viewpoint*; MIT Press: Massachusetts, MA, USA, 1993.

23. Usamentiaga, R.; Molleda, J.; García, D.F. Structured-light sensor using two laser stripes for 3D reconstruction without vibrations. *Sensors* **2014**, *14*, 20041–20063. [CrossRef] [PubMed]

24. Sturm, P.; Ramalingam, S.; Tardif, J.P.; Gasparini, S.; Barreto, J. Camera models and fundamental concepts used in geometric computer vision. *Found. Trends Comput. Gr. Vis.* **2011**, *6*, 1–183. [CrossRef]

25. Hanning, T. *High Precision Camera Calibration*; Springer: Wiesbaden, Germany, 2011.

26. Usamentiaga, R.; Molleda, J.; García, D.F.; Pérez, L.; Vecino, G. Real-time line scan extraction from infrared images using the wedge method in industrial environments. *J. Electron. Imaging* **2010**, *19*, 043017. [CrossRef]

27. Wilson, A. Choosing a 3D Vision System for Automated Robotics Applications. *Vis. Syst. Des.* **2014**, *19*. Available online: http://www.vision-systems.com/articles/print/volume-19/issue-11/features/choosing-a-3d-vision-system-for-automated-robotics-applications.html (accessed on 11 December 2014).

28. Ramos, B.; Santos, E. Comparative study of different digitization techniques and their accuracy. *Comput.-Aided Des.* **2011**, *43*, 188–206.

29. Yilmaz, H.M.; Yakar, M.; Gilec, S.A.; Dulgerler, O.N. Importance of digital close-range photogrammetry in documentation of cultural heritage. *J. Cult. Herit.* **2007**, *8*, 428–433. [CrossRef]

30. Werner, T.; Zisserman, A. New techniques for automated architecture reconstruction from photographs. In Proceedings of the Seventh European Conference on Computer Vision (ECCV 2002), Copenhagen, Denmark, 28–31 May 2002; pp. 541–555.

31. Werner, T.; Zisserman, A. Model selection for automated architectural reconstruction from multiple views. In Proceedings of the Thirteenth British Machine Vision Conference (BMVC 2002), Cardiff, UK, 2–5 September 2002; pp. 53–62.

32. Werner, T.; Schaffalitzky, F.; Zisserman, A. Automated architecture reconstruction from close-range photogrammetry. *Int. Arch. Photogramm. Remote Sens. Spat. Inf. Sci.* **2002**, *34*, 352–359.

33. Lane, S.N.; James, T.D.; Crowell, M.D. Application of digital photogrammetry to complex topography for geomorphological research. *Photogramm. Rec.* **2000**, *16*, 793–821. [CrossRef]

34. Fonstad, M.A.; Dietrich, J.T.; Corville, B.C.; Jensen, J.L.; Carbonneau, P.E. Topographic structure from motion: A new development in photogrammetric measurement. *Earth Surf. Process. Landf.* **2013**, *38*, 421–430. [CrossRef]

35. Dueholm, K.S.; Garde, A.A.; Pedersen, A.K. Preparation of accurate geological and structural maps, cross-sections or block diagrams from colour slides, using multi-model photogrammetry. *J. Struct. Geol.* **1993**, *15*, 933–937. [CrossRef]

36. Eisenbeiss, H.; Lambers, K.; Sauerbier, M.; Li, Z. Photogrammetric documentation of an archaeological site using and autonomous model helicopter. *Int. Arch. Photogramm. Remote Sens. Spat. Inf. Sci.* **2005**, *36*, 238–243.

37. Granshaw, S.I. Bundle adjustment methods in engineering photogrammetry. *Photogramm. Rec.* **1980**, *10*, 181–207. [CrossRef]

38. Fraser, C.S.; Brown, D.C. Industrial photogrammetry: New developments and recent applications. *Photogramm. Rec.* **1986**, *12*, 197–217. [CrossRef]

39. Goldan, M.; Kroon, R. As-built products modelling and reverse engineering in shipbuilding through combined digital photogrammetry and CAD/CAM technology. *J. Sh. Prod.* **2003**, *19*, 98–104.

40. Luhmann, T.; Robson, S.; Kyle, S.; Harley, I. *Close Range Photogrammetry. Principles, techniques and applications*; Whittles Publishing: Caithness, UK, 2006.

41. Canny, J. A computational approach to edge detection. *IEEE Trans. Pattern Anal. Mach. Intell.* **1986**, *8*, 679–698. [CrossRef] [PubMed]

42. Harris, C.; Stephens, M. A combined corner and edge detector. In Proceedings of the Alvey vision Conference, Manchester, UK, 2 September 1998.

43. Lucas, B.D.; Kanade, T. An iterative image registration technique with an application to stereo vision. In Proceedings of the Seventh International Joint Conference on Artificial Intelligence (IJCAI 1981), Vancouver, BC, Canada, 24–28 August 1981; pp. 674–679.

44. Lowe, D.G. Distinctive image features from scale-invariant keypoints. *Int. J. Comput. Vis.* **2004**, *60*, 91–110. [CrossRef]

45. Bay, H.; Ess, A.; Tuytelaars, T.; Van Gool, L. Speeded-up robust features (SURF). *Comput. Vis. Image Underst.* **2008**, *110*, 346–359. [CrossRef]

46. Kimmel, R.; Zhang, C.; Bronstein, A.; Bronstein, M. Are MSER features really interesting? *IEEE Trans. Pattern Anal. Mach. Intell.* **2011**, *33*, 2316–2320. [CrossRef] [PubMed]

47. IDS Whitepapers. Available online: https://en.ids-imaging.com/whitepaper.html (accessed on 30 July 2015).

48. Ensenso—Stereo 3D Cameras. Available online: http://www.ensenso.com (accessed on 13 December 2015).

49. Lim, J. Optimized projection pattern supplementing stereo systems. In Proceedings of the 2009 IEEE International Conference on Robotics and Automation (ICRA 2009), Kobe, Japan, 12–17 May 2009; pp. 2823–2829.

50. Konolige, K. Projected texture stereo. In Proceedings of the 2010 IEEE International Conference on Robotics and Automation (ICRA 2010), Anchorage, KY, USA, 3–7 May 2010; pp. 148–155.

51. Time of Flight-Heptagon. Available online: http://hptg.com/technology/time-of-flight/ (accessed on 13 December 2015).

52. Kolb, A.; Barth, E.; Koch, R.; Larsen, R. Time-of-flight sensors in computer graphics (state-of-the-art report). In Proceedings of Eurographics 2009-State of the Art Reports, Munich, Germany, 30 March–2 April 2009.

53. Chen, F.; Brown, G.; Song, M. Overview of three dimensional shape measurement using optical methods. *Opt. Eng.* **2000**, *39*, 10–22. [CrossRef]

54. Pagès, J.; Collewet, C.; Chaumette, F.; Salvi, J. A camera-projector system for robot positioning by visual servoing. In Proceedings of the 2006 Conference on Computer Vision and Pattern Recognition Workshop (CVPRW 2006), New York, NY, USA, 17–22 June 2006.

55. Salvi, J. An Approach to Coded Structured Light to Obtain Three Dimensional Information. Ph.D. Thesis, University of Gerona, Gerona, Spain, 1997.

56. Bernal, C.; Agustina, B.; Marín, M.M.; Camacho, A.M. Performance evaluation of optical scanner based on blue led structured light. *Proc. Eng.* **2013**, *63*, 591–598. [CrossRef]

57. LMI Technologies. Available online: http://lmi3d.com/products (accessed on 30 October 2015).

58. Viager, M. *Analysis of Kinect for Mobile Robots*; Individual course report; Technical University of Denmark: Kongens Lyngby, Denmark, 2011.

59. Khoshelham, K.; Elberink, S.O. Accuracy and resolution of kinect depth data for indoor mapping applications. *Sensors* **2012**, *12*, 1437–1454. [CrossRef] [PubMed]

60. Mahmud, M.; Joannic, D.; Roy, M.; Isheil, A.; Fontaine, J.F. 3D part inspection path planning of a laser scanner with control on the uncertainty. *Comput.-Aided Des.* **2011**, *43*, 345–355. [CrossRef]

61. Salmi, T.; Väätäinen, O.; Malm, T.; Montonen, J.; Marstio, I. Meeting new challenges and possibilities with modern robot safety technologies. In *Enabling Manufacturing Competitiveness and Economic Sustainability*; Springer International Publishing: Montreal, QC, Canada, 2014; pp. 183–188.

62. Point Cloud Library (PCL). Available online: http://pointclouds.org/documentation/tutorials/ (accessed on 29 July 2015).

63. Luhmann, T. Close range photogrammetry for industrial applications. *ISPRS J. Photogramm. Remote Sens.* **2010**, *65*, 558–569. [CrossRef]

64. Rieke-Zapp, D.; Tecklenburg, W.; Peipe, J.; Hastedt, H.; Haig, C. Evaluation of the geometric stability and the accuracy potential of digital cameras-Comparing mechanical stabilisation versus parametrisation. *ISPRS J. Photogramm. Remote Sens.* **2009**, *64*, 248–258. [CrossRef]

65. Broers, H.; Jansing, N. How precise is navigation for minimally invasive surgery? *Int. Orthop.* **2007**, *31*, 39–42. [CrossRef] [PubMed]

66. Laser Tracker Systems—Leica Geosystems. Available online: http://www.leica-geosystems.com/en/Laser-Tracker-Systems_69045.htm (accessed on 2 November 2015).

67. Shirinzadeh, B.; Teoh, P.L.; Tian, Y.; Dalvand, M.M.; Zhong, Y.; Liaw, H.C. Laser interferometry-based guidance methodology for high precision positioning of mechanisms and robots. *Robotics Computer-Integrated Manuf.* **2010**, *26*, 74–82. [CrossRef]

68. Vincze, M.; Prenninger, J.P.; Gander, H. A laser tracking system to measure position and orientation of robot end effectors under motion. *Int. J. Robotics Res.* **1994**, *13*, 305–314. [CrossRef]

69. Dumas, C.; Caro, S.; Mehdi, C.; Garnier, S.; Furet, B. Joint stiffness identification of industrial serial robots. *Robotica* **2012**, *30*, 649–659. [CrossRef]

70. Qu, W.W.; Dong, H.Y.; Ke, Y.L. Pose accuracy compensation technology in robot-aided aircraft assembly drilling process. *Acta Aeronaut. Astronaut. Sinica* **2011**, *32*, 1951–1960.

71. Leikas, E. Robot guidance with a photogrammetric 3-D measuring system. *Ind. Robot* **1999**, *26*, 105–108. [CrossRef]

72. Hefele, J.; Brenner, C. Real-time photogrammetric algorithms for robot calibration. *Int. Archives Photogramm. Remote Sens. Spat. Inf. Sci.* **2002**, *34*, 33–38.

73. Liu, B.; Zhang, F.; Qu, X. A method for improving the pose accuracy of a robot manipulator based on multi-sensor combined measurement and data fusion. *Sensors.* **2015**, *15*, 7933–7952. [CrossRef] [PubMed]

74. Amdal, K. Single camera system for close range industrial photogrammetry. *Int. Archives Photogramm. Remote Sens.* **1992**, *29*, 6–10.

75. Luhmann, T. Precision potential of photogrammetric 6DOF pose estimation with a single camera. *ISPRS J. Photogramm. Remote Sens.* **2009**, *64*, 275–284. [CrossRef]

76. IDS Case Studies. Available online: https://en.ids-imaging.com/case-studies.html (accessed on 30 July 2015).

77. Carroll, J. 3D vision system assists in robotic bin picking. *Vis. Syst. Des.* **2014**, *19*. Available online: http://www.vision-systems.com/articles/2014/08/3d-vision-system-assists-in-robotic-bin-picking.html (accessed on 29 August 2014).

78. Sturm, J.; Konolige, K.; Stachniss, C.; Burgard, W. 3D pose estimation, tracking and model learning of articulated objects from dense depth video using projected texture stereo. In Proceedings of the 2010 RGB-D: Advanced Reasoning with Depth Cameras Workshop, Zaragoza, Spain, 27 June 2010.

79. Weingarten, J.W.; Gruener, G.; Siegwart, R. A state-of-the-art 3D sensor for robot navigation. In Proceedings of the 2004 IEEE/RSJ International Conference on Intelligent Robots and Systems (IROS 2004), Sendai, Japan, 28 September–2 October 2004; pp. 2155–2160.

80. May, S.; Werner, B.; Surmann, H.; Pervolz, K. 3D time-of-flight cameras for mobile robotics. In Proceedings of the 2006 IEEE/RSJ International Conference on Intelligent Robots and Systems (IROS 2006), Beijing, China, 9–15 October 2006; pp. 790–795.

81. May, S.; Droeschel, D.; Holz, D.; Wiesen, C.; Fuchs, S. 3D pose estimation and mapping with time-of-flight cameras. In Proceedings of the 2008 IEEE/RSJ International Conference on Intelligent Robots and Systems (IROS 2008), 3D Mapping Workshop, Nice, France, 22–26 September 2008.

82. Hedge, G.; Ye, C. Extraction of planar features from Swissranger SR-3000 range images by a clustering method using normalized cuts. In Proceedings of the 2009 IEEE/RSJ International Conference on Intelligent Robots and Systems (IROS 2009), St. Louis, MO, USA, 10–15 October 2009; pp. 4034–4039.

83. Ohno, K.; Nomura, T.; Tadokoro, S. Real-time robot trajectory estimation and 3D map construction using 3D camera. In Proceedings of the 2006 IEEE/RSJ International Conference on Intelligent Robots and Systems (IROS 2006), Beijing, China, 9–15 October 2006; pp. 5279–5285.

84. Stipes, J.A.; Cole, J.G.P.; Humphreys, J. 4D scan registration with the SR-3000 LIDAR. In Proceedings of the 2008 International Conference on Robotics and Automation (ICRA 2008), Pasadena, CA, USA, 19–23 May 2008; pp. 2988–2993.

85. Arbeiter, G.; Fischer, J.; Verl, A. 3-D-Environment reconstruction for mobile robots using fast-SLAM and feature extraction. In Proceedings of the Forty-first International Symposium on Robotics (ISR 2010), Munich, Germany, 7–9 June 2010.

86. Kuhnert, K.D.; Stommel, M. Fusion of stereo-camera and PMD-camera data for realtime suited precise 3D environment reconstruction. In Proceedings of the 2006 IEEE/RSJ International Conference on Intelligent Robots and Systems (IROS 2006), Beijing, China, 9–15 October 2006; pp. 4780–4785.

87. Netramai, C.; Oleksandr, M.; Joochim, C.; Roth, H. Motion estimation of a mobile robot using different types of 3D sensors. In Proceedings of the Fourth International Conference on Autonomic and Autonomous Systems (ICAS 2008), Gosier, France, 16–21 March 2008; pp. 148–153.

88. Dellen, B.; Alenyà, G.; Foix, S.; Torras, C. 3D object reconstruction from swissranger sensors data using a spring-mass model. In Proceedings of the Fourth International Conference on Computer Vision Theory Applications, Lisbon, Portugal, 5–8 February 2009; pp. 368–372.

89. Foix, S.; Alenyà, G.; Andrade-Cetto, J.; Torras, C. Object modelling using a ToF camera under an uncertainty reduction approach. In Proceedings of the 2010 IEEE International Conference on Robotics Automation (ICRA 2010), Anchorage, KY, USA, 3–7 May 2010; pp. 1306–1312.

90. Haddadin, S.; Suppa, M.; Fuchs, S.; Bodenmüller, T.; Albu-Schäffer, A.; Hirzinger, G. Towards the robotic co-worker. *Robotics Res.* **2011**, *70*, 261–282.

91. Fuchs, S.; May, S. Calibration and registration for precise surface reconstruction with time of flight cameras. *Int. J. Intell. Syst. Technol. Appl.* **2008**, *5*, 274–284. [CrossRef]

92. Kuehnle, J.U.; Xue, Z.; Stotz, M.; Zoellner, J.M.; Verl, A.; Dillmann, R. Grasping in depth maps of time-of-flight cameras. In Proceedings of the 2008 IEEE International Workshop on Robotic and Sensors Environments (ROSE 2008), Ottawa, ON, Canada, 17–18 October 2008; pp. 132–137.

93. Saxena, A.; Wong, L.; Ng, A.Y. Learning grasp strategies with partial shape information. In Proceedings of the Twenty-third AAAI Conference on Artificial Intelligence, Chicago, IL, USA, 13–17 July 2008; pp. 1491–1494.

94. Maldonado, A.; Klank, U.; Beetz, M. Robotic grasping of unmodeled objects using time-of-flight range data and finger torque information. In Proceedings of the 2010 IEEE/RSJ International Conference on Intelligent Robots and Systems (IROS 2010), Taipei, Taiwan, 18–22 October 2010; pp. 2586–2591.

95. Foix, S.; Alenyà, G.; Torras, C. *Exploitation of Time-of-Flight (ToF) Cameras*; Technical Report; CSIC-UPC: Barcelona, Spain; December; 2010.

96. Claes, K.; Bruyninckx, H. Robot positioning using structured light patterns suitable for self calibration and 3D tracking. In Proceedings of the Thirteenth International Conference on Advanced Robotics, Daegu, Korea (South), 22–25 August 2007.

97. Le Moigne, J.J.; Waxman, A.M. Structured light patterns for robot mobility. *J. Robotics Autom.* **1988**, *4*, 541–548. [CrossRef]

98. Patra, S.; Bhowmick, B.; Banerjee, S.; Kalra, P. High resolution point cloud generation from kinect and HD cameras using graph cut. In Proceedings of the 2012 International Joint Conference on Computer Vision, Imaging and Computer Graphics Theory and Applications (VISAPP 2012), Rome, Italy, 24–26 February 2012.

99. Susperregi, L.; Sierra, B.; Castrillón, M.; Lorenzo, J.; Martínez-Otzeta, J.M.; Lazkano, E. On the use of a low-cost thermal sensor to improve kinect people detection in a mobile robot. *Sensors.* **2013**, *13*, 14687–14713. [CrossRef] [PubMed]

100. Wang, L.; Schmidt, B.; Nee, A.Y.C. Vision-guided active collision avoidance for human-robot collaborations. *Manuf. Lett.* **2013**, *1*, 5–8. [CrossRef]

101. FlexSort Project. Available online: http://www.prodintec.es/prodintec/g_noticias?accion=detalleNoticia&id =159 (accessed on 15 November 2015).

102. Brosed, F.J.; Santolaria, J.; Aguilar, J.J.; Guillomía, D. Laser triangulation sensor and six axes anthropomorphic robot manipulator modelling for the measurement of complex geometry products. *Robot. Comput.-Integr. Manuf.* **2012**, *2*, 660–671. [CrossRef]

103. Nguyen, V.; Martinelli, A.; Tomatis, N.; Siegwart, R. A comparison of line extraction algorithms using 2D laser rangefinder for indoor mobile robotics. In Proceedings of the 2005 IEEE/RSJ International Conference on Intelligent Robots and Systems (IROS 2005), Edmonton, AB, Canada, 2–6 August 2005; pp. 1929–1934.

104. Borges, G.A.; Aldon, M.J. line extraction in 2D range images for mobile robotics. *J. Intell. Robotic Syst.* **2004**, *40*, 267–297. [CrossRef]

105. Agapakis, J.E.; Katz, J.M.; Friedman, J.M.; Epstein, G.N. Vision-aided robotic welding: An approach and a flexible implementation. *Int. J. Robotics Res.* **1990**, *9*, 17–34. [CrossRef]

106. Fernández, A.; Acevedo, R.G.; Alvarez, E.A.; López, A.C.; García, D.F.; Usamentiaga, R.; Sánchez, J. Low-cost system for weld tracking based on artificial vision. *IEEE Trans. Ind. Appl.* **2011**, *47*, 1159–1167. [CrossRef]

107. Khadraoui, D.; Motyl, G.; Martinet, P.; Gallice, J.; Chaumette, F. Visual servoing in robotics scheme using a camera/laser-stripe sensor. In Proceedings of the 1996 IEEE International Conference on Robotics and Automation (ICRA 1996), Minneapolis, MN, USA, 22–28 April 1996; pp. 743–750.

108. Pears, N.; Probert, P. An optical range sensor for mobile robot guidance. In Proceedings of the 1993 IEEE International Conference on Robotics and Automation (ICRA 1993), Atlanta, GA, USA, 2–6 May 1993; pp. 659–664.

109. Einsele, T. Real-time self-localization in unknown indoor environments using a panorama laser range finder. In Proceedings of the 1997 IEEE/RSJ International Conference on Intelligent Robots and Systems (IROS 1997), Grenoble, France, 11 September 1997; pp. 697–702.

110. Jensfelt, P.; Christensen, H. Laser based position acquisition and tracking in an indoor environment. In Proceedings of the 1998 International Symposium on Robotics and Automation (ISRA 1998), Leuven, Belgium, 20 May 1998.

111. Zhang, L.; Ghosh, B.K. Line segment based map building and localization using 2D laser rangefinder. In Proceedings of the 2000 IEEE International Conference on Robotics and Automation (ICRA 2000), San Francisco, CA, USA, 24–28 April 2000; pp. 2538–2543.

112. Gühring, J. Dense 3D surface acquisition by structured light using off-the-shelf components. In Proceedings of Videometrics and Optical Methods for 3D Shape Measurement, San Jose, CA, USA, 20 January 2001. SPIE: 2001.

113. Hashimoto, K. A Review on vision-based control of robot manipulators. *Adv. Robotics* **2003**, *17*, 969–991.

114. Lazaros, N.; Sirakoulis, G.C.; Gasteratos, A. review of stereo vision algorithms: from software to hardware. *Int. J. Optomechatronics* **2008**, *2*, 435–462. [CrossRef]

 *sensors*

Article

Vision-Based SLAM System for Unmanned Aerial Vehicles

Rodrigo Munguía [1,2,*], Sarquis Urzua [2], Yolanda Bolea [1] and Antoni Grau [1,*]

1   Department of Automatic Control, Technical University of Catalonia UPC, Barcelona 08036, Spain; yolanda.bolea@upc.edu
2   Department of Computer Science, CUCEI, University of Guadalajara, Guadalajara 44430, Mexico; isi.sarquis@gmail.com
*   Correspondence: rodrigo.munguia@upc.edu (R.M.); antoni.grau@upc.edu (A.G.);
    Tel.: +52-33-1015-0602 (R.M.); +34-934-016-975 (A.G.)

Academic Editor: Gonzalo Pajares Martinsanz
Received: 8 December 2015; Accepted: 9 March 2016; Published: 15 March 2016

Abstract: The present paper describes a vision-based simultaneous localization and mapping system to be applied to Unmanned Aerial Vehicles (UAVs). The main contribution of this work is to propose a novel estimator relying on an Extended Kalman Filter. The estimator is designed in order to fuse the measurements obtained from: (i) an orientation sensor (AHRS); (ii) a position sensor (GPS); and (iii) a monocular camera. The estimated state consists of the full state of the vehicle: position and orientation and their first derivatives, as well as the location of the landmarks observed by the camera. The position sensor will be used only during the initialization period in order to recover the metric scale of the world. Afterwards, the estimated map of landmarks will be used to perform a fully vision-based navigation when the position sensor is not available. Experimental results obtained with simulations and real data show the benefits of the inclusion of camera measurements into the system. In this sense the estimation of the trajectory of the vehicle is considerably improved, compared with the estimates obtained using only the measurements from the position sensor, which are commonly low-rated and highly noisy.

Keywords: state estimation; unmanned aerial vehicle; monocular vision; localization; mapping

---

1. Introduction

In recent years, many researchers have addressed the issue of making Unmanned Aerial Vehicles (UAVs) more autonomous. In this context, the state estimation of the six degrees of freedom (6-DoF) of a vehicle (*i.e.*, its attitude and position) is a fundamental necessity for any application involving autonomy.

Outdoors, this problem is seemingly solved with on-board Global Positioning System (GPS) and Inertial Measurements Units (IMU) with their integrated version, the Inertial Navigation Systems (INS). In fact, unknown, cluttered, and GPS-denied environments still pose a considerable challenge. While attitude estimation is well-handled with available systems [1], GPS-based position estimation has some drawbacks. Specifically GPS is not a reliable service as its availability can be limited in urban canyons and is completely unavailable in indoor environments.

Moreover, even when GPS signal is available, the problem of position estimation could not be solved in different scenarios. For instance, aerial inspection of industrial plants is an application that requires performing precision manoeuvres in a complex environment. In this case, and due to the several sources of error, the position obtained with a GPS can vary with an error of several meters in just a few seconds for a static location [2]. In such a scenario, the use of GPS readings, smoothed or not, as the feedback signal of a control system can be unreliable because the control system cannot distinguish between sensor noise and actual small movements of the vehicle. Therefore, some additional sensory information (e.g., visual information) should be integrated into the system in order to improve accuracy.

The aforementioned issues have motivated the move of recent works towards the use of cameras to perform visual-based navigation in periods or circumstances when the position sensor is not available, when it is partially available, or when a local navigation application requires high precision. Cameras are well adapted for embedded systems because they are cheap, lightweight, and power-saving. In this way, a combination of vision and inertial measurements is often chosen as means to estimate the vehicle attitude and position. This combination can be performed with different approaches, as in [3], where the vision measurement is provided by an external trajectometry system, directly yielding the position and orientation of the robot. In [4], an external CCD camera provides the measurements. Other on-board techniques were proposed by [5,6], where an embedded camera uses different markers to provide a good estimation of position and orientation as well. This estimation was obtained using the specific geometry of different markers and assuming that the marker's position was known. The same idea was exploited by [7], implemented with the low-cost Wii remote visual sensor. Finally, visual sensing is sometimes provided by optical flow sensors to estimate the attitude, the position, and the velocity, as in [8]. In these different approaches, position estimation is obtained by computer vision and the attitude is either obtained by vision (see [3,6]) or by IMU sensors. In [9], even a single angular measurement could significantly improve attitude and position estimation.

Another family of approaches (for instance [10,11]) relies on visual SLAM (Simultaneous Localization and Mapping) methods. In this case, the mobile robot operates in a *priori* unknown environment using only on-board sensors to simultaneously build a map of its surroundings and locate itself inside this map.

Robot sensors have a large impact on the algorithm used in SLAM. Early SLAM approaches focused on the use of range sensors, such as sonar rings and lasers, see [12–15]. Nevertheless, some disadvantages appear when using range sensors in SLAM: correspondence or data association becomes difficult, the sensors are expensive and have a limited working range, and some of them are limited to 2D maps. For small unmanned aerial vehicles, there exist several limitations regarding the design of the platform, mobility, and payload capacity that impose considerable restrictions. Once again, cameras appear as a good option to be used in SLAM systems applied to UAVs.

In this work, the authors propose the use of a monocular camera looking downwards, integrated into the aerial vehicle, in order to provide visual information of the ground. With such information, the proposed visual-based SLAM system will be using visual information, attitude, and position measurements in order to accurately estimate the full state of the vehicle as well as the position of the landmarks observed by the camera.

Compared with another kind of visual configurations (e.g., stereo vision), the use of monocular vision has some advantages in terms of weight, space, power consumption, and scalability. For example, in stereo rigs, the fixed base-line between cameras can limit the operation range. On the other hand, the use of monocular vision introduces some technical challenges. First, depth information cannot be retrieved in a single frame, and hence, robust techniques to recover features depth are required. In this work, a novel method is developed following the research initiated in [16]. The proposed approach is based on a stochastic technique of triangulation to estimate features depth.

In this novel research, a new difficulty appears: the metric scale of the world cannot be retrieved if monocular vision is used as the unique sensory input to the system. For example, in the experiments presented in [17], the first ten measurements are aligned with the ground-truth in order to obtain the scale of the environment. In [18], the monocular scale factor is retrieved from a feature pattern with known dimensions. On the other hand, in many real scenarios GPS signal is available, at least for some periods. For this reason, in this work it is assumed that the GPS signal is known during a short period (for some seconds) at the beginning of the trajectory. Those GPS readings will be integrated into the system in order to recover the metric scale of the world. This period of time is what authors consider the initialization period. After this period, the system can rely only on visual information to estimate the position of the aerial vehicle.

The integration of GPS readings with visual information is not new in the literature, see [19]. In this sense, one of the contributions of this work is to demonstrate that the integration of very noisy GPS measurements into the system for an initial short period is enough to recover the metric scale of the world. Furthermore, the experiments demonstrate that for flight trajectories performed near the origin of the navigation reference frame, it is better to avoid the integration of such GPS measurements after the initialization period.

This paper is organized as follows: Section 2 states the problem description and assumptions. Section 3 describes the proposed method in detail. Section 4 shows the experimental results, and finally in Section 5, the conclusions of this work are presented.

2. System Specification

2.1. Assumptions

The platform that the authors consider in this work is a quadrotor freely moving in any direction in $\mathbb{R}^3 \times SO(3)$, as shown in Figure 1. The quadrotor is equipped with a monocular camera, an attitude and heading reference system (AHRS) and a position sensor (GPS). It is important to remark that the proposed visual-based SLAM approach can be applied to another kind of platforms.

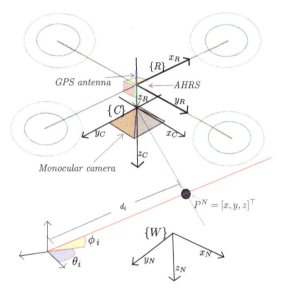

Figure 1. Coordinate systems: the local tangent frame is used as the navigation reference frame ^{N}. AHRS: Attitude and Heading Reference System.

The proposed system is mainly intended for local autonomous vehicle navigation. Hence, the local tangent frame is used as the navigation reference frame. The initial position of the vehicle defines the origin of the navigation coordinates frame. The navigation system follows the NED (North, East, Down) convention. In this work, the magnitudes expressed in the navigation, vehicle (robot), and camera frame are denoted respectively by the superscripts ^{N}, ^{R}, and ^{C}. All the coordinate systems are right-handed defined.

In this research, the sensors that have been taken into account are described and modelled in the following subsections.

2.2. Monocular Camera

As a vision system, a standard monocular camera has been considered. In this case, a central-projection camera model is assumed. The image plane is located in front of the camera's origin where a non-inverted image is formed. The camera frame ^{C} is right-handed with the z-axis pointing to the field of view.

The $\mathbb{R}^3 \Rightarrow \mathbb{R}^2$ projection of a 3D point located at $p^N = (x, y, z)^T$ to the image plane (u, v) is defined by:

$$u = \frac{x'}{z'} \quad v = \frac{y'}{z'} \tag{1}$$

where u and v are the coordinates of the image point p expressed in pixel units, and:

$$\begin{bmatrix} x' \\ y' \\ z' \end{bmatrix} = \begin{bmatrix} f & 0 & u_0 \\ 0 & f & v_0 \\ 0 & 0 & 1 \end{bmatrix} p^C \tag{2}$$

being p^C the same 3D point p^N, but expressed in the camera frame ^{C} by $p^C = R^{NC}(p^N - t_c^N)$. In this case, it is assumed that the intrinsic parameters of the camera are already known: (i) focal length f; (ii) principal point u_0, v_0; and (iii) radial lens distortion $k_1, ..., k_n$.

Let $R^{NC} = (R^{RN} R^{CR})^T$ be the rotation matrix that transforms the navigation frame ^{N} to the camera frame ^{C}. Let R^{CR} be a known value, and R^{RN} is computed from the current robot quaternion q^{NR}. Let $t_c^N = r^N + R^{RN} t_c^R$ be the position of the camera's optical center position expressed in the navigation frame.

Inversely, a directional vector $h^C = [h_x^C, h_y^C, h_z^C]^T$ can be computed from the image point coordinates u and v.

$$h^C(u, v) = \left[\frac{u_0 - u}{f}, \frac{v_0 - v}{f}, 1 \right]^T \tag{3}$$

The vector h^C points from the camera optical center position to the 3D point location. h^C can be expressed in the navigation frame by $h^N = R^{CN} h^C$, where $R^{CN} = R^{RN} R^{CR}$ is the camera-to-navigation rotation matrix. Note that for the $\mathbb{R}^2 \Rightarrow \mathbb{R}^3$ mapping case, defined in Equation (3), depth information is lost.

The distortion caused by the camera lens is considered through the model described in [20]. Using this model (and its inverse form), undistorted pixel coordinates (u, v) can be obtained from the distorted pixel (u_d, v_d), and conversely.

2.3. Attitude and Heading Reference System

An attitude and heading reference system (AHRS) is a combination of instruments capable of maintaining an accurate estimation of the vehicle attitude while it is manoeuvring. Recent manufacturing of solid-state or MEMS gyroscopes, accelerometers, magnetometers, and powerful microcontrollers as well, have made possible the development of small, low-cost, and reliable AHRS devices (e.g., [1,21,22]). For these reasons, in this work a loosely-coupled approach is considered. In this case, the information of orientation provided by the AHRS is explicitly fused into the system. Hence, the availability of high-rated (typically 50–100 Hz) attitude measurements provided by a decoupled AHRS device are assumed.

Attitude measurements y_a^N are modelled by:

$$y_a^N = a^N + v_a \tag{4}$$

where $a^N = [\phi_v, \theta_v, \psi_v]^T$, being ϕ_v, θ_v, and ψ_v Euler angles denoting respectively the roll, pitch, and yaw of the vehicle. Let v_a be a Gaussian white noise with power spectral density (PSD) σ_a^2.

2.4. GPS

The Global Positioning System (GPS) is a satellite-based navigation system that provides 3D position information for objects on or near the Earth's surface. The GPS system and global navigation satellite systems have been described in detail in numerous studies (e.g., [2,23]). Several sources of error affect the accuracy of GPS position measurements. The cumulative effect of each of these error sources is called the user-equivalent range error (UERE). In [2], these errors are characterized as a combination of slowly varying biases and random noise. In [24] it is stated that the total UERE is approximately 4.0 m (σ), from which 0.4 m (σ) correspond to random noise. In Figure 2, a comparison between the trajectory obtained with GPS and the actual one, flying in a small area, is shown.

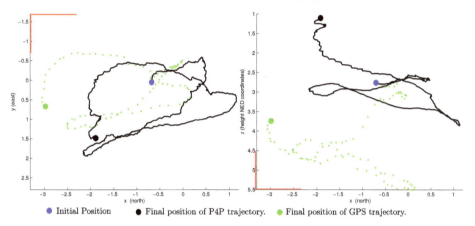

Figure 2. Example of GPS position measurements obtained for a flight performed by the aerial vehicle. Top view (**left plot**) and lateral view (**right plot**) are shown for clarity. Flight trajectory has been computed using the perspective on 4-point (P4P) method described in Section 4. Error drift in GPS readings is noticeable. NED: North, East, Down.

In this work, it is assumed that position measurements y_r can be obtained from the GPS unit, at least at the beginning of the trajectory, and they are modelled by:

$$y_r = r^N + v_r \tag{5}$$

where v_r is a Gaussian white noise with PSD σ_r^2, and r^N is the position of the vehicle.

Commonly, position measurements are obtained from GPS devices in geodetic coordinates (*latitude*, *longitude*, and *height*). Therefore, in Equation (5) it is assumed that GPS position measurements have been previously transformed to their corresponding local tangent frame coordinates. It is also assumed that the offset between the GPS antenna and the vehicle frame has been taken into account in the previous transformation.

2.5. Sensor Fusion Approach

The estimator proposed in this work is designed in order to estimate the full state of the vehicle, which will contain the position and orientation of the vehicle and their first derivatives, as well as the location of the landmarks observed by the camera.

Attitude estimation can be well-handled by the available systems in the vehicle, as has been mentioned in the above subsections. Typically, the output of the AHRS is directly used as a feedback to the control system for stabilizing the flying vehicle. On the other hand, the proposed method requires the camera–vehicle to know its orientation in order to estimate its position, as will be discussed later in the paper. In order to account for the uncertainties associated with the estimation provided by the

AHRS, the orientation is included into the state vector (see Section 3.1) and is explicitly fused into the system (see Section 3.4).

Regarding the problem of position estimation, it cannot be solved for applications that require performing precise manoeuvres, even when GPS signal is available, as it can be inferred from the example presented in Section 2.4. Therefore, some additional sensory information (e.g., monocular vision) should be integrated into the system in order to improve its accuracy. On the other hand, one of the most challenging aspects of working with monocular sensors has to do with the impossibility of directly recovering the metric scale of the world. If no additional information is used, and a single camera is used as the sole source of data to the system, the map and trajectory can only be recovered without metric information [25].

Monocular vision and GPS are not suitable to be used separately for navigation purposes in some scenarios. For this reason, the noisy data obtained from the GPS is added during the initialization period in order to incorporate metric information into the system. Hence, after an initial period of convergence, where the system is considered to be in the initialization mode, the system can operate relying only on visual information to estimate the vehicle position.

3. Method Description

3.1. Problem Description

The main goal of the proposed method is to estimate the following system state x:

$$x = [x_v, y_1^N, y_2^N, ..., y_n^N]^T \tag{6}$$

where x_v represents the state of the unmanned aerial vehicle, and y_i^N represents the location of the *i-th* feature point in the environment. At the same time, x_v is composed of:

$$x_v = [q^{NR}, \omega^R, r^N, v^N]^T \tag{7}$$

where $q^{NR} = [q_1, q_2, q_3, q_4]$ represents the orientation of the vehicle respect to the world (navigation) frame by a unit quaternion. Let $\omega^R = [\omega_x, \omega_y, \omega_z]$ be the angular velocity of the robot expressed in the same frame of reference. Let $r^N = [p_x, p_y, p_z]$ represent the position of the vehicle (robot) expressed in the navigation frame. Let $v^N = [v_x, v_y, v_z]$ denote the linear velocity of the robot expressed in the navigation frame. The location of a feature y_i^N is parametrized in its euclidean form:

$$y_i^N = [p_{x_i}, p_{y_i}, p_{z_i}]^T \tag{8}$$

In the remainder of the paper, the superscript N will be dropped from y_i^N to avoid confusion.

The architecture of the system is defined by the a classical loop of prediction-update steps in the Extended Kalman Filter (EKF) in its direct configuration. In this case, the vehicle state as well as the feature estimates are propagated by the filter, see Figure 3.

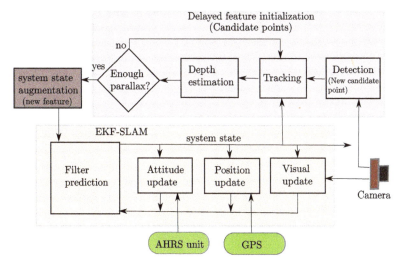

Figure 3. Block diagram showing the architecture of the system. EKF-SLAM: Extended Kalman Filter Simultaneous Localization and Mapping.

3.2. Prediction

At the same frequency of the AHRS operation, the vehicle system state x_v takes a step forward through the following unconstrained constant-acceleration (discrete) model:

$$
\begin{cases}
q_{k+1}^{NC} = \left(\cos \|w\| I_{4\times4} + \dfrac{\sin \|w\|}{\|w\|} W \right) q_k^{NC} \\
\omega_{k+1}^{C} = \omega_k^{C} + \Omega^{C} \\
r_{k+1}^{N} = r_k^{N} + v_k^{N} \Delta t \\
v_{k+1}^{N} = v_k^{N} + V^{N}
\end{cases}
\tag{9}
$$

In the model represented by Equation (9), a closed form solution of $\dot{q} = 1/2(W)q$ is used to integrate the current velocity rotation ω^C over the quaternion q^{NC}. In this case $w = [\omega_{k+1}^{C}\Delta t/2]^T$ and:

$$
W = \begin{bmatrix}
0 & -w_1 & -w_2 & -w_3 \\
w_1 & 0 & -w_3 & w_2 \\
w_2 & w_3 & 0 & -w_1 \\
w_3 & -w_2 & w_1 & 0
\end{bmatrix}
\tag{10}
$$

At every step, it is assumed that there is an unknown linear and angular velocity with acceleration zero-mean and known-covariance Gaussian processes σ_v and σ_w, producing an impulse of linear and angular velocity: $V^N = \sigma_v^2 \Delta t$ and $\Omega^C = \sigma_w^2 \Delta t$.

It is assumed that the map features y_i remain static (rigid scene assumption) so $x_{k+1} = [x_{v(k+1)}, y_{1(k)}, y_{2(k)}, ..., y_{n(k)}]^T$.

The state covariance matrix P takes a step forward by:

$$
P_{k+1} = \nabla F_x P_k \nabla F_x^T + \nabla F_u Q \nabla F_u^T
\tag{11}
$$

where Q and the Jacobians ∇F_x, ∇F_u are defined as:

$$
\nabla F_x = \begin{bmatrix} \dfrac{\partial f_v}{\partial \hat{x}_v} & 0_{13\times n} \\ 0_{n\times 13} & I_{n\times n} \end{bmatrix}, \nabla F_u = \begin{bmatrix} \dfrac{\partial f_v}{\partial u} & 0_{13\times n} \\ 0_{n\times 6} & 0_{n\times n} \end{bmatrix}, Q = \begin{bmatrix} U & 0_{6\times n} \\ 0_{n\times 6} & 0_{n\times n} \end{bmatrix}
\tag{12}
$$

Let $\frac{\partial f_v}{\partial x_v}$ be the derivatives of the equations of the nonlinear prediction model (Equation (9)) with respect to the robot state x_v. Let $\frac{\partial f_v}{\partial u}$ be the derivatives of the nonlinear prediction model with respect to the unknown linear and angular velocity. The Jacobian calculation is a complicated but tractable matter of differentiation, hence, no results are presented here. Uncertainties are incorporated into the system by means of the process noise covariance matrix $U = diag[\sigma_a^2 I_{3\times3}, \sigma_\omega^2 I_{3\times3}]$, through parameters σ_a^2 and σ_ω^2.

3.3. Visual Aid

Depth information cannot be obtained in a single measurement when bearing sensors (e.g., a projective camera) are used. To infer the depth of a feature, the sensor must observe this feature repeatedly as the sensor freely moves through its environment, estimating the angle from the feature to the sensor center. The difference between those angle measurements is the parallax angle. Actually, parallax is the key that allows the estimation of features depth. In the case of indoor sequences, a displacement of centimeters could be enough to produce parallax; on the other hand, the more distant the feature, the more the sensor has to travel to produce parallax.

In monocular-based systems, the treatment of the features in the stochastic map (initialization, measurement, *etc.*) is an important problem to address with direct implications in the robustness of the system. In this work, a novel method is proposed in order to incorporate new features into the system. In this approach, a single hypothesis is computed for the initial depth of features by means of a stochastic technique of triangulation. The method is based on previous authors' work [16], and new contributions have been introduced in this research.

3.3.1. Detection of Candidate Points

The proposed method states that a minimum number of features y_i is considered to be predicted appearing in the image, otherwise new features should be added to the map. In this latter case, new points are detected in the image through a random search. For this purpose, Shi-Tomasi corner detector [26] is applied, but other detectors could be also used. These points in the image, which are not yet added to the map, are called candidate points, see Figure 4. Only image areas free of both candidate points and mapped features are considered to detect new points with the saliency operator.

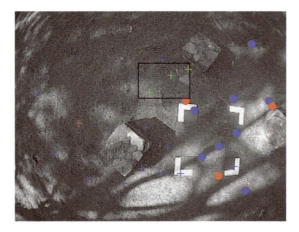

Figure 4. Candidate points are detected randomly in image regions without map features or candidate points. In this frame, the black rectangle indicates the current search region. Three new candidate points have been detected (green cross-marks). Candidate points being tracked are indicated by blue cross-marks. Visual features already mapped are indicated by circles. Red marks indicate unsuccessful matches.

At the k frame, when a visual feature is detected for the first time, the following entry c_l is stored in a table:

$$c_l = \left[(t_{c0}^N)^T, \theta_0, \phi_0, P_{y_{c_i}}, z_{uv} \right] \tag{13}$$

where $z_{uv} = [u, v]$ is the location in the image of the candidate point. Let $y_{c_i} = [t_{c0}^N, \theta_0, \phi_0]^T = h(\hat{x}, z_{uv})$ be a variable that models a 3D semi-line, defined on one side by the vertex t_{c0}^N, corresponding to the current optical center coordinates of the camera expressed in the navigation frame, and pointing to infinity on the other side, with azimuth and elevation θ_0 and ϕ_0, respectively, and:

$$\theta_0 = \text{atan2}(h_y^N, h_x^N)$$
$$\phi_0 = \text{acos} \left(\frac{h_z^N}{\sqrt{(h_x^N)^2+(h_y^N)^2+(h_z^N)^2}} \right) \tag{14}$$

where $h^N = [h_x^N, h_y^N, h_z^N]^T$ is computed as indicated in Section 2.2. P_{y_i} is a 5×5 covariance matrix which models the uncertainty of y_{c_i} in the form of $P_{y_{c_i}} = \nabla Y_{c_i} P \nabla Y_{c_i}^T$, where P is the system covariance matrix and ∇Y_{c_i} is the Jacobian matrix formed by the partial derivatives of the function $y_{c_i} = h(\hat{x}, z_{uv})$ with respect to $[\hat{x}, z_{uv}]^T$.

Also, a $p \times p$ pixel window, centered in $[u, v]$ is extracted and related to the corresponding candidate point.

3.3.2. Tracking of Candidate Points

To infer the depth of a feature, the sensor must observe this feature repeatedly until a minimum parallax is reached. For this reason, it is necessary to have a method to track the location in the image of candidate points whose initial depth must be computed. For feature points that have already been included into the system state, there is enough information (e.g., depth) to define probability regions where these points must lie based on the statistical information available in the system state (see [27]). On the other hand, for candidate points, there is not yet information about depth nor statistical correlations with other elements of the system. In this sense, one alternative is to use a general-purpose decoupled tracking method that works on the images and does not need assumptions about the system dynamics (e.g., [26]). Due to the lack of information about system dynamics, these kinds of methods usually define regions of search with symmetric geometry and fixed size. This factor can add some extra computational cost.

In this work, a novel technique to track candidate points is proposed. The idea is to take advantage of the knowledge about the direction of the movement of the camera in order to define regions of search defined by very thin ellipses. The ellipses are aligned with the epipolar lines where the candidate points must lie.

At every subsequent frame $k+1, k+2...k+n$, the location of candidate points is tracked. In this case, a candidate point is predicted to be appearing inside an elliptical region S_c centered in the point $[u, v]$, taken from c_l, see Figure 5. In order to optimize the speed of the search, the major axis of the ellipse is aligned with the epipolar line defined by image points e_1 and e_2.

The epipole e_1 is computed by projecting t_{c0}^N, which is stored in c_l, to the current image plane by Equations (1) and (2). Because there is not depth information of the candidate point, an hypothetical depth equal to one ($d = 1$) is chosen in order to determine a virtual 3D point p^N which lies in the semi-ray defined by c_l. The epipole e_2 is then computed by projecting this virtual 3D point p^N through Equations (1) and (2).

In this case, p^N will model a 3D point located at:

$$p^N = t_{c0}^N + m(\theta_0, \phi_0) d \tag{15}$$

where $m(\theta_0, \phi_0)$ is a directional unitary vector defined by: $m(\theta_0, \phi_0) = (\cos\theta_0 \sin\phi_0, \sin\theta_0 \sin\phi_0, \cos\phi_0)^T$.

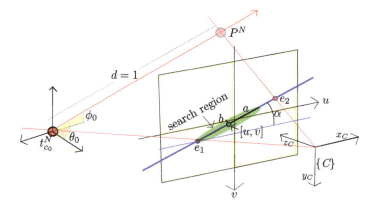

Figure 5. The established search region to match candidate points is constrained to ellipses aligned with the epipolar line.

The orientation of the ellipse S_c is determined by $\alpha_c = \text{atan2}(e_y, e_x)$, where e_y, e_x represents the y and x coordinates, respectively, of e, and $e = e_2 - e_1$. The size of the ellipse S_c is determined by its major and minor axis, respectively a and b.

The ellipse S_c is represented in its matrix form by:

$$S_c = R_c \begin{bmatrix} a & 0 \\ 0 & b \end{bmatrix} R_c^T$$

$$R_c = \begin{bmatrix} \cos \alpha_c & -\sin \alpha_c \\ \sin \alpha_c & \cos \alpha_c \end{bmatrix}$$

(16)

The ellipse S_c represents a probability region where the candidate point must lie in the current frame. The proposed tracking method is intended to be used during an initial short period of time. During this period, some information will be gathered in order to compute a depth hypothesis for each candidate point, prior to its initialization as a new map feature. For this reason, there is no extra effort to obtain more robust variations in scale or a rotations descriptor. In this case, direct patch cross-correlation is applied over all the image locations $[u_i, v_i] \in S_c$. If the score of a location $[u_i, v_i]$, determined by the best cross-correlation between the candidate patch and the n patches defined by the region of search, is higher than a specific threshold, then this pixel location $[u_i, v_i]$ is considered as the current candidate point location. Thus, c_l is updated with $z_{uv} = [u_i, v_i]$.

Unfortunately, because there is not yet reliable information about the depth of candidate points, it is difficult to determine an optimal and adaptive size of the ellipse. In this case, a is left as a free parameter to be chosen empirically as a function of the particularities of the application (e.g., maximum velocity of the vehicle, video frame rate). For the application presented in this work, good results were found with a value of $a = 20$ pixels.

On the other hand, it is possible to investigate the effects obtained by the variation of the relation of (b/a) which determines the proportion of the ellipse. In Figure 6, it can be noted that the time required to track a candidate point increases considerably as the ellipse tends to be a circle (left plot). On the other hand, the number of candidate points being tracked is lower when the ellipse tends to be a circle (middle plot). This is because some candidate points are lost when the ellipse is too thin, and new candidate points must be detected. Even so, the total time required for the whole tracking process of candidate points is much lower when the parameter b is chosen in order to define a very thin ellipse (right plot). For the foregoing reason the value of parameter b is recommended to be ten times lower than a.

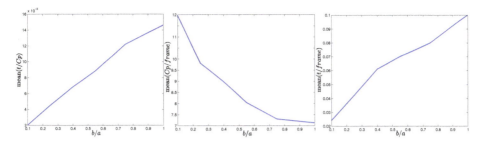

Figure 6. Results obtained by means of the variation of the relation between ellipse S_c axes (b/a). (**left plot**): average tracking time for a candidate point; (**middle plot**): average number of candidate points being tracked at each frame; (**right plot**): average total time per frame. These results were obtained using the same methodology described in Section 4.2.

3.3.3. Estimating Candidate Points Depth

Every time that a new image location $z_{uv} = [u, v]$ is obtained for a candidate point c_l, an hypothesis of depth d_i is computed by:

$$d_i = \frac{\|e_l\| \sin \gamma}{\sin \alpha_i} \qquad (17)$$

Let $\alpha_i = \pi - (\beta + \gamma)$ be the parallax. Let $e_l = t_{c_0}^N - t_c^N$ indicate the displacement of the camera from the first observation position to its current position, with:

$$\beta = \cos^{-1}\left(\frac{h_1 \cdot e_l}{\|h_1\| \|e_l\|}\right) \quad \gamma = \cos^{-1}\left(\frac{-h_2 \cdot e_l}{\|h_2\| \|e_l\|}\right) \qquad (18)$$

Let β be the angle defined by h_1 and e_l. Let h_1 be the normalized directional vector $m(\theta_i, \phi_i) = (\cos \theta_i \sin \phi_i, \sin \theta_i \sin \phi_i, \cos \phi_i)^T$ computed taking θ_i, ϕ_i from c_l, and where γ is the angle defined by h_2 and $-e_l$. Let $h_2 = h^N$ be the directional vector pointing from the current camera optical center to the feature location, computed as indicated in Section 2.2 from the current measurement z_{uv}, see Figure 7.

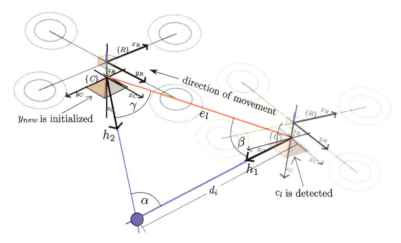

Figure 7. An hypothesis d_i for the depth of a candidate point is computed by triangulating between the first location when the point was detected and the current location of the vehicle.

At each step, there may be a considerable variation in depth computed by triangulation, specially for low parallax. In previous authors' work [28], it is shown that estimates are greatly improved by filtering the hypotheses of depth d_i with a simple low-pass filter. Moreover, in this work it is demonstrated that only a few degrees of parallax is enough to reduce the uncertainty in the depth estimation. When parallax α_i is greater than a specific threshold ($\alpha_i > \alpha_{min}$) a new feature $y_{new} = [p_{x_i}, p_{y_i}, p_{z_i}]^T = h(c_l, d)$ is added to the system state vector x:

$$x_{new} = [x_{old}; y_{new}]^T \tag{19}$$

where

$$y_{new} = t_{c_0}^N + m(\theta_i, \phi_i)d_i \tag{20}$$

The system state covariance matrix P is updated by:

$$P_{new} = \begin{bmatrix} P_{old} & 0 \\ 0 & P_{y_{new}} \end{bmatrix} \tag{21}$$

where $P_{y_{new}}$ is the 3 × 3 covariance matrix which models the uncertainty of the new feature y_{new}, and:

$$P_{y_{new}} = \nabla Y \begin{bmatrix} P_{y_i} & 0 \\ 0 & \sigma_d^2 \end{bmatrix} \nabla Y^T \tag{22}$$

In Equation (22), P_{y_i} is taken from c_l (Equation (13)). Let σ_d^2 be a parameter modelling the uncertainty of process of depth estimation. Let ∇Y be the Jacobian matrix formed by the partial derivatives of the function $y_{new} = h(c_l, d)$ with respect to $[(t_{c_0}^N)^T, \theta_0, \phi_0, d]^T$.

3.3.4. Visual Updates and Map Management

The process of tracking visual features y_i is conducted by means of an active search technique [27]. In this case, and in a different way from the tracking method described in Section 3.3.2, the search region is defined by the innovation covariance matrix S_i, where $S_i = \nabla H_i P_{k+1} \nabla H_i^T + \xi_i$.

Assuming that for the current frame, n visual measurements are available for features $y_1, y_2, ..., y_n$, then the filter is updated with the Kalman update equations as follows:

$$\begin{cases} x_k = x_{k+1} + K(z - h) \\ P_k = P_{k+1} - KSK^T \\ K = P_{k+1} \nabla H^T S^{-1} \\ S = \nabla H P_{k+1} \nabla H^T + \xi \end{cases} \tag{23}$$

where $z = [z_{uv_1}, z_{uv_2}, ..., z_{uv_n}]^T$ is the current measurement vector. Let $h = [h_1, h_2, ..., h_n]^T$ be the current prediction measurement vector. The measurement prediction model $h_i = (u, v) = h(x_v, y_i)$ has been defined in Section 2.2. Let K be the Kalman gain. Let S be the innovation covariance matrix. Let $\nabla H = [\nabla H_1, \nabla H_2, ... \nabla H_n]^T$ be the Jacobian formed by the partial derivatives of the measurement prediction model $h(x)$ with respect to the state x.

$$\nabla H_i = \left[\frac{\partial h_i}{\partial x_v}, ...0_{2\times3}... \frac{\partial h_i}{\partial y_i}, ...0_{2\times3}... \right] \tag{24}$$

Let $\frac{\partial h_i}{\partial x_v}$ be the partial derivatives of the equations of the measurement prediction model h_i with respect to the robot state x_v. Let $\frac{\partial h_i}{\partial y_i}$ be the partial derivatives of h_i with respect to feature y_i. Note that $\frac{\partial h_i}{\partial y_i}$ has only a nonzero value at the location (indexes) of the observed feature y_i. Let $\xi = (I_{2n\times2n})\sigma_{uv}^2$

be the measurement noise covariance matrix. Let σ_{uv}^2 be the variance modelling the uncertainty in visual measurements.

A SLAM framework that works reliably in a local way can easily be applied to large-scale problems using different methods, such as sub-mapping, graph-based global optimization [29], or global mapping [30]. Therefore, in this work, large-scale SLAM and loop-closing are not considered. However, these problems have been intensively studied in the past. Candidate points whose tracking process is failing are pruned from the system. Furthermore, visual features with high percentage of mismatching are removed from the system state and covariance matrix. The removal process is carried out using the approach described in [31].

3.4. Attitude and Position Updates

When an attitude measurement y_a^N is available, the system state is updated. Since most low-cost AHRS devices provide their output in Euler angles format, the following measurement prediction model $h_a = h(\hat{x}_v)$ is used:

$$\begin{bmatrix} \theta_v \\ \phi_v \\ \psi_v \end{bmatrix} = \begin{bmatrix} \text{atan2}(2(q_3 q_4 - q_1 q_2), 1 - 2(q_2^2 + q_3^2)) \\ \text{asin}(-2(q_1 q_3 + q_2 q_4)) \\ \text{atan2}(2(q_2 q_3 - q_1 q_4), 1 - 2(q_3^2 + q_4^2)) \end{bmatrix} \tag{25}$$

During the initialization period, position measurements y_r are incorporated into the system using the simple measurement model $h_r = h(\hat{x}_v)$:

$$h_r = [p_x, p_y, p_z]^T \tag{26}$$

The regular Kalman update equations (Equation (23)) are used to update attitude and position whenever is required, but using the corresponding Jacobian ∇H and measurement noise covariance matrix R.

The metric scale of the world cannot be retrieved using only monocular vision, as mentioned previously, and thus additional information must be added to the system. For instance, the metric scale can be retrieved if the position of some landmarks are known *a priori* with low uncertainty [32]. In this work, it is assumed that the GPS signal is available for an initial period at least. This period is considered as an initialization period that must allow the convergence of depth for at least some features close to their actual values. These first features added to the map during the initialization period set a metric scale in estimations. Afterwards, the system can operate relying only on visual information to estimate the location of the vehicle.

For the proposed method, the initialization period will end when at least n features show a certain degree of convergence. It has been theoretically demonstrated (e.g., [33]) that knowledge about the position of three landmarks can be enough to make the metric scale observable. However, in practice, there is always the possibility that the tracking process of some features fails at any time. For this reason, in this work the initialization period will be ending when $n \geq 3$ features have converged. In experiments, good results have been found with $n = 5$.

In [34], the convergence of features is tested using the Kullback distance. However, the complexity of the sampling method proposed to evaluate this distance is quite high. In the present work, good results have been found with the following criteria:

$$\max(eig(P_{y_i})) < \frac{\|y_i - r^N\|}{100} \tag{27}$$

where P_{y_i} is the 3×3 sub-matrix extracted from the covariance matrix P corresponding to the y_i feature. In this case, if the greater eigenvalue of P_{y_i} is smaller than one percent of the distance between the camera and the feature, then it is considered that the uncertainty in this feature has been minimized enough to take it as an initial reference of metrics.

It is important to note that the origin of the local reference system of navigation is established at the end of the initialization period. The reason is because at the beginning of the movement the GPS errors can wrongly dominate the estimations.

Since the proposed method is not deterministic, the duration of the initialization period varies even for the same input dataset (see Figure 8). For this reason, in order to simplify the experimental methodology, a fixed initialization period was used for computing the results of comparative studies presented in Section 4. In this manner, it was easier to align (in time) the estimated trajectories in order to perform a Monte Carlo validation. The fixed initial period was empirically determined to allow a high percentage of initial convergence. In a real scenario, the duration of the initialization period should be determined by an adaptive criteria, as authors have proposed in this section.

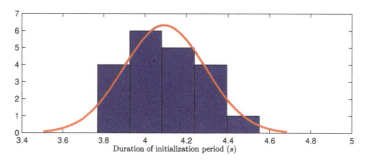

Figure 8. Histogram of the duration of the initialization period obtained after 20 runs of the proposed method. This particular case corresponds to the flight trajectory presented in Section 4.2.

4. Experimental Results

In this section, the results obtained using synthetic data from simulations are presented as well as the results obtained from experiments with real data. The experiments were performed in order to validate the performance of the proposed method. A MATLAB® implementation was used for this purpose.

4.1. Experiments with Simulations

In simulations, the model used to implement the vehicle dynamics was taken from [35]. To model the transient behaviour of the GPS error, the approach of [36] was followed. The monocular camera was simulated using the same parameter values of the camera used in the experiments with real data. The parameter values used to emulate the AHRS were taken from [1].

Figure 9 illustrates two cases of simulation: (a) The quadrotor was commanded to take off from the ground and then to follow a circular trajectory with constant altitude. The environment is composed by 3D points, uniformly distributed over the ground, which emulate visual landmarks; (b) The quadrotor was commanded to take off from the ground and then to follow a figure-eight-like trajectory with constant altitude. The environment is composed by 3D points, randomly distributed over the ground.

In simulations, it is assumed that the camera can detect and track visual features, avoiding the data association problem. Furthermore, the problem of the influence of the estimates on the control system was not considered. In other words, an almost perfect control over the vehicle is assumed.

Figure 10 shows the average mean absolute error (MAE) in position, obtained after 20 Monte Carlo runs of simulation. The MAE was computed for three scenarios: (i) using only GPS to estimate position; (ii) using GPS together with camera along all of the trajectory in order to estimate position and map; (iii) using GPS only during the initialization period, and then performing visual-based navigation and mapping.

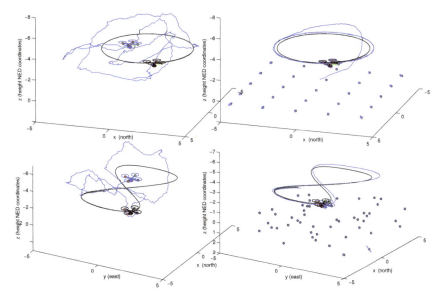

Figure 9. Comparison of the estimated trajectories obtained by filtering GPS data (**left plots**), and the estimated maps and trajectories obtained through visual-based navigation (**right plots**). Two different kind of trajectories and distributions of landmarks are simulated: (**upper plots**) a circular trajectory, (**lower plots**) a figure-eight-like trajectory. The GPS signal was used only during the initialization period. The actual trajectory is shown in black. The estimates are shown in blue.

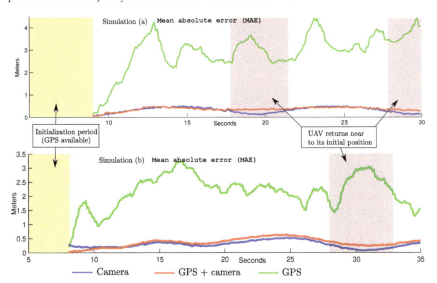

Figure 10. Mean absolute error (MAE) in position computed from two simulations (a and b) out of 20 Monte Carlo runs: (**upper plot**) simulation (a) results; (**lower plot**) simulation (b) results.

Figures 9 and 10 clearly show the benefits of incorporating visual information into the system. It is important to note that the trajectory obtained relying only on the GPS was computed by incorporating GPS readings into the filter, and do not denote raw measurements. In Figure 10 it is interesting to note that the computed MAE values for the trajectories obtained through visual-based navigation exhibit the classical SLAM behaviour when the quadrotor returns near to its initial position. In this case, the error is minimized close to zero. On the other hand, when the GPS is used all the time, the MAE remains more constant. In this case, it is seen that even when the vehicle is close to its trusted position, there is some influence of the GPS errors that affect the estimation. This behaviour suggests that for trajectories performed near to a local frame of reference, and even when the GPS signal is available, it is better to navigate having more confidence in visual information than in GPS data. On the other hand, in the case of trajectories moving far away from its initial frame of reference, the use of absolute referenced data obtained from the GPS imposes an upper bound on the ever growing error, contrary to what is expected with a pure vision-based SLAM approach.

In these experiments, it is important to note that the most relevant source of error comes from the slow-time varying bias part of the GPS. In this case, some of the effects of this bias can be tackled by the model in Equation (5) by means of increasing the measurement noise covariance matrix. On the other hand, it was found that increasing this measurement matrix too much can affect the convergence of initial features depth. A future work could be, for instance, to develop an adaptive criteria to fuse GPS data, or also to extend the method in order to explicitly estimate the slow-varying bias of the GPS.

4.2. Experiments with Real Data

A custom-built quadrotor is used to perform experiments with real data. The vehicle is equipped with an Ardupilot unit as flight controller [37], a NEO-M8N GPS unit, a radio telemetry unit 3 DR 915 Mhz, a DX201 DPS camera with wide angle lens, and a 5.8 GHz video transmitter. In experiments, the quadrotor has been manually radio-controlled (see Figure 11).

A custom-built C++ application running on a laptop has been used to capture data from the vehicle, which were received via MAVLINK protocol [38], as well as capturing the digitalized video signal transmitted from the vehicle. The data captured from the GPS, AHRS, and frames from the camera were synchronized and stored in a dataset. The frames with a resolution of 320×240 pixels, in gray scale, were captured at 26 fps. The flights of the quadrotor were conducted in a open area of a park surrounded by trees, see Figure 11. The surface of the field is mainly flat and composed by grass and dirt, but the experimental environment also included some small structures and plants. An average of 8–9 GPS satellites were visible at the same time.

Figure 11. A park was used as flight field. Data obtained from the sensors of a radio-controlled quadrotor has been used to test the proposed method. The eight year-old first author's son was in charge of piloting the flying vehicle.

In experiments, in order to have an external reference of the flight trajectory to evaluate the performance of the proposed method, four marks were placed in the floor, forming a square of known dimensions (see Figure 4). Then, a perspective on 4-point (P4P) technique [39] was applied to each frame in order to compute the relative position of the camera with respect to this known reference. It is important to note that the trajectory obtained by the above technique should not be considered as a perfect reference of ground-truth. However, this approach was very helpful to have a fully independent reference of flight for evaluation purposes. Finally, the MATLAB implementation of the proposed method has been executed offline for all the dataset in order to estimate the flight trajectory and the map of the environment.

An initial period of flight was considered for initialization purposes, as explained in Section 3.4. Figure 12 shows two different instances of a flight trajectory. For this test, the GPS readings were fused into the system only at the initialization period; after that, the position of the vehicle and the map of the environment were recovered using visual information. Since the beginning of the flight (left plots), it can be clearly appreciated how the GPS readings diverge from the actual trajectory. Several features have been included into the map just after a few seconds of flight (right plots).

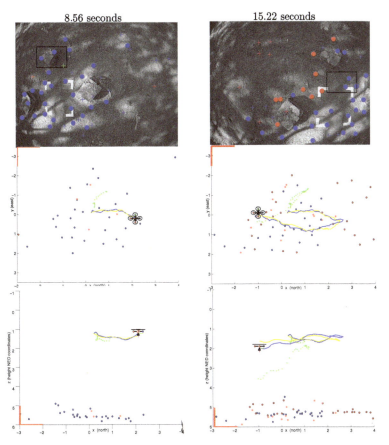

Figure 12. Estimated trajectory and map corresponding to two different instants of time during periods of visual-based navigation: (**upper plots**) real images at 8.56 s and 15.22 s of flight; (**middle plots**) zenital view of maps and estimated trajectories at 8.56 s and 15.22 s of flight; (**lower plots**) sectional view of maps and estimated trajectories at 8.56 s and 15.22 s of flight. The estimated trajectory is indicated in blue. The P4P visual reference is indicated in yellow. GPS position measurements are indicated in green. Comparing visual features with the estimated map, it can be appreciated that the physical structure of the environment is partially recovered.

Figure 13 shows a 3D perspective of the estimated map and trajectory after 30 s of flight. In this test, a good concordance between the estimated trajectory and the P4P visual reference were obtained, especially if it is compared with the GPS trajectory.

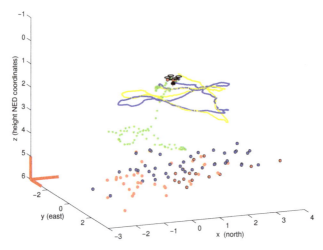

Figure 13. 3D plot of the estimated map and trajectory obtained in visual-based navigation mode. Considering the trajectory obtained by the P4P visual technique as a reference, it can be clearly appreciated that GPS is unreliable to estimate position when fine manoeuvres are performed.

In order to gain more insight about the performance of the proposed method, the same three experimental variants used in simulations were computed, but in this case with real data: (i) GPS; (ii) GPS + camera; (iii) camera (GPS only at the initialization). In this comparison, all the results were obtained averaging ten executions of each method. Is important to note that those averages are computed because the method is not deterministic since the search and detection of new candidate points is conducted in a random manner over the images (Section 3.3.1). The P4P visual reference was used as ground-truth. The number of visual features being tracked at each frame can affect the performance of monocular SLAM methods. For this reason, the methods were tested by setting two different values of minimum distance (M.D.) between the visual features being tracked. In this case, the bigger the value, the lesser the number of visual features that can be tracked.

Figure 14 shows the progression over time for each case. A separate plot for each coordinate (north, east, and down) is presented. Table 1 gives a numerical summary of the results obtained in this experimental comparison with real data. These results confirm the results obtained through simulations. For trajectories estimated using only GPS data, the high average MAE in position makes this approach not suitable for its use as feedback to control fine manoeuvres. In this particular case, it is easy to see that the major source of error comes from the altitude computed by the GPS (see Figure 14, lower plots). Additional sensors (e.g., a barometer) can be used to mitigate this particular error. However, the error in the horizontal plane (north–east) can be still critical for certain applications. In this sense, the benefits obtained by including visual information into the system are evident.

As it could be expected, the number of map features increases considerably as the minimum distance between visual points is decremented. However, it is interesting to remark that, at least for these experiments, there was no important improvement in error reduction. Regarding the use of the GPS altogether with monocular vision, a slightly better concordance was obtained between the P4P reference and trajectory estimated avoiding the GPS data (after the initialization). These results still suggest that, at least for small environments, it could be better to rely more on visual information than on GPS data after the initialization period.

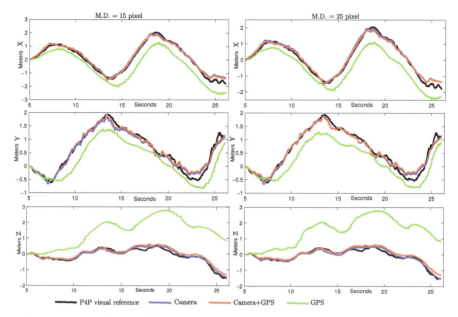

Figure 14. Estimated average of position expressed in coordinates for a minimum distance of 15 pixels: north (**left upper plot**), east (**left middle plot**), and down (**left lower plot**), and for a minimum distance of 25 pixels: north (**right upper plot**), east (**right middle plot**), and down (**right lower plot**). A period of 5 s of initialization was considered where the GPS was available.

Table 1. Numerical results in real data experiments; (i) M.D. stands for minimum distance between features (in pixels units); (ii) N.O.F. stands for average number of features maintained into the system state; (iii) aMAE stands for average mean absolute error (in meters).

	M.D. (15p)	M.D. (15p)	M.D. (25p)	M.D. (25p)
Method	N.O.F.	aMAE (m)	N.O.F. (s)	aMAE (m)
GPS	-	$1.70 \pm 0.77\sigma$	-	$1.70 \pm 0.77\sigma$
Camera + GPS	$56.4 \pm 10.2\sigma$	$0.21 \pm 0.11\sigma$	$30.9 \pm 4.9\sigma$	$0.22 \pm 0.10\sigma$
Camera	$57.9 \pm 9.3\sigma$	$0.20 \pm 0.09\sigma$	$30.9 \pm 5.6\sigma$	$0.20 \pm 0.08\sigma$

The feasibility to implement monocular SLAM methods in real-time has been widely studied in the past. In particular, since the work of Davison in 2003 [32], the feasibility for EKF-based methods was shown for maps composed of up to 100 features using standard hardware. Later, in [29], it was shown that filter-based methods might be beneficial if limited processing power is available. Even real-time performance has been demonstrated for relatively high computation demanding techniques as the optimization-based method proposed in [40]. In the application proposed in this work, it can be seen (Table 1) that the number of features that are maintained into the system state (even for the low M.D.) are considerably below an upper bound that should allow a real-time performance, for instance by implementing the algorithm in C or C++ languages.

5. Conclusions

In this work, a vision-based navigation and mapping system with application to unmanned aerial vehicles has been presented. The visual information is obtained with a camera integrated in the flying vehicle pointing to the ground. The proposed scheme is closely related to monocular SLAM systems where a unique camera is used to concurrently estimate a map of visual features as well as the trajectory of the camera. As a difference from the purely monocular SLAM approaches, in this work a multi-sensor scheme is followed in order to take advantage of the set of sensors commonly available in UAVs in order to overcome some technical difficulties associated with monocular SLAM systems.

When a monocular camera is used, depth information cannot be retrieved in a single frame. In this work, a novel method is developed with this purpose. The proposed approach is based on a stochastic technique of triangulation to estimate features depth. Another important challenge that arises with the use of monocular vision comes with the fact that the metric scale of the environment can be only retrieved with a known factor if no additional information is incorporated into the system. In this work, the GPS readings are used during an initial short period of time in order to set the metric scale of estimation. After this period, the system operates relying uniquely on visual information to estimate the location of the vehicle.

Due to the highly noisy nature of the GPS measurements, it is unreliable to work only with filtered GPS data in order to obtain an accurate estimation of position to perform fine manoeuvres. In this case, visual information is incorporated into the system in order to refine such estimations.

The experimental results obtained through simulations as well as with real data suggest the following and relevant conclusions: (i) the integration into the system of very noisy GPS measurements during an initial short period is enough to recover the metric scale of the world; (ii) for flight trajectories performed near to the origin of the navigation frame of reference it is better to avoid integration of GPS measurements after the initialization period.

Acknowledgments: This research has been funded with EU Project AEROARMS project reference H2020-ICT-2014-1-644271, http://www.aeroarms-project.eu/. First author also wants to thank his son Roderic Munguía for his contribution to this work.

Author Contributions: Rodrigo Munguía designed the algorithm and coordinated the research, Sarquis Urzua implementated the experiments with real data, Yolanda Bolea helped to design the experiments and implemented the simulation part, and Antoni Grau wrote the paper and also coordinated the research. All the authors contributed equally in the research that leads to this paper.

Conflicts of Interest: The authors declare no conflict of interest.

Nomenclature

p^N	3D point, defined in euclidean coordinates, expressed in frame N
p^C	3D point, defined in euclidean coordinates, expressed in frame C
u, v	Undistorted pixel coordinates of a visual feature
u_d, v_d	Distorted pixel coordinates of a visual feature
$f, u_0, v_0, k_1, ..k_n$	Intrinsic parameters of the camera
R^{NC}	Rotation matrix from navigation to camera frame
t_c^N	Position of the camera optical center expressed in the navigation frame
h^C	Vector pointing from t_c^N to P^C expressed in frame C
y_a^N	Attitude measured
a^N	Actual attitude
ϕ_v, θ_v, ψ_v	Roll, pitch, and yaw of the vehicle
v_a	Modelled Gaussian white noise in attitude
σ_a^2	Attitude measurement variance
y_r^N	GPS position measurement
v_r	Modelled Gaussian white noise in position

σ_r^2	Position measurement variance
x	Augmented system state
P	System state covariance matrix
x_v	State of the vehicle
q^{NR}	Quaternion representing the orientation of the vehicle
ω^R	Angular velocity of the vehicle
r^N	Vehicle position
v^N	Lineal velocity of the vehicle
y_i^N	Map feature
x_i, y_i, z_i	Euclidean coordinates of features
V^N	Linear velocity impulse
Ω^C	Angular velocity impulse
σ_v^2	Linear velocity impulse variance
σ_ω^2	Angular velocity impulse variance
Q	Process noise covariance matrix
∇F_x	Jacobian of the prediction model with respect to the system state
∇F_u	Jacobian of the prediction model with respect to the unknown inputs
c_l	Data stored for each candidate point
y_{c_i}	3D semi-line defined by a candidate point
$t_{c_0}^N$	Camera position when the candidate point was first observed
θ_0, ϕ_0	Azimuth and elevation of the candidate point when it was first observed
z_{uv}	Visual point location
h^N	Vector pointing from $t_{c_0}^N$ to P^N expressed in frame ^{C}
$P_{y_{c_i}}$	Covariance matrix of y_{c_i}
∇Y_{c_i}	Jacobian of the function y_{c_i} with respect to the system state and visual measurement
d	Feature depth
e	Epipolar point
S_c	Elliptical region of search of candidate points
α_c	Orientation of the ellipse S_c
a, b	Major and minor axis of the ellipse S_c
α	Parallax of the candidate point
e_l	Displacement of the camera from its first observation to its current position
y_{new}	New feature to be added to the system state
$P_{y_{new}}$	Covariance matrix which models the uncertainty of y_{new}
∇Y	Jacobian of the function y_{new} with respect to c_l and d
σ_d^2	Modelled uncertainty associated with the process of depth estimation
S	Innovation covariance matrix
K	Kalman gain
ξ	Measurement noise covariance matrix
σ_{uv}^2	Visual measurement variance
z	Measurement vector
h	Predicted measurement vector
h_i	Measurement prediction model for the i feature
∇H	Jacobian of the function h with respect to the system state x
h_a	Measurement prediction model of attitude
h_r	Measurement prediction model of position
P_{y_i}	sub-matrix of P corresponding to a feature y_i

References

1. Munguia, R.; Grau, A. A Practical Method for Implementing an Attitude and Heading Reference System. *Int. J. Adv. Robot. Syst.* **2014**, *11*, doi:10.5772/58463.
2. Parkinson, B. *Global Positionig System: Theory and Applications*; American Institute of Aeronautics and Astronautics: Washington, DC, USA, 1996.
3. Gurdan, D.; Stumpf, J.; Achtelik, M.; Doth, K.M.; Hirzinger, G.; Rus, D. Energy-Efficient Autonomous Four-Rotor Flying Robot Controlled at 1 kHz. In Proceedings of the 2007 IEEE International Conference on Robotics and Automation, Roma, Italy, 10–14 April 2007; pp. 361–366.
4. Kim, J.; Min-Sung Kang, S.P. Accurate Modeling and Robust Hovering Control for a Quad-Rotor VTOL Aircraft. *J. Intell. Robot. Syst.* **2010**, *57*, 9–26.
5. Mori, R.; Kenichi Hirata, T.K. Vision-Based Guidance Control of a Small-Scale Unmanned Helicopter. In Proceedings of the Conference on Intelligent Robots and Systems, San Diego, CA, USA, 29 October–2 November 2007.
6. Zhang, T.; Kang, Y.; Achtelik, M.; Kuehnlenz, K.; Buss, M. Autonomous Hovering of a Vision-IMU Guided Quadrotor. In Proceedings of the International Conference on Mechatronics and Automation, Changchun, China, 9–12 August 2009.
7. Wenzel, K.E.; Paul Rosset, A.Z. Low-Cost Visual Tracking of a Landing Place and Hovering Flight Control with a Microcontroller. In Proceedings of the Selected Papers from the 2nd International Symposium on UAV, Reno, NV, USA, 8–10 June 2009; pp. 297–311.
8. Luis-Rodolfo, E.R.; Garcia-Carrillo, I.F. Vision-Based Altitude, Position and Speed Regulation of a Quadrotor Rotorcraft. In Proceedings of the Conference on Intelligent Robots and Systems, Taipei, Taiwan, 18–22 October 2010.
9. Munguia, R.; Manecy, A. State estimation for a bio-inspired hovering robot equipped with an angular sensor. In Proceedings of the 2012 9th International Conference on Electrical Engineering, Computing Science and Automatic Control (CCE), Mexico City, Mexico, 26–28 September 2012; pp. 1–6.
10. Artieda, J.; Sebastian, J.; Campoy, P.; Correa, J.; Mondragon, I.; Martinez, C.; Olivares, M. Visual 3-D SLAM from UAVs. *J. Intell. Robot. Syst.* **2009**, *55*, 299–321.
11. Weiss, S.; Scaramuzza, D.; Siegwart, R. Monocular SLAM based navigation for autonomous micro helicopters in GPS-denied environments. *J. Field Robot.* **2011**, *28*, 854–874.
12. Zhao, H.; Chiba, M.; Shibasaki, R.; Shao, X.; Cui, J.; Zha, H. SLAM in a dynamic large outdoor environment using a laser scanner. In Proceedings of the IEEE International Conference on Robotics and Automation, ICRA 2008, Pasadena, CA, USA, 19–23 May 2008; pp. 1455–1462.
13. Bosse, M.; Roberts, J. Histogram Matching and Global Initialization for Laser-only SLAM in Large Unstructured Environments. In Proceedings of the 2007 IEEE International Conference on Robotics and Automation, Roma, Italy, 10–14 April 2007; pp. 4820–4826.
14. Fallon, M.; Folkesson, J.; McClelland, H.; Leonard, J. Relocating Underwater Features Autonomously Using Sonar-Based SLAM. *IEEE J. Ocean. Eng.* **2013**, *38*, 500–513.
15. Yap, T.; Shelton, C. SLAM in large indoor environments with low-cost, noisy, and sparse sonars. In Proceedings of the IEEE International Conference on Robotics and Automation, ICRA '09, Kobe, Japan, 12–17 May 2009; pp. 1395–1401.
16. Munguia, R.; Grau, A. Monocular SLAM for Visual Odometry: A Full Approach to the Delayed Inverse-Depth Feature Initialization Method. *Math. Probl. Eng.* **2012**, *2012*, 676385.
17. Forster, C.; Lynen, S.; Kneip, L.; Scaramuzza, D. Collaborative monocular SLAM with multiple Micro Aerial Vehicles. In Proceedings of the 2013 IEEE/RSJ International Conference on Intelligent Robots and Systems (IROS), Tokyo, Japan, 3–7 Novemebr 2013; pp. 3962–3970.
18. Mirzaei, F.; Roumeliotis, S. A Kalman Filter-Based Algorithm for IMU-Camera Calibration: Observability Analysis and Performance Evaluation. *IEEE Trans. Robot.* **2008**, *24*, 1143–1156.
19. Weiss, S.; Achtelik, M.; Chli, M.; Siegwart, R. Versatile distributed pose estimation and sensor self-calibration for an autonomous MAV. In Proceedings of the 2012 IEEE International Conference on Robotics and Automation (ICRA), Saint Paul, MN, USA, 14–18 May 2012; pp. 31–38.
20. Bouguet, J. Camera Calibration Toolbox for Matlab. Available online: http://www.vision.caltech.edu/bouguetj/calib_doc/ (accessed on 14 March 2016).

21. Jurman, D.; Jankovec, M.; Kamnik, R.; Topic, M. Calibration and data fusion solution for the miniature attitude and heading reference system. *Sens. Actuators A Phys.* **2007**, *138*, 411–420.

22. Wang, M.; Yang, Y.; Hatch, R.; Zhang, Y. Adaptive filter for a miniature MEMS based attitude and heading reference system. *Position Locat. Navig. Symp.* **2004**, 193–200, doi:10.1109/PLANS.2004.1308993.

23. Grewal, M.S.; Lawrence, R.; Weill, A.P.A. *Global Positioning Systems, Inertial Navigation, and Integration*; Wiley: Hoboken, NJ, USA, 2007.

24. Zogg, J-M. *Essentials of Sattelite Navigation*, Technical Report; u-blox AG: Thalwil, Switzerland, 2009.

25. Davison, A.; Reid, I.; Molton, N.; Stasse, O. MonoSLAM: Real-Time Single Camera SLAM. *IEEE Trans. Pattern Anal. Mach. Intell.* **2007**, *29*, 1052 –1067.

26. Shi, J.; Tomasi, C. Good features to track. In Proceedings of the 1994 IEEE Computer Society Conference on Computer Vision and Pattern Recognition, Proceedings CVPR '94, Seattle, WA, USA, 21–23 June 1994.

27. Davison, A.J.; Murray, D.W. Mobile robot localisation using active vision. In Proceedings of the 5th European Conference on Computer Vision (ECCV '98), Freiburg, Germany, 2–6 June 1998.

28. Munguia, R.; Grau, A. Concurrent Initialization for Bearing-Only SLAM. *Sensors* **2010**, *10*, 1511–1534.

29. Strasdat, H.; Montiel, J.; Davison, A. Real-time monocular SLAM: Why filter? In Proceedings of the 2010 IEEE International Conference on Robotics and Automation (ICRA), Anchorage, AK, USA, 3–7 May 2010; pp. 2657 –2664.

30. Munguia, R.; Grau, A. Closing Loops With a Virtual Sensor Based on Monocular SLAM. *Instrum. Meas. IEEE Trans.* **2009**, *58*, 2377 –2384.

31. Munguia, R.; Grau, A. Monocular SLAM for Visual Odometry. In Proceedings of the IEEE International Symposium on Intelligent Signal Processing, WISP 2007, Alcala de Henares, Spain, 3–5 October 2007; pp. 1–6.

32. Davison, A. Real-time simultaneous localisation and mapping with a single camera. In Proceedings of the Ninth IEEE International Conference on Computer Vision, Nice, France, 13–16 October 2003; Volume 2, pp. 1403–1410.

33. Belo, F.A.W.; Salaris, P.; Fontanelli, D.; Bicchi., A. A Complete Observability Analysis of the Planar Bearing Localization and Mapping for Visual Servoing with Known Camera Velocities. *Int. J. Adv. Robot. Syst.* **2013**, doi:10.5772/54603.

34. Bailey, T. Constrained initialisation for bearing-only SLAM. In Proceedings of the IEEE International Conference on Robotics and Automation, ICRA '03, Taipei, Taiwan, 14–19 September 2003; Volume 2, pp. 1966–1971.

35. Corke, P.I. *Robotics, Vision & Control: Fundamental Algorithms in Matlab*; Springer: Berlin, Germany, 2011.

36. Rankin, J. An error model for sensor simulation GPS and differential GPS. In Proceedings of the Position Location and Navigation Symposium, Las Vegas, NV, USA, 11–15 April 1994; pp. 260–266.

37. Community, O.S. Ardupilot. 2015. Available online: http://ardupilot.com (accessed on 14 March 2016).

38. Mavlink Communication Protocol. Available online: http://qgroundcontrol.org/mavlink/start (accessed on 14 March 2016).

39. Chatterjee, C.; Roychowdhury, V.P. Algorithms for coplanar camera calibration. *Mach. Vis. Appl.* **2000**, *12*, 84–97.

40. Klein, G.; Murray, D. Parallel Tracking and Mapping for Small AR Workspaces. In Proceedings of the 6th IEEE and ACM International Symposium on Mixed and Augmented Reality, ISMAR 2007, Santa Barbara, CA, USA, 14–17 November 2007; pp. 225 –234.

Article

Machine Learning and Computer Vision System for Phenotype Data Acquisition and Analysis in Plants

Pedro J. Navarro [1,*,†], **Fernando Pérez** [2,†], **Julia Weiss** [2,†] **and Marcos Egea-Cortines** [2,†]

[1] DSIE, Universidad Politécnica de Cartagena, Campus Muralla del Mar, s/n. Cartagena 30202, Spain
[2] Genética, Instituto de Biotecnología Vegetal, Universidad Politécnica de Cartagena, Cartagena 30202, Spain; fernando.perez8@um.es (F.P.); Julia.weiss@upct.es (J.W.); marcos.egea@upct.es (M.E.-C.)
*   Correspondence: pedroj.navrro@upct.es; Tel.: +34-968-32-6546
†   These authors contributed equally to this work.

Academic Editor: Gonzalo Pajares Martinsanz
Received: 3 March 2016; Accepted: 26 April 2016; Published: 5 May 2016

Abstract: Phenomics is a technology-driven approach with promising future to obtain unbiased data of biological systems. Image acquisition is relatively simple. However data handling and analysis are not as developed compared to the sampling capacities. We present a system based on machine learning (ML) algorithms and computer vision intended to solve the automatic phenotype data analysis in plant material. We developed a growth-chamber able to accommodate species of various sizes. Night image acquisition requires near infrared lightning. For the ML process, we tested three different algorithms: *k*-nearest neighbour (kNN), Naive Bayes Classifier (NBC), and Support Vector Machine. Each ML algorithm was executed with different kernel functions and they were trained with raw data and two types of data normalisation. Different metrics were computed to determine the optimal configuration of the machine learning algorithms. We obtained a performance of 99.31% in kNN for RGB images and a 99.34% in SVM for NIR. Our results show that ML techniques can speed up phenomic data analysis. Furthermore, both RGB and NIR images can be segmented successfully but may require different ML algorithms for segmentation.

Keywords: computer vision; image segmentation; machine learning; data normalisation; circadian clock

1. Introduction

The advent of the so-called omics technologies has been a major change in the way experiments are designed and has driven new ways to approach biology. One common aspect to these technology-driven approaches is the continuous decrease in price in order to achieve high throughput. As a result biology has become a field where big data accumulates, and which requires analytical tools [1]. The latest newcomer in the field of automatic sampling is the so-called phenomics. It comprises any tool that will help acquire quantitative data of phenotypes. Plant growth and development can be considered as a combination of a default program that interacts with biotic and abiotic stresses, light and temperature to give external phenotypes. And measuring, not only the outcome or end point, but also kinetics and their changes is becoming increasingly important to understand plants as a whole and become more precise at experimental designs. One of the newest developments is automatic image acquisition [2].

One of the fields where automatic image acquisition has defined its development is circadian clock analysis as promoters driving reporter genes such as luciferase or Green Fluorescent Protein allowed the identification of mutants and further characterization of the gene network at the transcriptional level [3,4]. Artificial vision systems have been used to study different aspects of plant growth and development such as root development [5], leaf growth [6], flowers and shoots [7] or seedling [8]. An important challenge of image acquisition in plant biology is the signalling effect of different light wavelengths including blue light, red and far red. As a result image acquisition in the dark requires infrared lightning [7,9].

Phenotyping of small plants such as *Arabidopsis thaliana* can be performed with a vertical camera taking pictures of rosettes at time intervals [10]. Larger plants or the parallel phenotyping of several traits require image acquisition from lateral positions [11]. Thus obtaining lateral images or the reconstruction of 3-dimensional images is performed by combination of cameras or moving them to acquire images [12].

Although hardware development requires a multidisciplinary approach, the bottleneck lies in image analysis. Ideally images should be analysed in an automatic fashion. The number of images to be processed when screening populations or studying kinetics can easily go into the thousands. The partition of digital images into segments, known as segmentation is a basic process allowing the acquisition of quantitative data that may be a number of pixels of a bidimensional field, determining the boundaries of interest in an object [13]. Segmentation discriminates between background and defines the region under study and is the basis for further data acquision.

The development of artificial intelligence processes based on machine learning (ML) has been an important step in the development of software for omic analysis and modelling [14]. Examples include support vector machines (SVM) for Illumina base calling [15], *k*-nearest neighbour (kNN) classification for protein localization [16] or Naïve Bayes Classifiers for phylogenetic reconstructions [17]. Furthermore, ML approaches have been used extensively in image analysis applied to plant biology and agriculture [18,19].

Plant growth occurs in a gated manner *i.e.*, it has a major peak during the late night in hypocotyls, stems or large leaves [11,20,21]. This is the result of circadian clock regulation of genes involved in auxin and gibberellin signalling and cell expansion [22]. One of the inputs to the circadian clock is blue light transmitted through proteins that act as receptors such as ZEITLUPE/FLAVIN-BINDING, KELCH REPEAT, F-BOX and LOV KELCH PROTEIN2 [23,24]. Phytochromes absorb red and far red light such as PHYTOCHROME A [25,26]. As a result night image acquisition has to be done with near infrared (NIR) light giving the so-called extended night signal [27]. The aim of this work was to develop the corresponding algorithms to obtain data from day and night imaging. We used machine learning to analyse a set of images taken from different species during day and night. We used the aforementioned SVM, NBC and kNN to obtain image segmentations. Our results demonstrate that ML has great potential to tackle complex problems of image segmentation.

2. Materials and Methods

Figure 1a shows a schematic of the system. The data acquisition system is composed of four modules which we describe below.

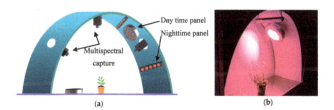

(a) (b)

Figure 1. Growth chamber: (**a**) Functional schematic of the system; (**b**) Experiment with *Petunia x hybrida* during daytime in the real system.

2.1. Ilumination Subsystem

We pursued two goals with the illumination subsystem. First we wanted to grow plants under conditions close to their natural environments and second we wanted to acquire pictures during the night-time without interfering with the behaviour of the plant. For this purpose, we have established two illumination periods: daytime and night-time. The illumination subsystem is composed of two LED (light-emitting diode) panels which, allows to carry-out the capture image process and the same time it allows to supply the precise combination of the wavelengths for growing up correctly.

The daytime LED panel is formed by a combination of five types of LEDs emitting wavelengths with peaks in UV light (290 nm), blue light (450 and 460 nm) and red light (630 and 660 nm). The LED panel has a power of fifty watts. It is usually used for indoor growing of crop plants. The merging of wavelengths produces an illumination with a pink-red appearance (Figure 2a).

The night-time LED panel is composed by a bar of 132 NIR LEDs (three rows of forty four LEDs) with a wavelength of 850 nm (Figure 2b).

We programmed a system that would give a day/night timing whereby day light was created by turning on the daytime LED. In order to capture night images, the night-time LED panel was turned on for a period between 3 and 5 s coupled to an image capture trigger. The system can be programmed by the user for different periods of day and night lengths and time course of picture acquisition. The minimal period is one picture every 6 s and the maximal is one picture in 24 h.

(a) (b)

Figure 2. Illumination subsystem (a) Daytime LED panel; (b) Nightime LED panel.

2.2. Capture Subsystem

The capture module is in charge of image capture during day and night and the control of the illumination subsystem. The main capture subsystem element is a multispectral 2-channel Charge-Coupled Device (CCD) camera. A prism placed in the same optical path between the lens and CCDs allows a simultaneous capture the visible (or RGB) and NIR image (see Figure 3a). This feature has reduced the amount of cameras being used by the system and has avoided the construction of a mechanical system to move the lenses or the cameras in front of the plants. The camera has a resolution of 1024 (h) × 768 (v) active pixels per channel. During day and night a resolution of 8 bit per pixel was used in all the channels (R-G-B-NIR). Figure 3b,c shows the response of the NIR-CCD and RGB-CCD of the multispectral camera.

Capture and illumination subsystems are controlled via a GUI developed in C/C++ (Figure 4a,b). It comprises eight digital input/output channels and six analog ones in an USB-GPIO module (Figure 5a). The system had 10 bit resolution. It was configured using the *termios* Linux library in C/C++.

The second component was the optocoupler relay module. It had four optocoupled outputs, optocoupled to a relay triggering at voltages between 3 and 24 V. Both day light and night light LEDs were connected to two relays (Figure 5b), in such a way that the configuration via the control software dictates the beginning of image acquisition, triggers light turning on or off coordinating the light pulses with the camera during the day and night.

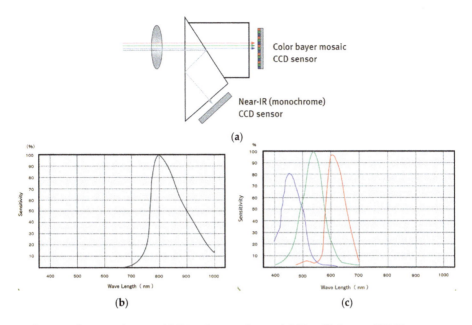

Figure 3. Capture subsystem (**a**) Prism between lens and CCDs; (**b**) Camera NIR-IR response; (**c**) Camera RGB response.

Figure 4. Graphical User Interface for capture subsystem: (**a**) Image control tap; (**b**) Time control tab.

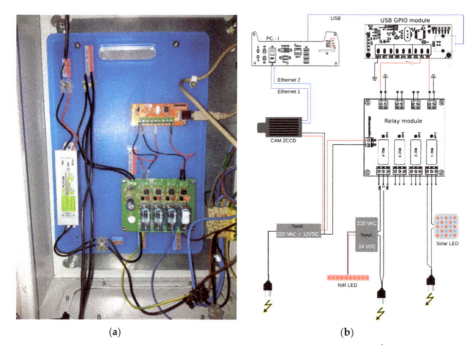

Figure 5. Hardware of the capture subsystem. (**a**) USB-GPIO module (red-board) and opto-coupler relay module (green-board); (**b**) Electric connections between all hardware modules in the growth-chamber.

2.3. Image Processing Module

Each experiment generates two types of images: one NIR image during the night-time and another RGB image during the daytime. In order to obtain an automatic image segmentation, we designed an algorithm to classify the objects from the images of the experiment in two groups: organs and background. The algorithm developed is divided in three stages.

2.3.1. Extraction of Samples of Images Representative from the Different Classes

During the first stage of the algorithm we have selected a set of representative samples formed by n matrix with size of $k \times k$ pixels of each class. The size of the regions can be of 1×1, 16×16, 32×32, 64×64 or 128×128 pixels. This will depend of size and morphology of the organ to be classified and of the period of the daytime or night-time involved. During the day we used a single pixel per channel while we used the larger pixel regions to increase the information obtained in the IR channel.

2.3.2. Features Vector

The feature vector is usually composed by a wide variety of different types of features. The most utilized features are related to: intensity of image pixels [28], geometries [29] and textures (first and second-order statistical features) [30,31]. In addition the feature vector is computed over image transformations such as Fourier, Gabor, and Wavelet [32]. Colour images comprise three channels for R, G and B. As a result the amount of information is multiplied by three and the number of possible combinations and image transformations are incremented.

We have applied two types of features vector techniques depending whether the image was captured during daytime (RGB image) or night-time (NIR images). We tested several colour spaces to construct the features vector of daytime: RGB primary space, HSV perceptual space and CIE L*a*b* luminance-chrominance space [33]. The use of a single colour space produced poor results. In order to improve the performance we increased the number of feature vectors. We used cross combinations of the RGB, CIE L*a*b*, HSV and found that the best performance was with RGB and CIE L*a*b*. Thus we constructed the features vector formed by the pixel the corresponding pixel values. In the NIR images, we used a features vector computed over two decomposition levels of the Haar wavelet transform.

Colour Images

The features vector of the colour images is composed of six elements extracted from the pixel values of two colour spaces: RGB and CIE L*a*b*. We selected a large set of random pixels of each class to construct the features vector Figure 6a (organs-class1-green and background-class2-white). It was necessary to convert RGB colour space of the original image to CIE L*a*b* colour space. Figure 6b shows the twenty values of the features vector of class 1.

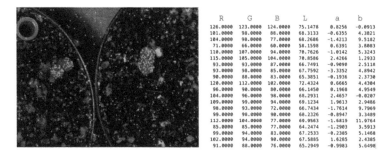

R	G	B	L	a	b
126.0000	123.0000	124.0000	75.1478	0.8256	-0.0913
101.0000	98.0000	88.0000	68.3133	-0.6355	4.3821
104.0000	98.0000	77.0000	68.2686	-1.4213	9.5182
71.0000	66.0000	60.0000	58.1598	0.6391	3.8083
110.0000	107.0000	94.0000	70.7626	-1.0142	5.3243
115.0000	105.0000	104.0000	70.8586	2.4266	1.2933
93.0000	93.0000	87.0000	66.7491	-0.9090	2.5118
93.0000	98.0000	85.0000	67.7592	-3.3352	4.8942
90.0000	88.0000	83.0000	65.3851	-0.1936	2.3730
120.0000	112.0000	102.0000	72.4324	0.6665	4.4304
96.0000	90.0000	80.0000	66.1450	0.1968	4.9549
104.0000	96.0000	98.0000	68.2931	2.4657	-0.0207
109.0000	99.0000	94.0000	69.1234	1.9613	2.9486
98.0000	93.0000	72.0000	66.7434	-1.7614	9.7969
99.0000	98.0000	90.0000	68.2326	-0.8947	3.3489
112.0000	104.0000	77.0000	69.9563	-1.6819	11.9764
85.0000	85.0000	77.0000	64.2474	-1.2903	3.5913
99.0000	94.0000	83.0000	67.2533	-0.2385	5.1468
102.0000	94.0000	90.0000	67.5885	1.6285	2.4385
91.0000	88.0000	76.0000	65.2949	-0.9983	5.6498

Figure 6. Colour images features vector construction. A matrix of different pixel values corresponding to R, G, B, and L*a*b* colour spaces.

NIR Images

Discrete Wavelet Transformation (DWT) generates a set of values formed by the "wavelet coefficients". Being $f(x, y)$ an image of $M \times N$ size, each level of wavelet decomposition is formed by the convolution of the image $f(x, y)$ with two filters: a low-pass filter (LPF) and a high-pass filter (HPF). The different combinations of these filters result in four images here described as LL, LH, HL and HH. In the first decomposition level four subimages or bands are produced: one smooth image, also called approximation, $f_{LL}^{(1)}(x, y)$, that represents an approximation of the original image $f(x, y)$ and three detail subimages $f_{LH}^{(1)}(x, y)$, $f_{HL}^{(1)}(x, y)$ and $f_{HH}^{(1)}(x, y)$, which represent the horizontal, vertical and diagonal details respectively. There are several wavelet mother functions that can be employed, like Haar, Daubechies, Coiflet, Meyer, Morlet, and Bior, depending on the specific problem to be identified [34,35]. Figure 7 shows the pyramid algorithm of wavelet transform in the first decomposition level.

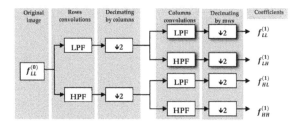

Figure 7. First level of direct 2D-DWT decomposition.

In this work we have computed a features vector based on the wavelet transform with basis Haar [36]. The features vector is formed of four elements: maximum, minimum, mean and Shannon entropy of coefficients wavelets calculated in the horizontal, vertical and diagonal subimages in two decomposition levels (see Equations (1)–(5)): We have eliminated the approximation subimage due to it contains a representation decimated of the original image:

$$f_{1,\,..,\,6} = \max \left\{ f_{LH}^{(l)}(x,y),\, f_{LH}^{(l)}(x,y),\, f_{LH}^{(l)}(x,y) \right\},\, \forall l = 1, 2 \tag{1}$$

$$f_{7,..,12} = \min \left\{ f_{LH}^{(l)}(x,y),\, f_{LH}^{(l)}(x,y),\, f_{LH}^{(l)}(x,y) \right\},\, \forall l = 1, 2 \tag{2}$$

$$f_{13,...,18} = \text{mean} \left\{ f_{LH}^{(l)}(x,y),\, f_{LH}^{(l)}(x,y),\, f_{LH}^{(l)}(x,y) \right\},\, \forall l = 1, 2 \tag{3}$$

$$f_{19,\,..,24} = \text{shannon\_entropy} \left\{ f_{LH}^{(l)}(x,y),\, f_{LH}^{(l)}(x,y),\, f_{LH}^{(l)}(x,y) \right\},\, \forall l = 1, 2 \tag{4}$$

Shannon Entropy is calculated as the Equation (5):

$$\text{shannon\_entropy} \left(f_s^{(l)} \right) = - \sum_{i=1}^{M/2^l} \sum_{j=1}^{N/2^l} p\left(w_{ij}\right) \, log_2 \left(p\left(w_{ij}\right)\right) \tag{5}$$

The letter l represents the value of the wavelet decomposition level, s the subimages (LL, HL, LH, HH) created in the wavelet decomposition, and w_{ij} represents the wavelet coefficient (i, j), located in the s-subimage, at l-decomposition level. p represents the occurrence probability of the wavelet coefficient w_{ij}.

Feature vector has been obtained applying the Equations (1)–(5) to each region in two wavelet decomposition levels with Haar basis. The result was a feature vector of twenty-four elements ($f_{1,..,24}$) per region of size $k \times k$.

2.3.3. Classification Process

We have tested three machine-learning algorithms: (1) *k*-nearest neighbour (kNN); (2) naive Bayes classifier (NBC), and Support Vector Machine (SVM). The algorithms selected belong to the type of supervised classification. These type of algorithms require of a training stage before performing the classification process.

kNN classifier is a non-parametric method for classifying objects in a multi-dimensional space. After being trained, kNN assigns a specific class to a new object depending on the majority of votes from its neighbours. This measure is based in metrics such as Euclidean, Hamming or Mahalanobis distances. In the implementation of kNN algorithm it is necessary to assign an integer value to k. This parameter represents the *k*-neighbors used to carry-out the voting classification. A k optimal determination will allow that the good model adjusts to future data [37]. It is recommendable to use data normalisation coupled to kNN classifiers in order to avoid the predominance of big values over small values in the features vector.

NBC uses a probabilistic learning classification. Classifiers based on Bayesian methods utilize training data to calculate an observed probability of each class based on feature values. When the classifier is used later on unlabeled data, it uses the observed probabilities to predict the most likely class for the new features. As NBC works with probabilities it does not need data normalization.

SVM is a supervised learning algorithm where given labeled training data, it outputs a boundary which divides data by categories and categorizes new examples. The goal of a SVM is to create a boundary, called hyperplane, which leads to homogeneous partitions of data on either side. SVMs can also be extended to problems were the data are not linearly separable. SVMs can be adapted for use with nearly any type of learning task, including both classification and numeric prediction. SVM classifier tend to perform better after data normalisation.

In this work we have used raw data and two types of normalisation procedures which have been computed over features space: dn0: without normalisation; dn1: mean and standard-deviation normalization and dn2: mode and standard-deviation normalisation. Features space of each class is composed by a $m \times n$ matrix (see Equation (6)):

$$f_{ij}^{nC} = \begin{bmatrix} f_{11}^1 & \cdots & f_{1n}^1 \\ \vdots & \ddots & \vdots \\ f_{1m}^1 & \cdots & f_{mn}^1 \\ & & \cdot \\ & & \cdot \\ f_{11}^{nC} & \cdots & f_{1n}^{nC} \\ \vdots & \ddots & \vdots \\ f_{1m}^{nC} & \cdots & f_{mn}^{nC} \end{bmatrix} \quad \nabla\, i = 1,\ldots,m\,;\, j = 1,\ldots,n\,;\, nC = 1,2 \tag{6}$$

Being i-th row, the vector of features i-th of features space formed by n features. m represents the number of vectors in the features space and nC represents the number of classes in th space.

The normalised features space, F_{ij}^{nc}, depending on the normalisation types (dn0, dn1, dn2) is computed as is shown in the Equation (7):

$$F_{ij}^{nc} = \begin{bmatrix} \frac{f_{11}^1 - st_1}{st_2} & \cdots & \frac{f_{1n}^1 - st_1}{st_2} \\ \vdots & \ddots & \vdots \\ \frac{f_{1m}^1 - st_1}{st_2} & \cdots & \frac{f_{mn}^1 - st_1}{st_2} \\ & & \cdot \\ & & \cdot \\ \frac{f_{11}^{nC} - st_1}{st_2} & \cdots & \frac{f_{1n}^{nC} - st_1}{st_2} \\ \vdots & \ddots & \vdots \\ \frac{f_{1m}^{nC} - st_1}{st_2} & \cdots & \frac{f_{mn}^{nC} - st_1}{st_2} \end{bmatrix} \rightarrow \begin{cases} dn0 \begin{cases} st_1 = 0 \\ st_2 = 1 \end{cases} raw \quad data \\ dn1 \begin{cases} st_1 = mean(f_{ij}^{nC}) \\ st_2 = StdDes(f_{ij}^{nC}) \end{cases} \\ dn2 \begin{cases} st_1 = mode(f_{ij}^{nC}) \\ st_2 = StdDes(f_{ij}^{nC}) \end{cases} \end{cases} \tag{7}$$

To obtain the best result in classification process, the ML algorithms were tested with different configuration parameters. kNN was tested with Euclidean and Minkowski distances with three type of data normalisation, NBC was tested with Gauss and Kernel Smoothing Functions (KSF) without data normalisation, and SVM was tested with linear and quadratic functions, on three types of data normalisation. The ML algorithms used two classes of objects. One for the plant organs and a second one for the background. In all of them we applied the leave-out cross validation (LOOCV) method to measure of the error of the classifier. Basically LOOCV method extracts a sample of the training set and it constructs the classifier with the remaining of the training samples. Then it evaluates the classification error and the process is repeated for all the training samples. At end the LOOCV method computes the mean of the errors and it obtains a measure of how model is adjusted to data. This method allows comparing the results of the different ML algorithms, provided that they will be applied to same sample data. Table 1 shows a summary of parameters used for the classification process.

Table 1. kNN, NBC and SVM configuration parameters.

Configuration	kNN	NBC	SVM
method	Euclidean, Minkowski	Gauss, KSF	Linear, quadratic
data normalisation	dn0, dn1, dn2	dn0	dn1, dn2
metrics	LOOCV, ROC	LOOCV, ROC	LOOCV, ROC
classes	2	2	2

2.4. Experimental Validation

In order to test and validate the functioning of the system and ML methods, acquired pictures of *Antirrhinum majus* and *Antirrhinum linkianum* were used to analyse growth kinetics. The camera was positioned above the plants under study. The daylight LED panel was above the plants while the night-time LED was at a 45°. Data acquisition was performed for a total of six days. We obtained one image every 10 min during day and night. Day night cycles were set to 12:12 h and triggering of the NIR LED for image acquisition during the night was done for a total of 6 s.

In the experiment we obtained 864 colour images from the RGB sensor which were transformed to CIE L*a*b* colour space and 864 gray scale images from the NIR sensor. From each group (RGB and NIR images) we obtained fifty ground-truth images which were segmented manually by human experts. From the fifty ground-truth colour images, we selected 1200 samples of 1 pixel, which we used to train the RGB image processing ML algorithms. From the second fifty ground-truth NIR images we took 1200 regions of 32×32 pixels which were to train the NIR image processing ML algorithms. In both cases we selected 600 samples belonging to organs class and 600 samples belonging to background class.

Figure 8 shows two images from each day of the experiment for different capture periods. We can distinguish easily the growth stages of the two species during the daytime and night-time.

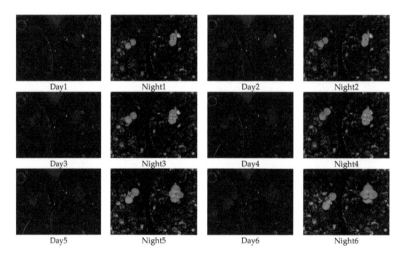

Figure 8. Images of the experiment captured every 12 h.

3. Results and Discussion

We evaluated the results of training stage of the ML algorithms with a leave-one-out cross-validation method (LOOCV) and with the Receiver Operating Characteristic (ROC) curve over data training sets obtained from RGB and NIR images. LOOCV and ROC curves have been applied under the different ML configurations shown in the Table 1. This allowed to select the optimal ML algorithm to be applied to each type of image depending on when it was captured: during daytime or night-time. Once we determined the optimal ML algorithm for each image set, we used the metric miss-classification [37] to evaluate the final performance of the implemented ML algorithms.

3.1. LOOCV

Table 2 shown the errors obtained after to apply LOOCV to the two images groups respectively. In both images groups the minimum error in data model adjust is produced with kNN classifier. Data normalisation based on in the mean (dn1) produced the best result in both cases, too. The maximum error is produced by the NBC classifier, with KSF and Gauss kernel, respectively.

Table 2. Colour images and NIR images. LOOCV error for kNN, NBC, SVM.

Classifier		kNN		NBC		SVM	
Configuration		*Euclidean*	*Minkowski*	*Gauss*	*KSF*	*Linear*	*Quadratic*
Colour	dn0	0.0283	0.0433	0.0750	0.0758	-	-
	dn1	**0.0242**	0.0467	-	-	0.0533	0.0383
	dn2	0.0283	0.0433	-	-	0.0667	0.0450
NIR	dn0	0.0288	0.0394,	0.0356	0.0319	-	-
	dn1	**0.0169**	0.0281	-	-	0.0326	0.0319
	dn2	0.0288	0.0394	-	-	0.0344	0.0325

3.2. ROC Curves

We evaluated the performance of the ML algorithms using Receiver Operating Characteristic (ROC) curve. The ROC curve is created by comparing the sensitivity (the rate of true positives TP, see Equation (8)), *versus* 1-specificity (the rate of false positives FP see Equation (9)), at various threshold levels [38]. The ROC curves, shown in Figures 8 and 9 allow comparing the results between of ML algorithms per each groups of images:

$$Sensitivity = \frac{TP}{TP + FN} \tag{8}$$

$$Specificity = \frac{TN}{TN + FP} \tag{9}$$

The Area Under the Curve (AUC) is usually used by ML to compare statistical models. AUC can be interpreted as the probability that the classifier will assign a higher score to a randomly chosen positive example than to a randomly chosen negative example [39,40].

Figure 9 shows ROC curves computed over the set of colour images training. The higher values of AUC were obtained by kNN classifier with Euclidean distance. Concerning the data normalisation, we achieved similar results with raw data and normalisation based on the mode and standard deviation (dn0 and dn1).

Figure 10 shows ROC curves computed over the set of NIR images training. We can observe that the best results were obtained with the SVM classifier with quadratic functions and using a normalised data based on the mean and standard deviation (dn1). Table 3 shows AUC values obtained from ROC curves of the Figures 9 and 10.

In both cases after classification stages, we performed a postprocessing stage composed of morphological operations and an area filter to eliminate noise and small particles. The segmented images were merged with the original images (Figures 11 and 12).

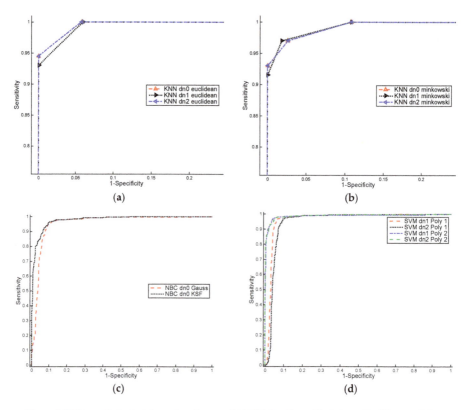

Figure 9. ROC results for training colour images. (**a**) kNN classifier with distance Euclidean and data normalisation: dn0, dn1 and dn2; (**b**) kNN classifier with distance Minkowski and data normalisation: dn0, dn1 and dn2; (**c**) BN classifier with Gauss and KSF kernels and data normalisation dn0; (**d**) SVM classifier with lineal and quadratic polynomial functions and data normalisation: dn1 and dn2.

Table 3. Colour images and NIR images. AUC for kNN, NBC, SVM.

Classifier		kNN		NBC		SVM	
Configuration		*Euclidean*	*Minkowski*	*Gauss*	*KSF*	*Linear*	*Quadratic*
Colour	dn0	**0.9984**	0.9974	0.9542	0.9778	-	-
	dn1	0.9979	0.9976	-	-	0.9622	0.9875
	dn2	**0.9984**	0.9974	-	-	0.9496	0.9886
NIR	dn0	0.9987	0.9979	0.9877	0.9963	-	-
	dn1	0.9975	0.9993	-	-	0.9867	**1.000**
	dn2	0.9987	0.9979	-	-	0.9868	0.9932

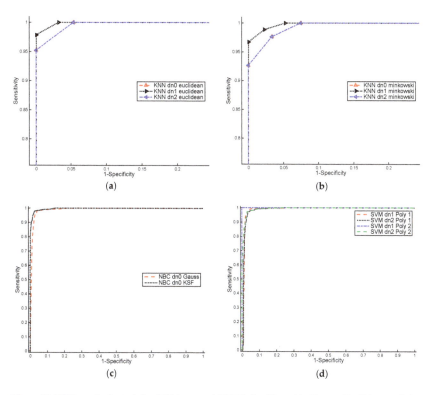

Figure 10. ROC results for training NIR images. (**a**) kNN classifier with distance Euclidean and data normalisation: dn0, dn1 and dn2; (**b**) kNN classifier with distance Minkowski and data normalisation: dn0, dn1 and dn2; (**c**) BN classifier with Gauss and KSF kernels and data normalisation dn0; (**d**) SVM classifier with lineal and quadratic polynomial functions and data normalisation: dn1 and dn2.

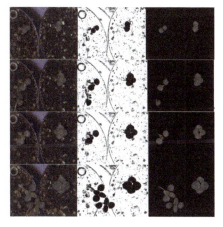

Figure 11. Results of the colour image segmentation based on kNN classifier in different growing stages. Left shows four colour images selected at different growth stages. The second column presents the result of segmentation using kNN classifier with Euclidean distance and without normalisation data (dn0). The third column shows the results after post-processing stage.

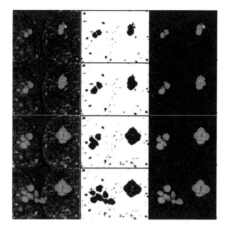

Figure 12. Results of the NIR image segmentation based on SVM. Left shows four NIR images chosen at different growth stages growing during night-time. The second column shows the segmentation results after application of the SVM classifier with quadratic function and with data normalisation dn1. The third column shows the results after post-processing stage.

3.3. Error Segmentation

The misclassification error (ME) represents the percentage of the background pixels that are incorrectly allocated to the object (*i.e.*, to the foreground) or *vice versa*. The error can be calculated by means of Equation (10), where B_{GT} (Background Ground-Truth image) and O_{GT} (Object Ground-Truth image) represent the ground-truth image of the background and of the object taken as reference, and B_T (Background Test) and O_T (Object Test) represent the image to be assessed. In the event that the test image coincides with the pattern image, the classification error will be zero and therefore the performance of the segmentation will be the maximum [41].

$$ME = \frac{|B_{GT} \cap B_T| + |O_{GT} \cap O_T|}{|B_{GT}| + |O_{GT}|} \tag{10}$$

The performance of the implemented algorithms is assessed according to the Equation (11):

$$\eta = 100 \cdot (1 - ME) \tag{11}$$

Table 4 shows mean values computed after to segment the fifty ground-truth images of each group with optimal ML algorithm (kNN and SVM) selected in the previous subsection.

Table 4. Performance in the image segmentation.

Classifier	kNN	SVM
Performance	99.311%	99.342%

The performance of the image segmentation calculated shows excellent results in both groups (Table 4). It has been necessary to increase the complexity of the features vector in the NIR images (using Wavelet transform) to obtain a similar results of performance. This is probably an expected result as RGB images had three times more of information than a NIR image.

4. Conclusions and Future Work

In this work we have developed a system based on ML algorithms and computer vision intended to solve the automatic phenotype data analysis. The system is composed by a growth-chamber with capacities to perform experiments with numerous species. The design of the growth-chamber has allowed easy positioning of different cameras and illuminations. The system can take thousands of images through the capture subsystem, capture spectral images and it creates time-lapse series of the specie during the experiment. One of the main goals of this work has been to capture images during the night-time without affecting plant growth.

We have used three different ML algorithms for image segmentation: k-nearest neighbour (kNN), Naive Bayes classifier (NBC), and Support Vector Machine. Each ML algorithm was executed with different kernel functions: kNN with Euclidean and Minkowski distances, NBC with Gauss and KSF functions and SVM with linear and quadratic functions. Furthermore ML algorithms have been trained with two types of data normalisation: dn0 (raw data), dn1 (mean and standard deviation) and dn2 (mode & standard deviation). Our results show that RGB images are better classified with the kNN classifier, Euclidean distance and without data normalisation. In contrast, NIR images performed better with SVM classifier with quadratic function and with data normalisation dn1.

In the last stage we have applied ME metrics to measure the image segmentation performance. We have achieved a performance of 99.3% in both ground-truth colour images and ground-truth NIR images. Currently the algorithms are being used in an automatic image segmentation processing to study circadian rhythm in wild type lines, transposon-tagged mutants and transgenic lines with modifications in genes involved in the control of growth and the circadian clock. Regarding future work, we consider important to identify a new feature vector which produces better performance rates, improve the illumination subsystem, reduce the computation time of the windowing segmentation in NIR images.

Acknowledgments: The work has been partially supported and funded by the Spanish Ministerio de Economía y Competitividad (MINECO) under the projects ViSelTR (TIN2012-39279), BFU2013-45148-R and cDrone (TIN2013-45920-R).

Author Contributions: Pedro J. Navarro, Fernándo Pérez, Julia Weiss and Marcos Egea-Cortines conceived and designed the experiments; Pedro J. Navarro, Fernándo Pérez, Julia Weiss, and Marcos Egea-Cortines performed the experiments; Pedro J. Navarro and Fernándo Pérez analyzed the data; Julia Weiss contributed reagents/materials/analysis tools; Pedro J. Navarro, Fernándo Pérez and Marcos Egea-Cortines wrote the paper. Pedro J. Navarro, Fernándo Pérez, Julia Weiss and Marcos Egea-Cortines corrected the draft and approved the final version.

Conflicts of Interest: The authors declare no conflict of interest.

Abbreviations

The following abbreviations are used in this manuscript:

LM	Machine learning
KSF	Kernel smoothing function
kNN	k-nearest neighbour
NBC	Naïve Bayes classifier
SVM	Support vector machines
LOOCV	Leave-one-out cross validation method
AUC	Area under curve
ME	Misclassification error

References

1. Fahlgren, N.; Gehan, M.; Baxter, I. Lights, camera, action: High-throughput plant phenotyping is ready for a close-up. *Curr. Opin. Plant Biol.* **2015**, *24*, 93–99. [CrossRef] [PubMed]
2. Deligiannidis, L.; Arabnia, H. *Emerging Trends in Image Processing, Computer Vision and Pattern Recognition*; Elsevier: Boston, MA, USA, 2014.

3.   Dee, H.; French, A. From image processing to computer vision: Plant imaging grows up. *Funct. Plant Biol.* **2015**, *42*, iii–v. [CrossRef]

4.   Furbank, R.T.; Tester, M. Phenomics—Technologies to relieve the phenotyping bottleneck. *Trends Plant Sci.* **2011**, *16*, 635–644. [CrossRef] [PubMed]

5.   Dhondt, S.; Wuyts, N.; Inzé, D. Cell to whole-plant phenotyping: The best is yet to come. *Trends Plant Sci.* **2013**, *18*, 428–439. [CrossRef] [PubMed]

6.   Tisné, S.; Serrand, Y.; Bach, L.; Gilbault, E.; Ben Ameur, R.; Balasse, H.; Voisin, R.; Bouchez, D.; Durand-Tardif, M.; Guerche, P.; *et al.* Phenoscope: An automated large-scale phenotyping platform offering high spatial homogeneity. *Plant J.* **2013**, *74*, 534–544. [CrossRef] [PubMed]

7.   Li, L.; Zhang, Q.; Huang, D. A review of imaging techniques for plant phenotyping. *Sensors* **2014**, *14*, 20078–20111. [CrossRef] [PubMed]

8.   Honsdorf, N.; March, T.J.; Berger, B.; Tester, M.; Pillen, K. High-throughput phenotyping to detect drought tolerance QTL in wild barley introgression lines. *PLoS ONE* **2014**, *9*, 1–13. [CrossRef] [PubMed]

9.   Barron, J.; Liptay, A. Measuring 3-D plant growth using optical flow. *Bioimaging* **1997**, *5*, 82–86. [CrossRef]

10. Aboelela, A.; Liptay, A.; Barron, J.L. Plant growth measurement techniques using near-infrared imagery. *Int. J. Robot. Autom.* **2005**, *20*, 42–49. [CrossRef]

11. Navarro, P.J.; Fernández, C.; Weiss, J.; Egea-Cortines, M. Development of a configurable growth chamber with a computer vision system to study circadian rhythm in plants. *Sensors* **2012**, *12*, 15356–15375. [CrossRef] [PubMed]

12. Nguyen, T.; Slaughter, D.; Max, N.; Maloof, J.; Sinha, N. Structured light-based 3d reconstruction system for plants. *Sensors* **2015**, *15*, 18587–18612. [CrossRef] [PubMed]

13. Spalding, E.P.; Miller, N.D. Image analysis is driving a renaissance in growth measurement. *Curr. Opin. Plant Biol.* **2013**, *16*, 100–104. [CrossRef] [PubMed]

14. Navlakha, S.; Bar-joseph, Z. Algorithms in nature: The convergence of systems biology and computational thinking. *Mol. Syst. Biol.* **2011**, *7*. [CrossRef] [PubMed]

15. Kircher, M.; Stenzel, U.; Kelso, J. Improved base calling for the Illumina Genome Analyzer using machine learning strategies. *Genome Biol.* **2009**, *10*, 1–9. [CrossRef] [PubMed]

16. Horton, P.; Nakai, K. Better prediction of protein cellular localization sites with the *k* nearest neighbors classifier. *Proc. Int. Conf. Intell. Syst. Mol. Biol.* **1997**, *5*, 147–152. [PubMed]

17. Yousef, M.; Nebozhyn, M.; Shatkay, H.; Kanterakis, S.; Showe, L.C.; Showe, M.K. Combining multi-species genomic data for microRNA identification using a Naive Bayes classifier. *Bioinformatics* **2006**, *22*, 1325–1334. [CrossRef] [PubMed]

18. Tellaeche, A.; Pajares, G.; Burgos-Artizzu, X.P.; Ribeiro, A. A computer vision approach for weeds identification through Support Vector Machines. *Appl. Soft Comput. J.* **2011**, *11*, 908–915. [CrossRef]

19. Guerrero, J.M.; Pajares, G.; Montalvo, M.; Romeo, J.; Guijarro, M. Support Vector Machines for crop/weeds identification in maize fields. *Expert Syst. Appl.* **2012**, *39*, 11149–11155. [CrossRef]

20. Covington, M.F.; Harmer, S.L. The circadian clock regulates auxin signaling and responses in Arabidopsis. *PLoS Biol.* **2007**, *5*, 1773–1784. [CrossRef] [PubMed]

21. Nusinow, D.A.; Helfer, A.; Hamilton, E.E.; King, J.J.; Imaizumi, T.; Schultz, T.F.; Farré, E.M.; Kay, S.A.; Farre, E.M. The ELF4-ELF3-LUX complex links the circadian clock to diurnal control of hypocotyl growth. *Nature* **2011**, *475*, 398–402. [CrossRef] [PubMed]

22. De Montaigu, A.; Toth, R.; Coupland, G. Plant development goes like clockwork. *Trends Genet.* **2010**, *26*, 296–306. [CrossRef] [PubMed]

23. Baudry, A.; Ito, S.; Song, Y.H.; Strait, A.A.; Kiba, T.; Lu, S.; Henriques, R.; Pruneda-Paz, J.L.; Chua, N.H.; Tobin, E.M.; *et al.* F-box proteins FKF1 and LKP2 Act in concert with ZEITLUPE to control arabidopsis clock progression. *Plant Cell* **2010**, *22*, 606–622. [CrossRef] [PubMed]

24. Kim, W.-Y.Y.; Fujiwara, S.; Suh, S.-S.S.; Kim, J.; Kim, Y.; Han, L.Q.; David, K.; Putterill, J.; Nam, H.G.; Somers, D.E. ZEITLUPE is a circadian photoreceptor stabilized by GIGANTEA in blue light. *Nature* **2007**, *449*, 356–360. [CrossRef] [PubMed]

25. Khanna, R.; Kikis, E.A.; Quail, P.H. EARLY FLOWERING 4 functions in phytochrome B-regulated seedling de-etiolation. *Plant Physiol.* **2003**, *133*, 1530–1538. [CrossRef] [PubMed]

26. Wenden, B.; Kozma-Bognar, L.; Edwards, K.D.; Hall, A.J.W.; Locke, J.C.W.; Millar, A.J. Light inputs shape the Arabidopsis circadian system. *Plant J.* **2011**, *66*, 480–491. [CrossRef] [PubMed]

27. Nozue, K.; Covington, M.F.; Duek, P.D.; Lorrain, S.; Fankhauser, C.; Harmer, S.L.; Maloof, J.N. Rhythmic growth explained by coincidence between internal and external cues. *Nature* **2007**, *448*, 358–363. [CrossRef] [PubMed]

28. Fernandez, C.; Suardiaz, J.; Jimenez, C.; Navarro, P.J.; Toledo, A.; Iborra, A. Automated visual inspection system for the classification of preserved vegetables. In Proceedings of the 2002 IEEE International Symposium on Industrial Electronics, ISIE 2002, Roma, Italy, 8–11 May 2002; pp. 265–269.

29. Chen, Y.Q.; Nixon, M.S.; Thomas, D.W. Statistical geometrical features for texture classification. *Pattern Recognit.* **1995**, *28*, 537–552. [CrossRef]

30. Haralick, R.M.; Shanmugam, K.; Dinstein, I.H. Textural features for image classification. *IEEE Trans. Syst. Man Cybern.* **1973**, 610–621. [CrossRef]

31. Zucker, S.W.; Terzopoulos, D. Finding structure in co-occurrence matrices for texture analysis. *Comput. Graph. Image Process.* **1980**, *12*, 286–308. [CrossRef]

32. Bharati, M.H.; Liu, J.J.; MacGregor, J.F. Image texture analysis: Methods and comparisons. *Chemom. Intell. Lab. Syst.* **2004**, *72*, 57–71. [CrossRef]

33. Navarro, P.J.; Alonso, D.; Stathis, K. Automatic detection of microaneurysms in diabetic retinopathy fundus images using the L*a*b color space. *J. Opt. Soc. Am. A Opt. Image Sci. Vis.* **2016**, *33*, 74–83. [CrossRef] [PubMed]

34. Mallat, S.G. A theory for multiresolution signal decomposition: The wavelet representation. *IEEE Trans. Pattern Anal. Mach. Intell.* **1989**, *11*, 674–693. [CrossRef]

35. Ghazali, K.H.; Mansor, M.F.; Mustafa, M.M.; Hussain, A. Feature extraction technique using discrete wavelet transform for image classification. In Proceedings of the 2007 5th Student Conference on Research and Development, Selangor, Malaysia, 11–12 December 2007; pp. 1–4.

36. Arivazhagan, S.; Ganesan, L. Texture classification using wavelet transform. *Pattern Recognit. Lett.* **2003**, *24*, 1513–1521. [CrossRef]

37. Lantz, B. *Machine Learning with R*; Packt Publishing Ltd.: Birmingham, UK, 2013.

38. Hastie, T.; Tibshirani, R.; Friedman, J. The elements of statistical learning. *Elements* **2009**, *1*, 337–387.

39. Bradley, A. The use of the area under the ROC curve in the evaluation of machine learning algorithms. *Pattern Recognit.* **1997**, *30*, 1145–1159. [CrossRef]

40. Hand, D.J. Measuring classifier performance: A coherent alternative to the area under the ROC curve. *Mach. Learn.* **2009**, *77*, 103–123. [CrossRef]

41. Sezgin, M.; Sankur, B. Survey over image thresholding techniques and quantitative performance evaluation. *J. Electron. Imaging* **2004**, *13*, 146–168.

Article

An Approach to the Use of Depth Cameras for Weed Volume Estimation

Dionisio Andújar [1,*], José Dorado [2], César Fernández-Quintanilla [2] and Angela Ribeiro [1]

[1]   Center for Automation and Robotics, Spanish National Research Council, CSIC-UPM, Arganda del Rey, Madrid 28500, Spain; angela.ribeiro@csic.es
[2]   Institute of Agricultural Sciences, Spanish National Research Council, CSIC, Madrid 28006, Spain; jose.dorado@ica.csic.es (J.D.); cesar@ica.csic.es (C.F.-Q.)
*   Correspondence: dionisioandujar@hotmail.com; Tel.: +34-91-745-2500

Academic Editor: Gonzalo Pajares Martinsanz
Received: 4 May 2016; Accepted: 22 June 2016; Published: 25 June 2016

Abstract: The use of depth cameras in precision agriculture is increasing day by day. This type of sensor has been used for the plant structure characterization of several crops. However, the discrimination of small plants, such as weeds, is still a challenge within agricultural fields. Improvements in the new Microsoft Kinect v2 sensor can capture the details of plants. The use of a dual methodology using height selection and RGB (Red, Green, Blue) segmentation can separate crops, weeds, and soil. This paper explores the possibilities of this sensor by using Kinect Fusion algorithms to reconstruct 3D point clouds of weed-infested maize crops under real field conditions. The processed models showed good consistency among the 3D depth images and soil measurements obtained from the actual structural parameters. Maize plants were identified in the samples by height selection of the connected faces and showed a correlation of 0.77 with maize biomass. The lower height of the weeds made RGB recognition necessary to separate them from the soil microrelief of the samples, achieving a good correlation of 0.83 with weed biomass. In addition, weed density showed good correlation with volumetric measurements. The canonical discriminant analysis showed promising results for classification into monocots and dictos. These results suggest that estimating volume using the Kinect methodology can be a highly accurate method for crop status determination and weed detection. It offers several possibilities for the automation of agricultural processes by the construction of a new system integrating these sensors and the development of algorithms to properly process the information provided by them.

Keywords: Kinect v2; weed/crop structure characterization; weed detection; plant volume estimation; maize

---

1. Introduction

New trends in agriculture allow for the precise management of the spatial occurrence of pests within agricultural fields. The requirement of agriculture to have a low impact on the environment and the challenge of feeding an increasing population provides precision agriculture (PA) with the opportunity to face this challenge. PA uses knowledge of spatial and temporal variations in crops. The management of the spatio-temporal information needs methods and technologies that improve day by day to fulfill new requirements of food sustainability processes that optimize available resources [1]. Pests significantly curtail agricultural production and are responsible for a decrease of approximately 40% in potential global crop yields because they transmit disease, feed on crops, and compete with crop plants [2]. A major issue in Europe is the current reliance on chemical methods of pest control (CE 1107/2009 and 2009/128/CE). Thus, site-specific crop management could be the solution for a lower environmental impact while maximizing yields. The case of weed management is of high importance because herbicides are the most used pesticides in the world. The introduction of the concept of site-specific weed management (SSWM) is an attempt to manage the heterogeneity of fields though the use of new sensors and machinery to precisely treat weed patches only, so that the use of herbicides can be drastically reduced. The use of SSWM is conditioned by crop value, the proportion of the field infested by weeds, the shape and number of the patches, and the technologies for sampling and spraying [3]. The use of sensing technologies can separate weed from crops, identifying patches within the field through weed characteristics such as color or height. The quantification of these values is needed to determine correct management. Young et al. [4] noted a 53% reduction in the applications of grass herbicides in wheat through using SSWM. Similarly, Gerhards and Christensen [5] showed a herbicide reduction greater than 75% in grass weeds in a wheat crop over four years. Andújar et al. [6] put forward that SSWM was the most profitable strategy with normal levels of infestation and showed herbicide savings of more than 80% in maize crops. High-value crops such as horticultural plants or fruit trees, where only a small proportion of the field is infested by weeds, are the ideal targets for SSWM because the benefits can highly justify the associated cost of detection [7]. The increasing adoption of SSWM techniques is driven by the increasing knowledge of new farmers and the related economic benefits of using new technologies. In addition, some directives of the European Union lead to a reduction of the inputs used for pest control (European Council 2009, Brussels, Belgium), indirectly promoting the use of precision agriculture technologies, which allow for the reduction of pesticide use.

Currently, new tools for managing the heterogeneity within agricultural fields are emerging from new sensors and combined systems. When applied to crops, these systems allow cost optimization by new cost-efficient management and minimization of the environmental impact of tillage and chemical products. They are mainly based on different types of sensors that are able to discriminate weeds and crops or reconstruct plant shapes by phenotyping methods. Phenotyping techniques characterize plant structure by using non-invasive methodologies with new sensors that have been recently developed. The geometrical characterization of plant structures leads to an understanding of internal processes and establishes differences among plant species for good identification. The evaluation and modeling of plant architecture using phenotyping requires high precision and accuracy for measuring protocols [8]. An accurate model improves the decisions about the use of the information taken by phenotyping. Plant phenotyping models are created though different types of sensors, from imaging to non-imaging techniques, that recreate plant shapes [9]. Different systems can be used for this purpose. Weed identification using machine vision by portable imaging and analysis software is the most investigated technique [10]. However, under outdoor agricultural environments, different problems appear. The problem of variable and uncontrolled illumination that, among other things, produces shadows and leaf overlapping is a major challenge [11]. Spectral reflectance sensors [12,13] and fluorescence sensors [14] are also used. Regarding distance sensors, ultrasonic sensors and LiDAR are sensors available on the market. These sensors can be used for plant height and volume estimation [14,15]. In addition, other systems have been explored for plant characterization, including radar systems, stereovision, magnetic resonance, and thermal imaging.

Structured-light scanners have opened a new door in 3D modeling. They are automatic 3D acquisition devices that create high fidelity models of real 3D objects in a highly time-effective way and at low cost. There are two major manufacturers in the market: Microsoft Kinect 1.0 and 2.0 (Microsoft, Redmond, WA, USA) and the Asus Xtion (Asus, Taipei, Taiwan). Asus Xtion and Kinect 1.0 sensors combine structure light with computer vision techniques: depth from focus and depth from stereo [16]. Chéné et al. [17] reconstructed the geometric shape of rosebush, yucca, and apple tree and introduced an automatic algorithm for leaf segmentation from a single top view image acquisition. The algorithm could be applied to automate processes for plant disease or stress detection because leaf morphology can be related to internal processes in the plant. Wang & Li [18] applied the RGB-depth Kinect sensor to estimate onion fruit diameter and volume. The comparison between the measurements taken with RGB images and those taken with RGB-D showed a higher average accuracy for RGB-D. Plant structure characterization is a widely known technique that uses similar sensors based on time-of-flight (ToF) cameras [19,20], and is used for the estimation of the foliar density of fruit trees to control a spraying system [21]. Wang et al. [22] developed a picking robot that imitated human behavior for fruit recollection. Agrawal et al. [23] designed an inexpensive robotic system that was affordable for growers and was compatible with the existing hydroponic infrastructure for growing crops such as lettuce. Paulus et al. [24] compared a low-cost 3D imaging system with a high-precision close-up laser scanner for phenotyping purposes. The measured parameters from the volumetric structure of sugar beet taproots, the leaves of sugar beet plants, and the shape of wheat ears were compared. Although the study showed myriad possibilities for using depth cameras in plant phenotyping and the potential of using these sensors in automated application procedures, their reliability was lower than that of laser scanners. However, the low cost and additional information provided by RGB makes it a good alternative to high-cost tools. Although light radiation impedes its use in outdoor experiments, the use of the new Kinect v2 sensor allows for its use under high illumination conditions because the measurement principle is based on ToF methodology [25,26]. It also has a higher resolution capacity and can process a higher volume of data with a wider field of view. This paper presents a novel solution to separate weeds and maize plants through depth images taken in real outdoor conditions of maize fields, demonstrating its capabilities and limitations in discriminating weeds from crops using color segmentation and plant height measurements.

2. Material and Methods

2.1. Data Collection System

The Kinect v2 sensor has a depth camera, an RGB camera of 1080p, and an infrared emitter. The depth sensor is based on an indirect ToF measurement principle. An infrared light is emitted and reflected by the impacted object. The time it takes for light to travel from the infrared illuminator to the impacted object and back to the infrared camera is stored and transformed by wave modulation and phase detection to calculate the distance between the emitter and the object. The RGB camera has a resolution of 1920×1080 pixels, giving it an array of pixel RGB values that need to be converted into a WPF representation. The IR (infrared) camera used for the acquisition of depth data, as the depth camera, has a resolution of 512×424 pixels that allows the sensor to work in darkness and allows for the tracking of IR reflective objects while filtering out IR lights. When sensing depths of 70 degrees horizontally and 60 degrees vertically, the sensor can take in information from a wider field of view than the previous version of the sensor. Thus, objects can be closer to the sensor and still in its field of view, and the camera is also effective at longer distances, covering a larger total area. Although the technical characteristics provided by Microsoft state that the operative measurement range works from 0.5 m to 4.5 m, some tools enable the reconstruction of bigger 3D meshes by using ICP (Iterative Closest Point) algorithms, by moving the Kinect sensor around the scanned object. These meshes can be processed and edited as unstructured 3D triangular meshes in different software applications. The sensor readings are composed of a high number of frames per second with overlapping areas in the continuous measurements. The overlapped zones allow for the creation of better models and the automatic removal of outliers in the mesh. The software merges the points into a common point cloud taken by the relative position of the sensor, which is relative to itself at the initial point.

Meshes were acquired from the Kinect v2 sensor running Software Development Kit (SDK) on an Intel desktop computer with Windows 8 (Microsoft, Redmond, WA, USA). The data acquisition software was developed starting from the official Microsoft SDK 2.0 (Microsoft, Redmond, WA, USA)), which includes the drivers for the Kinect sensor. It allows the necessary drivers for the sensor to be obtained, as well as the use of a set of customizable sample functions that were implemented for the measurements, combined with some OpenCV functions. The aim was to store both depth and RGB information at a rate of at least 25 frames per second. The data were processed using a volumetric reconstruction based on a memory and speed efficient data structure [27]. The chosen storage format was compatible with the open software (Meshlab®, University of Pisa, Pisa, Italy) that was subsequently used to process the acquired raw data. The system was supplied with electric power by a portable gasoline generator of 6 kw. The Kinect sensor and additional devices were mounted in an ATV (All Terrain Vehicle, Yamaha, Shizuoka, Japan). The Kinect sensor was positioned in the middle of the sampling area and the ATV was stopped. Then, measurements were taken for 1 to 2 s, moving the sensor mechanically with a rotator motor from 45° to −45°, avoiding the effect of leaf overlapping by displacing the sensor through different point of views (Figure 1).

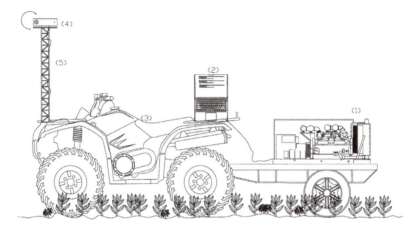

Figure 1. Schematic design of the system with the components integrated for maize-weed detection. (1) Portable gasoline generator; (2) laptop; (3) ATV; (4) Kinect v2 sensor; (5) support structure.

2.2. Field Measurements

The experimental set up was conducted under field conditions at La Poveda Research Farm (Arganda del Rey, Madrid, Spain). The soil was conventionally tilled. Maize was planted on the 6 April 2015 with 0.75-m row spacing and a population of 90,000 plants/ha. The experimental field was infested with *Sorghum halepense* (L.) Pers., *Datura ferox* L., *Salsola kali* L., *Polygonum aviculare* L., and *Xanthium strumarium* L., with some other minor weeds. Weeds were always shorter than maize plants. Weed control treatments were based on pre-emergence and post-emergence herbicide applications. Weed assessments were made in May when maize was at stage 14 to 16 of BBCH scale [28] and weeds BBCH 12 to BBCH 22 [29] (Figure 2).

(a)   (b)

Figure 2. (**a**) Some frames located in the experimental field; (**b**) example of a mesh obtained with the Kinect v2 sensor.

The sample locations were chosen to search for different weed compositions of grass, broad-leaved weeds and crops, as well as mixtures of them. The samples were from pure samples of the different species, as well as an equitable distribution of different mixtures and compositions. Weeds and crops were sampled using an 1 m × 1 m frame, which was divided into four quadrats. A total of ten 1 m × 1 m frames were sampled. Then, samples were distributed in quadrats of 0.5 × 0.5 m², with a total of 40 samples. Half of the quadrats were positioned in the inter-row area, and half of them were located in the maize row. Thus, the samples were equally distributed in areas with crops and areas occupied only by weeds. Some of the samples were weed-free or maize-free due to their positions. Because weed detection needs to be fast due to the narrow window for weed treatment, a full 3D model for an entire field would not be realistic. Thus, single 3D models were constructed from the top position. Every reading was taken in direct sunlight under real field conditions at midday on a sunny day (average of 40,000 lux) and not using any shading structure. Following the readings, the density and biomass of the weeds and maize were assessed within the quadrats (0.5 m × 0.5 m) located in the row and inter-row positions. Weed emergences were counted in each sample quadrat by species. Plants were cut and processed in the lab to determine the dry weights of the different species.

2.3. Data Processing

The meshes were processed to calculate maize and weed volume (Figure 2b). The offline processing of the samples was conducted using the open software Meshlab® (3D-CoForm, Brighton, UK). The initial point cloud was processed in three steps [30,31] (Figure 3): (a) Data outliers and noise in the point cloud were filtered out. The filter identifies the points with no connection (1 cm out of the grid) and removes them. (b) The areas out of the 1 m × 1 m frame were removed and the mesh was divided into samples of 0.5 m × 0.5 m. The division and removal of areas out of the frame were automatically done based on the color segmentation of the white frame. (c) Plant volumes were calculated from the mesh by computation of polyhedral mass properties [32]. Firstly, the maize parts of the mesh were isolated by height (Figure 4). Because the maize plants were considerably taller than the weeds, the upper part and the connected faces of the mesh were selected. From this selection, a new mesh was created, and maize volume was extracted. Then, after the selection and removal of the maize plants in the mesh, an RGB filter was applied to select the reaming green parts corresponding to weeds. The mesh volume of weeds was then obtained. The calculated volumes extracted from the meshes were statistically compared with the actual weight values, which were manually assessed.

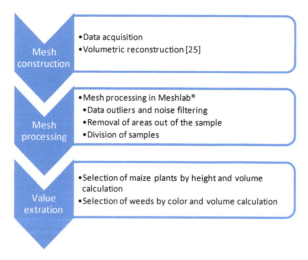

Figure 3. Data processing structure of the Kinect sensor information.

Figure 4. An example of maize isolation and removal by height selection, and an example of color selection for weed extraction.

2.4. Statistical Analysis

The data of maize and weed volume obtained with Kinect v2 sensor were analyzed and compared with the actual values of dry biomass weight using regression methods. Pearson's correlation coefficients were calculated to evaluate the simple linear relationships between the actual parameters and those obtained with the sensor. A correlation analysis was made prior to the regression analysis to provide initial information regarding the bivariate relationships among the parameters.

A canonical discriminant analysis (CDA) was used to predict weed classification of the system using weed height. This methodology is based on pattern recognition to find an optimal linear combination of features for the separation into groups. The dependent variable is expressed as functions of linear combinations of the input features [33]. The method tried to show the capabilities of the method to separate (a) infested samples by broad-leaved weeds; (b) infested samples by grasses; and (c) infested by mixture of both weed types. The analyses were performed using the SPSS statistical software (v. 20, IBM Corporation, Armonk, NY, USA).

3. Results and Discussion

The constructed models obtained by scanning maize samples using the new Kinect v2 device showed a high correlation for the calculated structural parameters between maize and weed soil measurements. The maize plants that were present in the sample always demonstrated a greater height than the weeds, which allowed for the identification of maize plants in the sample. Maize plants ranged from 40 cm to 55 cm, while weeds varied depending on weed species. *Sorhgum halepense* was the highest weed with a maximum height of 20 cm, while broad-leaf weeds were always shorter than 10 cm. The height of maize plants allowed for the selection of the upper part of the mesh and the connected faces of the mesh for the separation of maize and weeds. In addition, the difference in height between grasses and broad-leaf weeds identified both types of weeds. Maize plants had a higher volume in the sample than weeds. The shorter height of weeds led to a lower volume and necessitated RGB recognition to separate them from the microrelief of the samples. High correlations between the constructed models and the ground-truth variables resulted in significance at a level of 1%. The total volume of the model was highly correlated with the total vegetation weight, with an r of 0.93. Maize volume was well correlated with maize weight, with a Pearson's r of 0.77. The correlation between weed weight and weed volume reached an r-value of 0.83. These values show the potential of the new Kinect sensor in crop identification and weed detection. Figure 5a shows the linear regression for total biomass and total volume, with a high relation between values. In addition, once the maize volume was isolated from the sample, the maize biomass weight and maize volume showed a good relationship (Figure 5b). Thus, Kinect v2 volume estimation is an accurate method for crop status determination. This information could be useful for fertilizer applications using the same principles but with other sensors such as NVDI, ultrasonic, and LiDAR sensors. Thus, those areas with a higher or lower volume could receive a differential dose of fertilizers according to the growing state. Similar studies have shown the capability of Kinect sensors to characterize plant volume for several uses. The authors agree that the characterization of plant structures with Kinect leads to an overestimation of the actual plant values because the sensors cannot reconstruct the end details. Nock et al. [34] could not reconstruct *Salix* branches below 6.5 mm in diameter using an Asus Xtion, whereas a Kinect v1 worked with branches ranging from 2 to 13 mm in diameter at a distance of 0.5–1.25 m from the target object. However, they agree that these low-cost 3D imaging devices for plant phenotyping are reliable and can be integrated into automated applications of agricultural tasks [24]. Although the measurements tend to overestimate the actual dimensions of the plants, as in our experiments, plant reconstruction is of high fidelity compared with RGB images, and 3D modeling using this sensor constitutes a highly valuable method for plant phenotyping. In addition, the method worked properly at very early weed stages, which is when most of the weed treatments should be applied. Thus, this technology may be applicable for weed identification at various crop growth stages. Although some other methodologies, such as LiDAR, measure distance with a higher resolution, its inability to reproduce RGB limits the capability for weed and soil separation.

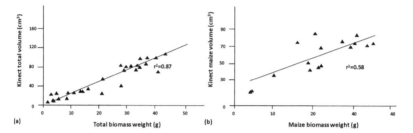

Figure 5. Linear regression between both the total volume that was estimated using the Kinect v2 sensor and the total biomass (**a**) and the maize volume with the maize biomass (**b**). The R^2 denotes the correlation coefficient of the simple regression.

In addition, the discrimination of the color of the weeds opens a new window for weed control. The Kinect v1 sensor had a lower resolution and an inability to work in outdoor conditions due to lighting conditions, which were the main deterrents to its usage. The improvements in the Kinect v2 sensor allowed for the acquisition of meshes outdoors with real and uncontrolled lighting. The separation of weed areas from the mesh and the calculated volume resulted in a linear correlation between the volumes estimated from the models and the actual parameters of weed biomass. The regression showed an r^2 of 0.7, indicating a good relationship with the volume data obtained with the Kinect device. Figure 6 shows the simple regression for weed biomass and weed volume, with a good relation between values, which demonstrate the suitability of the sensor for weed detection. Although RGB recognition is a good complement for weed isolation in 3D models, using color-image processing by itself for volume calculation results in worse predictions than depth images [18].

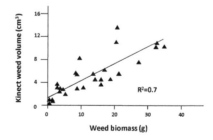

Figure 6. Simple linear regression between the total weed volume that was calculated with the depth and color images and the weed biomass weight.

When the plant densities were analyzed, the number of maize plants did not show a significant correlation with the calculated model. This fact was mainly due to the almost constant number of maize plants per sample. On the contrary, weed density showed good correlations with the volumetric measurements. They resulted in a significant difference: $p < 0.01$. The total volume of the model was correlated with weed density, with an R^2 of 0.59. Once the weed volume was isolated from the rest of sample elements, a similar R^2 value of 0.6 was obtained.

The CDA showed promising results for classification into three predefined groups (Table 1). All of the samples infested exclusively by broad-leaf weeds were properly classified (100%). In addition, 92.3% of the cases infested with presence of grasses were classified correctly. More than a half (53.8%) were properly classified as pure samples of grass weeds. The rest of the cases (38.5%) were classified as mixture of grasses and broad-leaf grasses. Samples composed of mixtures were classified as mixtures and grasses in the same proportion. Incorrect classification of mixtures could be due to the fact that the maximum height obtained is based on grasses present in the sample. Although the method did not allow for the discrimination of grasses and mixture, this does not suppose a major practical problem if a mixture of herbicides would be used on SSWM when a higher height is detected.

Table 1. Confusion matrix of the canonical discriminant classification showing a percentage of correct group classifications for the three predefined groups.

	Predicted		
	Monocots	**Dicots**	**Mixture**
Monocots	53.8	7.7	38.5
Dicots	0.0	100.0	0.0
Mixture	50.0	0.0	50.0

The system has shown large potential to work in outdoor conditions. Although further studies in software development are necessary to improve the results of the data processing stage, the sensor shows several potential applications that could not be considered for the previous version of Kinect. The sensor is able to work under natural high-lighting conditions, collecting a high number of points with accurate resolution. RGB detection, which is also improved, allowed for the separation of weeds from soil in the sample, creating an opportunity for real-time weed control applications. The possible applications of 3D models open a new door to continue the development and optimization of precision agriculture processes. Although the use of 3D models for geometric characterization of vegetation is not new, the sensors fulfill some needs regarding processing time, robustness, and information quality to adapt to the requirements of real-time agricultural applications.

Similar to this study, Chen et al. [35] derived some structural parameters of maize using the Kinect v1 version. The authors stated that leaf area index and leaf angle can be properly described with this sensor. Thus, considering this information and the plant volume extracted in the current study, the sensor can help to manage the information coming from the field to gain a better understanding of plant growth. Previous studies have shown that structure and size calculations from meshes are similar to manual measurements [31]. Concerning the use of RBG information, the camera in the Kinect v2 sensor has been improved and is in fact even better at identifying weeds in the samples. Yamamoto et al. [36] described a procedure for acquiring plant information on strawberry plants. The information on height, width, volume, leaf area, leaf inclination, and leaf color was assessed using a Kinect v1 sensor and compared with the actual results. The results agreed well with experimental evidence, and similar values of volume agreement to those of the current study were shown. In addition, motion information allowed for the segmentation of videos, avoiding the effect of shadows and overlap, one of the major problems in 2D imaging for weed classification. Thus, the impact of depth cameras on new trends in SSWM is high. The potential possibilities of greater automation of agricultural processes should be studied and improved by the construction of a new system integrating these sensors. The development of better algorithms for information processing will also be necessary. These innovations can increase productivity and reduce herbicide use within agricultural fields using the current sensing technology that is able to detect weeds where they are present, combined with an automatic system to treat weeds site-specifically.

4. Conclusions

The possibilities and disadvantages of depth sensors regarding their practical use in weed detection were assessed based on results obtained from field experiments. The dual methodology using height selection and RGB segmentation properly separated crops, weeds, and soil. Depth data provided geometrical information to the model and allowed for the discrimination of maize plants due to their higher height compared with weeds. A similar methodology could be used in other row crops, where crops have different heights compared with weeds or where weeds are located out of the crop row. In addition, the separation between broad leaf weeds and grasses was achievable using the same principle as height selection. The RGB data provided enough information for the separation of vegetation from soil. Thus, depth data provided geometrical information for classifying features according to their location and RGB according to its color. The constructed models show the potential of depth sensors to collect and fuse the spatial and color information of crops and to extract information for several purposes. The application of selective treatments using the proposed methodology would allow for a higher reduction of herbicide use in maize fields. In addition, the low cost and high frame rate makes this sensor a promising tool for site-specific weed management.

The proposed system offers opportunities for automation of agricultural processes by integrating depth cameras with fast algorithms. The possibilities for on-the-go operations and the development of new algorithms to automatically and more quickly build 3D models need to be further explored. These sensors and algorithms need to be used and integrated to properly process the captured information.

Acknowledgments: The Spanish Ministry of Economy and Competitiveness has provided support for this research via projects AGL2014-52465-C4-3-R and AGL2014-52465-C4-1-R, and Bosch Foundation.

Author Contributions: José Dorado and César Fernández-Quintanilla helped to design the experiments and provided material to the experiment; Angela Ribeiro contributed by providing material to the experiment, analyzing data and revising the document; Dionisio Andújar analyzed the data and wrote the paper.

Conflicts of Interest: The authors declare no conflict of interest.

References

1. Zhang, Q. *Precision Agriculture Technology for Crop Farming*; CRC Press: Washington, DC, USA, 2015.
2. Oerke, E.C.; Dehne, H.W.; Schnbeck, F.; Weber, A. *Crop Production and Crop Protection: Estimated Losses in Major Food and Cash Crops*; Elsevier: Amsterdam, The Netherlands, 1999.
3. Ruiz, D.; Escribano, C.; Fernandez-Quintanilla, C. Assessing the opportunity for site-specific management of Avena sterilis in winter barley fields in Spain. *Weed Res.* **2006**, *46*, 379–387. [CrossRef]
4. Young, D.L.; Kwon, T.J.; Smith, E.G.; Young, F.L. Site-specific herbicide decision model to maximize profit in winter wheat. *Precis. Agric.* **2003**, *4*, 227–238. [CrossRef]
5. Gerhards, R.; Christensen, S. Real-time weed detection, decision making and patch spraying in maize, sugar beet, winter wheat and winter barley. *Weed Res.* **2003**, *43*, 385–392. [CrossRef]
6. Andújar, D.; Ribeiro, A.; Fernandez-Quintanilla, C.; Dorado, J. Herbicide savings and economic benefits of several strategies to control *Sorghum halepense* in maize crops. *Crop Prot.* **2003**, *50*, 17–23. [CrossRef]
7. Andújar, D.; Rueda-Ayala, V.; Jackenkroll, M.; Dorado, J.; Gerhards, R.; Fernández-Quintanilla, C. The Nature of Sorghum Halepense (L.) Pers. Spatial Distribution Patterns in Tomato Cropping Fields. *Gesunde Pflanz.* **2013**, *65*, 85–91. [CrossRef]
8. Dhondt, S.; Wuyts, N.; Inzé, D. Cell to whole-plant phenotyping: The best is yet to come. *Trends Plant Sci.* **2013**, *8*, 1–12. [CrossRef] [PubMed]
9. Li, L.; Zhang, Q.; Huang, D. A Review of Imaging Techniques for Plant Phenotyping. *Sensors* **2014**, *14*, 20078–20111. [CrossRef] [PubMed]

10. Lee, W.S.; Alchanatis, V.; Yang, C.; Hirafuji, M.; Moshou, D.; Li, C. Sensing technologies for precision specialty crop production. *Comput. Electron. Agric.* **2010**, *74*, 2–33. [CrossRef]
11. McCarthy, C.L.; Hancock, N.H.; Raine, S.R. Applied machine vision of plants: A review with implications for field deployment in automated farming operations. *Intel. Serv. Robot.* **2010**, *3*, 209–217. [CrossRef]
12. Sui, R.; Thomasson, J.A.; Hanks, J.; Wooten, J. Ground-based sensing system for weed mapping in cotton. *Comput. Electron. Agric.* **2008**, *60*, 31–38. [CrossRef]
13. Andújar, D.; Ribeiro, A.; Fernández-Quintanilla, C.; Dorado, J. Accuracy and feasibility of optoelectronic sensors for weed mapping in wide row crops. *Sensors* **2011**, *11*, 2304–2318. [CrossRef] [PubMed]
14. Andújar, D.; Escola, A.; Dorado, J.; Fernandez-Quintanilla, C. Weed discrimination using ultrasonic sensors. *Weed Res.* **2011**, *51*, 543–547. [CrossRef]
15. Andújar, D.; Rueda-Ayala, V.; Moreno, H.; Rosell-Polo, J.R.; Escolà, A.; Valero, C.; Gerhards, R.; Fernández-Quintanilla, C.; Dorado, J.; Giepentrog, H.W. Discriminating crop, weeds and soil surface with a terrestrial LIDAR sensor. *Sensors* **2013**, *13*, 14662–14675. [CrossRef] [PubMed]
16. Gonzalez-Jorge, H.; Riveiro, b.; Vazquez-Fernandez, E.; Martínez-Sánchez, J.; Arias, P. Metrological evaluation of Microsoft Kinect and Asus Xtion sensors. *Measurement* **2013**, *46*, 1800–1806. [CrossRef]
17. Chéné, Y.; Rousseau, D.; Lucidarme, P.; Bertheloot, J.; Caffier, V.; Morel, P.; Belin, E.; Chapeau-Blondeau, F. On the use of depth camera for 3D phenotyping of entire plants. *Comput. Electron. Agric.* **2012**, *82*, 122–127. [CrossRef]
18. Wang, W.; Li, C. Size estimation of sweet onions using consumer-grade RGB-depth sensor. *J. Food Eng.* **2014**, *142*, 153–162. [CrossRef]
19. Van der Heijden, G.; Song, Y.; Horgan, G.; Polder, G.; Dieleman, A.; Bink, M.; Palloix, A.; Van Eeuwijk, F.; Glasbey, C. SPICY: Towards automated phenotyping of large pepper plants in the greenhouse. *Funct. Plant Biol.* **2012**, *39*, 870–877. [CrossRef]
20. Busemeyer, L.; Mentrup, D.; Moller, K.; Wunder, E.; Alheit, K.; Hahn, V.; Maurer, H.P.; Reif, J.C.; Muller, J.; Rahe, F.; et al. BreedVision—A multi-sensor platform for non-destructive field-based phenotyping in plant breeding. *Sensors* **2013**, *13*, 2830–2847. [CrossRef] [PubMed]
21. Correa, C.; Valero, C.; Barreiro, P.; Ortiz-Cañavate, J.; Gil, J. Usando Kinect como sensor para pulverización inteligente. In *VII Congreso Ibérico de Agroingeniería y Ciencias Hortícolas*; UPM: Madrid, Spain, 2013; pp. 1–6. (In Spanish)
22. Wang, H.; Mao, W.; Liu, G.; Hu, X.; Li, S. Identification and location system of multi-operation apple robot based on vision combination. *Trans. Chin. Soc. Agric. Mach.* **2012**, *43*, 165–170.
23. Agrawal, D.; Long, G.A.; Tanke, N.; Kohanbash, D.; Kantor, G. Autonomous robot for small-scale NFT systems. In Proceedings of the 2012 ASABE Annual International Meeting, Dallas, TX, USA, 29 July–1 August 2012.
24. Paulus, S.; Behmann, J.; Mahlein, A.K.; Plümer, L.; Kuhlmann, H. Low-cost 3D systems: Suitable tools for plant phenotyping. *Sensors* **2014**, *14*, 3001–3018. [CrossRef] [PubMed]
25. Lachat, E.; Macher, H.; Mittet, M.A.; Landes, T.; Grussenmeye, P. First experiences with kinect v2 sensor for close range 3d modelling. In Proceedings of the International Archives of the Photogrammetry, Remote Sensing and Spatial Information Sciences (ISPRS Conference), Avila, Spain, 31 August–4 September 2015; pp. 93–100.
26. Fankhauser, P.; Bloesch, M.; Rodriguez, D.; Kaestner, R.; Hutter, M.; Siegwart, R. Kinect v2 for mobile robot navigation: Evaluation and modeling. In Proceedings of the 2015 IEEE International Advanced Robotics (ICAR), Istanbul, Turkey, 27–31 July 2015.
27. Nießner, M.; Zollhöfer, M.; Izadi, S.; Stamminger, M. Real-time 3d reconstruction at scale using voxel hashing. *ACM Trans. Graphics* **2013**, *32*. [CrossRef]
28. Lancashire, P.D.; Bleiholder, H.; Langeluddecke, P.; Stauss, R.; van den Boom, T.; Weber, E.; Witzen-Berger, A. A uniform decimal code for growth stages of crops and weeds. *Ann. Appl. Biol.* **1991**, *119*, 561–601. [CrossRef]
29. Hess, M.; Barralis, G.; Bleiholder, H.; Buhr, L.; Eggers, T.; Hack, H.; Stauss, R. Use of the extended BBCH scale-general for the descriptions of the growth stages of mono- and dicotyledonous weed species. *Weed Res.* **1997**, *37*, 433–441. [CrossRef]
30. Andújar, D.; Fernández-Quintanilla, C.; Dorado, J. Matching the Best Viewing Angle in Depth Cameras for Biomass Estimation Based on Poplar Seedling Geometry. *Sensors* **2015**, *15*, 12999–13011. [CrossRef] [PubMed]

31. Azzari, G.; Goulden, M.L.; Rusu, R.B. Rapid Characterization of Vegetation Structure with a Microsoft Kinect Sensor. *Sensors* **2013**, *13*, 2384–2398. [CrossRef] [PubMed]

32. Mirtich, B. Fast and Accurate Computation of Polyhedral Mass Properties, 2007. Available online: http://www.cs.berkeley.edu/~{}jfc/mirtich/massProps.html (accessed on 1 Dececember 2015).

33. Kenkel, N.C.; Derksen, D.A.; Thomas, A.G.; Watson, P.R. Review: Multivariate analysis in weed science research. *Weed Sci.* **2002**, *50*, 281–292. [CrossRef]

34. Nock, C.A.; Taugourdeau, O.; Delagrange, S.; Messier, C. Assessing the potential of low-cost 3D cameras for the rapid measurement of plant woody structure. *Sensors* **2013**, *13*, 16216–16233. [CrossRef] [PubMed]

35. Chen, Y.; Zhang, W.; Yan, K.; Li, X.; Zhou, G. Extracting corn geometric structural parameters using Kinect. In Proceedings of the 2012 IEEE International Geoscience and Remote Sensing Symposium (IGARSS), Munich, Germany, 22–27 July 2012.

36. Yamamoto, S.; Hayashi, S.; Saito, S.; Ochiai, Y. Measurement of growth information of a strawberry plant using a natural interaction device. In Proceedings of the American Society of Agricultural and Biological Engineers Annual International Meeting, Dallas, TX, USA, 29 July–1 August 2012; pp. 5547–5556.

Article

Design of a Computerised Flight Mill Device to Measure the Flight Potential of Different Insects

Antonio Martí-Campoy [1,*], Juan Antonio Ávalos [2], Antonia Soto [2],
Francisco Rodríguez-Ballester [1], Victoria Martínez-Blay [2] and Manuel Pérez Malumbres [3]

[1] Instituto de Tecnologías de la Información y Comunicaciones (ITACA), Universitat Politècnica de València, Camino de Vera s/n, 46022 Valencia, Spain; prodrig@disca.upv.es
[2] Instituto Agroforestal Mediterráneo (IAM), Universitat Politècnica de València, Camino de Vera s/n, 46022 Valencia, Spain; juavama@msn.com (J.A.Á); asoto@eaf.upv.es (A.S.); vicmarbl@etsia.upv.es (V.M.-B.)
[3] Department of Physics and Computer Science, Miguel Hernandez University, Ave. Universidad s/n-Ed. Alcudia, 03202 Elche, Spain; mels@umh.es
* Correspondence: amarti@disca.upv.es; Tel.: +34-96-387-7007; Fax: +34-96-387-7579

Academic Editor: Gonzalo Pajares Martinsanz
Received: 8 January 2016; Accepted: 30 March 2016; Published: 7 April 2016

Abstract: Several insect species pose a serious threat to different plant species, sometimes becoming a pest that produces significant damage to the landscape, biodiversity, and/or the economy. This is the case of *Rhynchophorus ferrugineus* Olivier (Coleoptera: Dryophthoridae), *Semanotus laurasii* Lucas (Coleoptera: Cerambycidae), and *Monochamus galloprovincialis* Olivier (Coleoptera: Cerambycidae), which have become serious threats to ornamental and productive trees all over the world such as palm trees, cypresses, and pines. Knowledge about their flight potential is very important for designing and applying measures targeted to reduce the negative effects from these pests. Studying the flight capability and behaviour of some insects is difficult due to their small size and the large area wherein they can fly, so we wondered how we could obtain information about their flight capabilities in a controlled environment. The answer came with the design of flight mills. Relevant data about the flight potential of these insects may be recorded and analysed by means of a flight mill. Once an insect is attached to the flight mill, it is able to fly in a circular direction without hitting walls or objects. By adding sensors to the flight mill, it is possible to record the number of revolutions and flight time. This paper presents a full description of a computer monitored flight mill. The description covers both the mechanical and the electronic parts in detail. The mill was designed to easily adapt to the anatomy of different insects and was successfully tested with individuals from three species *R. ferrugineus*, *S. laurasii*, and *M. galloprovincialis*.

Keywords: measurement; coleoptera; insect pests; flight study; flight mill; flight potential; behaviour; computer

---

1. Introduction

In the order of Coleoptera, there is a large number of species that constitute a serious danger to trees and plants, causing negative effects upon them during their feeding and reproduction. Reducing crop productivity or degrading the quality and good-looking appearance of trees and plants in gardens and landscapes are some of these effects. Moreover, in some cases, the negative effect is devastating, as it entails the death of the plant infested by such insect. Chemical or biological treatments have been developed to successfully fight against these kinds of insects. However, these treatments are not always completely effective. This is the case of insect borers that hide inside trees, making chemical and biological spraying difficult to reach them. For example, the Cerambycid beetle *Monochamus galloprovincialis* Olivier (Coleoptera: Cerambycidae) usually affects ill or even dead pine trees, but its real risk is that this insect is a vector for the nematode *Bursaphelenchus xylophilus* Steiner & Buhrer (Nematoda: Parasitaphelenchidae) which may kill a tree in just a few months.

Another example is *Semanotus laurasii* Lucas (Coleoptera: Cerambycidae) that directly attacks cypress trees, eventually producing the death of the plant. Furthermore, more than 20 years ago, *Rhynchophorus ferrugineus* Olivier (Coleoptera: Dryophthoridae) was introduced in Spain, and is today considered a global pest affecting different palm species on all continents [1]. Some important advances have been achieved, both in the knowledge about these insects and in the development of detection, prevention, and control mechanisms, but the pests are still out of control in several areas.

Knowledge about these insects, concerning all aspects of their biology and behaviour, is absolutely necessary if we are to find ways to defeat them. Researching aspects such as feeding habits, life span, host preferences, spread behaviour, and so on, involve several disciplines and collaborative efforts between researchers from different areas.

One of these factors is the flight skill of insects. Obtaining knowledge about parameters like maximum flight distance or specific flight patterns may help to develop strategies and procedures to prevent the spread of the pest such as pest outbreak management or proper trap locations [2]. Such studies may be accomplished indoors or outdoors.

In outdoor studies, the collected data is very close to reality, including environmental factors like wind and temperature. However, when insects are able to fly long distances, a huge effort is necessary to monitor their displacements. Several examples of outdoor study techniques exist. Traps are used in the technique of mark-release-recapture [3–5]. Recapture is avoided in [6] where insects are labelled with emitting radioisotopes that allow detecting them even if they are inside a tree. In recent years, radioisotopes were substituted by a radio-transmitter attached to the insect [7,8].

Indoor studies offer the possibility to obtain data about the potential behaviour of insect flight. Furthermore, they allow us to work with quarantine pests, because releasing these kinds of insects into the wild is not possible, like in the case of *R. ferrugineus*.

Another advantage of laboratory studies is the possibility to work with a large number of individuals; this may make the research easier and faster. However, the collected data must be carefully analysed before it can be translated to outdoor realities.

Indoor studies are usually performed by means of three techniques: static tethering, flight balances and pendulums, and flight mills [9].

In static tethered flight analysis, the insect is constrained in a fixed position, and flight behaviour is derived from the air displacement produced by wing movement [10] or by measuring the forces produced in the structure where the insect is attached [11].

In flight balances and pendulums [12], the insect is attached to the end of an arm, and the arm is attached to a pivot that allows it to ascend or descend as the result of insect wing beating. Pendulum-based monitoring presents a similar operation [13]. In both balances and pendulums, a scale is located behind the insect to measure the lift force produced.

The tethered flight mill method is considered a model system for laboratory analysis of insects flight behaviour [14], and has been successfully used to study the flight performance of a large number of species belonging to different orders.

However, the data and results obtained from the use of a mill must be accepted with some scepticism. The authors are convinced that the flight performance of an insect tethered to a mill is not the same as that obtained when an individual freely flies outdoors. Environmental parameters, like changes in daylight or the presence of sexual and feed attractants may affect the potential distance an insect is able to fly. A clear example is a tail wind helping the insect fly longer distances with less effort. Despite these constraints, knowledge coming from flight mills may be very useful for obtaining data about basic flight attributes [15]. Furthermore, it may also be useful for improving knowledge about the insect by other means, like outdoor experiments. In this case, data collected from the flight mill may help to arrange traps in a mark-release-recapture study, for example.

2. Rationale

There is a need for a device like a flight mill to carry out studies on insects that pose environmental threats or jeopardize agriculture. This mill needs to be adapted to the particular characteristics of such insects [15]. Therefore, it is important to show, with full detail, the design, construction, and use of this kind of device as a resource to help fight these pests.

This paper presents a detailed description of the construction, instrumentation, and use of a flight mill. The mill was designed with insects of different sizes and weights in mind, and one of its main goals is that the mill must easily adapt to a variety of insects, both in its hardware and software aspects. Furthermore, allowing for the reproduction of this design is an important goal.

The idea of tethering an insect to some structure and studying its flight is not new. In fact, it is several decades old. In 1952, Krogh and Weis-Fogh [16] presented a so-called roundabout for studying sustained flight of swarms of locusts, up to 32, using a stroboscope to measure the speed of the periphery and derive the flight intensity of each individual.

In [17], a flight mill was presented to allow semi-free flight of one individual of the Mexican fruit fly based on a vertical shaft with magnetic flotation. The insect is attached to an arm and the arm is attached to a magnet. This magnet is placed over another magnet facing like poles, therefore repelling it. To keep the floating magnet in place, a hypodermic syringe needle is inserted through both magnets, acting as an axle. Its authors claim null vertical friction, but some friction still remains between the axle and the floating magnet as it rotates that is reduced by covering part of the axle with Teflon.

In [18], the flight mill was built with magnetic bearings that vertically hold a steel entomological pin. A silica capillary builds the arm where individuals of leafhoppers (Hemiptera: Auchenorrhyncha) are attached. Magnetic floating or magnetic sustenance is commonly used to build flight mills [14,19,20].

However, the use of magnets presents some problems, mainly when the flight mill has to be adapted to different insects. In [21], the flight mill based on magnetic bearings was used to study the emerald ash borer (EAB), Agrilus planipennis Fairmaire. As stated in the aforementioned work, in the first experimental attempts, the larger size and greater strength of this beetle (as compared to leafhoppers) allowed them to wrest themselves free from the magnetic field after a few revolutions. More powerful magnets were then used, but that implied an increase of the torsional drag and the effort required to rotate the mill, possibly affecting the flight speed or its duration. Even more, some insects may be unable to start flying.

Thus, the strength of the magnetic field should be adjusted as a function of the insect being studied, and this tuning requires changing magnets, so significant modifications to the flight mill may be required in order to study different insects.

Some insects start to fly when tarsal contact with a substrate is removed [22]. For these kinds of insect, a mechanism to allow the insect to land may be necessary in the flight mill. This way, the individual may stop and restart flying at will, avoiding forcing the insect to fly until it becomes exhausted. With this goal in mind, [23] used a balance where an attached moth can take off and land freely. However, this balance allows measuring the time the moth is flying, but not the distance, an important parameter when studying pest insects.

The insects studied with the flight mill presented in this paper do not present tarsal reflex, and they start and resume flying freely even with no ground contact. During experimentation, some insects were observed with closed wings and moving legs like they were walking while hanging from the arm.

In order to build a flight mill with low friction to interfere as little as possible with the insects' flight, and easily adaptable to insects of different sizes, the authors propose building a flight mill with ball bearings, similar to the one used in [24]. The advantage of the ball bearing is that it offers a robust union for its use with large size and mass insects like the red palm weevil. But at the same time, the friction of the ball bearing is very low, so it is also suitable for small, low strength insects. Regarding the arm, most previous works focus on reducing friction in the arm-base joint, but they pay little attention to the total mass the insect needs to move when flying. In this work, we use carbon fibre rods to build a light but steady arm.

Another important aspect when designing and building a flight mill is the way to instrument it. There are several options for detecting movement and counting revolutions. [25] use a miniature lamp and a phototransistor to detect the interruption of the light received by the phototransistor. [19] use a photoelectric switch to detect the passage of the arm. The same function is performed using an infrared emitter and receiver in [21]. The advantage of using infrared light is to avoid interference from visible light, so visible wavelength light can be manipulated according to the behavioural requirements of the insects being studied. In [24], a bike milometer is used to directly acquire statistical data about speed and distance. The milometer magnet is attached to the arm and the milometer sensor is attached to the mill pivot. Although some milometers may be connected to a computer and therefore record data automatically, there is no possibility of obtaining raw data about arm revolutions and only overall information is gathered.

Finally, arm revolution events, together with the time they occur, should be recorded in a computer system in order to automatically post-process these data and extract derivative magnitudes about the insect flight. Several works, including [12,14,25,26] employ hardware designed ad hoc to connect the arm revolution detector to a computer.

The flight mill described in this paper uses an infrared emitter and receiver to detect the movement of the arm when it interrupts the infrared beam, the same approach followed in other works like [21]. The output of the detector is connected to a standard PCI local bus [27] data acquisition card (DAQ) installed inside a personal computer where infrared beam interruptions are recorded.

This paper presents details about the connection of the infrared detector with the DAQ, sampling frequency, and developed software in order to help other research groups reproduce the design. Any similar DAQ may be used with no changes in the hardware and minor changes in the associated software.

Finally, the software architecture is presented. The data acquisition and processing is carried out by two separate applications. The first one records events in the mill without regard to the attached insect, but considering the mill's mechanical characteristics, like the arm length and the number of infrared beam interruptions per revolution.

The second application processes the previous raw records, also taking into account insect-related parameters as described in the following sections. This second application provides a list of single flights including flight time and distance, and also a summary of the whole experiment.

Using two separate applications to obtain these results increases the amount of stored data, but it allows performing a single flight campaign, and carrying out several analyses afterwards, changing insect-related parameters without having the insects fly again.

3. The Flight Mill

The flight mill developed in this work may be divided into three main blocks, as shown in Figure 1. The first block is the mechanical mill where insects are attached and allowed to fly. The second block is the mill instrumentation for collecting information about mill rotations. This instrumentation is accomplished by means of an infrared sensor that detects arm revolutions. A computer with an off-the-self data acquisition board records the data coming from the sensor. The last block is a set of three software applications. The first is used to align the sensors. The second is devoted to interacting with the data acquisition card and storing raw data in the computer about arm rotations. And the third application processes the raw data according to user settings and produces results about several flight parameters, like distance and time flown, among others.

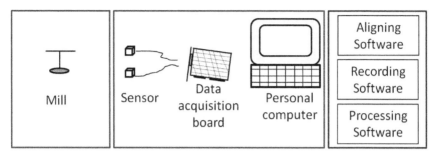

Figure 1. Functional blocks of the proposed flight mill.

3.1. The Mill

The mill is based on the design of [24] but incorporates some modifications. The main elements are the base, pivot, arm, and joint between the pivot and arm. Detailed descriptions of these elements and the way they are joined follows. Letters in brackets refers to elements in Figure 2.

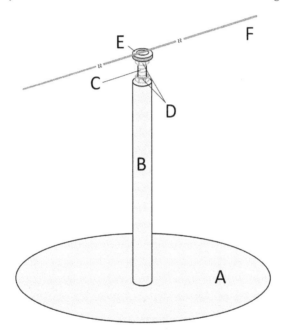

Figure 2. Mill design.

A heavy iron disc (A) ($\varnothing$ = 200 mm, 3.7 mm thick and weighing close to 900 gr) is used as the mill base, and it has been chosen to provide stability. The bottom side of this disc is covered with foam (thickness = 2 mm) to reduce the effect of vibrations that may come from the facility where the mill is installed. At the centre of the base, an iron rod is attached vertically (B) ($\varnothing$ = 12.2 mm, length = 150 mm) acting as the mill pivot. The pivot is attached by welding. An endless screw is at the end of the pivot (C) ($\varnothing$ = 4 mm, length = 150 mm), and inserted and screwed in. The insertion depth is about 30 mm and may be secured by a nut (D) located at the lower end of the screw. A second nut (D), close to the upper end of the screw is used to adjust and secure a miniature flanged ball bearing (E). Both nuts are hex nuts compliant with standard DIN-934.

The miniature ball bearing (E) is the mill's key mechanical component. Its purpose is to allow the mill arm to move with low friction and avoid any lever effect due to the insect weight. The ball bearing's main parameters are: manufacturer Minebea Co., Tokyo, Japan; width = 3 mm; internal diameter = 4 mm; external diameter = 8 mm; flange diameter = 9.2 mm. To this ball bearing, a washer (E) (internal ∅ range = [8 mm to 8.4 mm], external ∅ range = [14 mm to 15 mm]) is attached. Cyanoacrylate glue (Super Glue-3, Henkel Ibérica, Barcelona, Spain) is used to attach the washer to the ball bearing flange. Washer dimensions are critical because avoiding contact between it and the inner ring edge is mandatory, and at the same time the washer must offer the largest possible surface for attaching the mill arm. The weight of the union ball bearing and washer (E) is 2.04 gr.

Two options were used to build the mill arm for adapting it to different insect species. In both cases, the arm is a carbon fibre rod (F) (Eolo Sports Industrias, Gijón, Spain). For large insects, such as *R. ferrugineus*, the rod is 2 mm in diameter, 640 mm in length, and weighs 3 gr. For smaller insects, such as *S. laurasii*, the rod is 1 mm in diameter, 480 mm in length, and weighs 0.54 gr. The fibre rod attaches to the washer using hot silicone (Salki, Comersim S.A.U., Pamplona, Spain) as shown in Figure 3. One revolution of the thicker arm provides for a flight path of 2.01 m and 1.50 m for the thinner rod. Special care should be taken to join the arm to the ball bearing-washer system, centring the arm both lengthways and laterally to the rotating axis of the ball bearing to avoid deviation in the flight path length.

Figure 3. Detail of the union of the fibre carbon rod, washer, and ball bearing.

At one end of the arm, a plasticine counterweight (Jovi, S.A., Barcelona, Spain) is placed to compensate for the weight of the insect. At the other end of the arm, there are two parallel, horizontal directed pins for attaching the insect (see Figure 4).

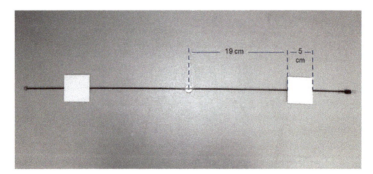

Figure 4. Detail of card stock, counterweight, and insect attachment pins.

Finally, the mill is assembled inserting the arm-washer-ball bearing (see Figure 5 and detail in Figure 6) at the top of the screw. For the assembly, the upper hex nut has been placed 3 mm from the top end of the screw. Then the ball bearing rests at the end of the screw, supported by the nut. In Figure 7, the dimensions of a DIN-934 hex nut and the ball bearing described previously are shown. For a nut with hole diameter (d) of 4 mm, the flat diameter (D) is 7 mm, the same as the inner diameter of the outer ring of the ball bearing outside diameter (D') minus the fillet radius (FR). This way, contact between the inner and outer rings of the ball bearing is avoided, and the arm can move freely, with only the very low friction of the ball bearing.

Figure 5. Arm assembly in the endless screw supported by the upper nut.

Figure 6. Detail of the washer-ball bearing assembly supported by the nut, without the arm.

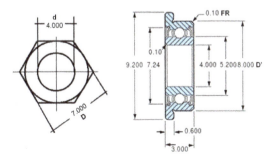

Figure 7. Dimensions of DIN-934 hex nut and ball bearing from Minebea Co., Tokyo, Japan.

The mill is installed on an open rack, composed of six chipboard shelves (length = 90 cm, width = 90 cm, thickness = 2 cm) with a space of 30 cm between shelves. A total of five mills have been installed in this rack, shown in Figures 8 and 9.

Figure 8. Rack composed of 5 flight mills.

Figure 9. Rack detail with an *R. ferrugineus* individual tethered to the mill.

3.2. Instrumenting the Mill

The flight mill's main objective is to collect data about the time and distance the attached insect is able to fly. Derivative magnitudes like speed, average flight distance, and maximum flight time are of interest also. Flight distance may be measured by simply counting the number of arm revolutions. But measuring the time is needed in order to obtain the remaining magnitudes.

We define an experiment as the process of attaching an insect to the mill, allowing it to fly for a predefined time (normally hours), and then removing it from the mill. To obtain the results mentioned before, taking timestamps of start and end flying events is not enough, since the insect may take breaks or perform irregular flight periods. In order to obtain accurate data about flight behaviour, accurately measuring the time of each arm revolution while the insect is attached to it is necessary.

To measure the time the arm takes to make a revolution, some electronic devices are used: first, an infrared emitter and receiver are coupled to produce and detect an infrared beam that is interrupted by the rotating arm; second, a data acquisition board with digital inputs collects the information from the infrared devices; and third, a personal computer populated with the data acquisition board and running Microsoft Windows XP stores and performs calculations on the collected data. Figure 10 shows a block diagram of this arrangement.

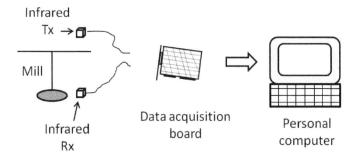

Figure 10. Block diagram of mill instrumentation.

The infrared emitter and receiver are used to detect the mill's arm passing. The receiver is located at the base of the mill, close to an edge. It is attached to the shelf with hot silicone. The emitter is attached to the bottom of the shelf above it and is also attached with hot silicone. Here are some considerations about the use of the infrared detector:

- The infrared beam produced is not very wide, about 6 mm in diameter, so the emitter and receiver alignment is important. A simple computer application, described later, is used to help the user align the beam.
- The receiver's sensing area has a 5.59 mm diameter. Taking into account the width of the mill's arm (1 or 2 mm), it may be possible that the arm is not wide enough to completely block the beam from the infrared detector. To ensure the beam is completely interrupted when the arm crosses it, a small (50 mm × 50 mm) rectangular piece of card stock is added to each arm leg 19 cm from the centre of the arm (see Figure 4).
- By adding card stock to the mill arm, we can guarantee that the infrared receiver has enough time to detect the beam interruption. Using some simple geometry calculus, it can be determined the beam is interrupted for 3.6 ms considering the insect is attached 24 cm (radius of the short fibre rod) from the centre of the arm, and assuming the insect flies at 50 Km/h. The lower the flight speed or the longer the arm, the longer the time the beam is interrupted and thus the detection is easier for the infrared receiver. Each complete revolution of the mill arm blocks the infrared beam twice, so actually half revolutions are counted.

Both the infrared emitter and receiver are connected to a data acquisition board (APCI-IB40; ARCOM CONTROL SYSTEMS, Cambridge, UK). This board allows the connection of up to 40 digital inputs/outputs (DIO), and is also able to power low consumption devices using a 5 volts output and a GND terminal. The power consumption of each emitter and receiver pair is about 10 mA; as the rack allows for the installation of five mills, the total power requirement for the installation is about 50 mA. This is low enough to directly power all five pairs of infrared emitters and receivers from the acquisition board.

Five DIO from the board are configured as inputs, and the output of the infrared receivers are connected to them. The APCI-IB40 board has an outer, female, 50 way, D-type connector, where digital inputs and outputs can be connected. Also, from this connector, a DC power supply of 12 V and 5 V, as well as ground, can be obtained. Direct wire plugging to this connector is not possible, so a male, aerial, 50 way, D-type connector with two unshielded multi-conductor cables (Alphawire; Sunbury-on-Thames, UK) was used. One end of the ribbon cable was soldered to terminals of the aerial D-type connector and the other end was stripped and screwed to a rack with terminal blocks (Sofamel; Barcelona, Spain). Terminal blocks were connected to a male, panel mounting, 3-way circular connector. Single wires were used to connect infrared emitter and receiver terminals to a female, 3-way aerial circular connector (Amphenol; Wallingford, CT, USA). This solution provides easy facility maintenance. Detail is shown in Figure 11.

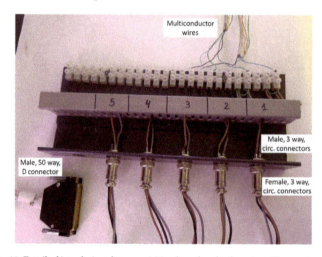

Figure 11. Detail of interfacing data acquisition board and infrared emitters and receivers.

The APCI-IB40 board also includes four 16-bit timer/counters nominated timer/counter 0 to timer/counter 3. Timer/counter 0 is used to trigger an interrupt signal at a fixed frequency based on a programmable divider from its clock source, which is selectable to be 1 MHz, 100 KHz, or 10 KHz. In order to not miss any activation of the five outputs coming from the infrared receivers, these signals must be sampled using a period shorter than the minimum beam disruption time, which in turn depends on the insect's speed. This way, timer/counter 0 is configured to raise an interruption every 5 ms, allowing measuring arm linear speeds up to 35 Km/h. To the authors' knowledge, this speed is greater than the maximum speed of any insect under study [28]. An interrupt is raised every 5 ms configuring timer/counter 0 in auto-reload mode, with a reload value of 500 counts and input clock frequency of 100 KHz.

3.3. Software

Three applications were developed to set up the system, to record raw flight information, and to process the recorded information. The partition of the software into three applications aims to improve the flexibility of the system. All three applications are programmed in the C++ language [29] using Borland C++ Builder version 6.0 [29].

3.3.1. Infrared Beam Alignment

A simple graphical application is used to help the user in the infrared beam alignment procedure. The application configures the APCI-IB40 board to generate an interrupt every 5 ms by means of the timer/counter 0. The interrupt service routine checks the output of the five infrared receivers and shows a coloured label indicating whether the receiver is detecting the beam (green label) or not (red label). A human operator may then adjust the receiver position until the label changes to green, indicating the emitter and receiver are aligned. Figure 12 shows a capture of the graphical interface of this application where only the infrared beams of mills 2 and 3 are correctly aligned.

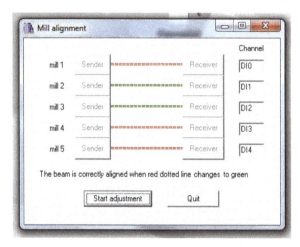

Figure 12. Interface of application for infrared beam alignment.

3.3.2. Capturing Events

This application is in charge of the detection of events originating from the five mills in the installation and recording the associated information into a file in the computer hard disk drive. The graphical interface of this application (see Figure 13) allows the user to customise several experiment parameters. What follows is the list of per-mill parameters used:

- Digital channel number of the acquisition board where the infrared receiver signal is connected.
- Number and name of the file where events were recorded. An automatic filename is provided by the application using the mill number, the experiment number, and the experiment date. The experiment number is automatically increased when the user ends an experiment. In all cases, the user may change the file name and choose a filename that best suits their needs.
- Sex, age, and sexual state of the insect.
- Arm length. This option provides flexibility to the mill and allows using different arms depending upon the insect under study.
- Arm type, indicating if the arm interrupts the infrared beam once or twice per revolution. This option is intended to deal with other ways of detecting the arm movement.
- Desired experiment time limit. Once this time is reached, the application stops recording events.

All these parameters work individually for each of the five mills.

The application offers two buttons for each mill, one to start capturing events and the other to manually stop the capture. When the start button is pressed, a new file is created and experiment parameters are stored in it. A simple plain text format is used to arrange the data inside the file. Also, the current date and experiment duration is written in the file.

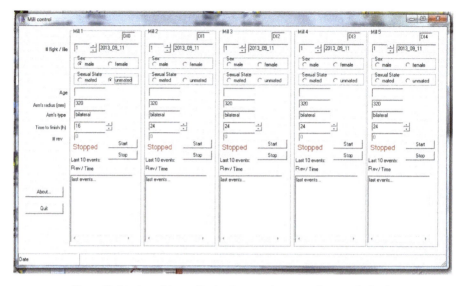

Figure 13. Interface of the application for capturing events (arm revolutions).

The application configures the timer/counter 0 in the acquisition board to generate an interrupt every 5 ms. This configuration is accomplished only once, the first time the user presses one of the five start buttons. Then, every 5 ms, the interrupt service routine checks the state of the infrared sensor for the active mills (those in which their start button has been pressed but neither of their stop buttons has been pressed nor the experiment time limit has been reached). Because the time the beam is interrupted by the card stock piece attached to each arm leg depends on the insect's flight speed, the time the infrared beam is broken may last quite longer than 5 ms, so simply checking the infrared receiver output does not work: a single beam interruption produced by a slow insect may be detected several times and lead to a false larger number of counts. To avoid this problem, the interrupt service routine detects the falling edge (a state change from a high to low logic level from two consecutive samples) in the receiver output using a Boolean variable to store the state of the receiver output in the previous sample and comparing it with the current receiver output state. Every falling edge is recorded in the file, coupled with the current system time (hours, minutes, seconds, and milliseconds provided by the operating system).

The interrupt service routine performs two more operations. First, it checks if any experiment has reached its time limit (the user-selected experiment time) in order to stop checking the sensor output, close the file of events, and increase the experiment number for the related mill; and second, for mills with a new event, updates the application graphical interface with some information about the recorded events: the number of times the beam has been disrupted and the list of the last ten events with their associated timestamps.

When the user presses the stop button or the experiment time limit is reached, the file containing all the information for the experiment is closed and the experiment number for the stopped mill is automatically increased. Then the graphical interface shows the mill is ready for a new experiment. From this moment forward, a new experiment may be configured and manually started.

3.3.3. Processing Recorded Files

This application is dedicated to process the data recorded by the *Capturing events* application described in the previous subsection. This last application is a purely software one and does not interact with the mill hardware. It would be easy to include all the required functionality in a single application, but the authors preferred to create separate applications because some parameters and options of this application depend on the insect under study, so more and frequent modifications are expected here than for the application that records raw events, an application that relies on hardware devices that are not easily modifiable. Even when the arm is changed to adapt it to a different kind of insect, all the program functions remain the same, and only some parameters, such as the length of the arm, need to be changed in the application devoted to recording events.

Having a separate application to process the raw recorded data is also advantageous in the sense the same raw data may be used to perform multiple different analyses with modified parameters without the need for insects to fly again. From a naive point of view, counting the half revolutions of the arm is enough to determine the distance the insect has flown for the whole experiment, and using the start and end time is enough to calculate the average flight speed. But there are several reasons for performing a detailed analysis of the recorded data:

- While flying for several hours, the insects may take several breaks, so stops and pauses have to be detected.
- When an insect stops flying, a backward movement may happen while the arm is near the infrared detection area, producing two breaks of the infrared beam with a very short in-between interval that leads to an erroneous flight speed calculation if this possibility is not taken into account
- When an insect needs or wants to stop flying, it is used to landing and stopping. However, while attached to the mill, an insect may stop flying but remain in movement until friction gently slows and stops it. Counting the revolutions the arm completes after the insect closes its wings leads to an overestimation of the distance flown.

The authors propose the use of single flights, like proposed in [30], to address the aforementioned issues. For the duration of an experiment, which may span from a few minutes to several hours to even longer than a day, the insect behaviour is divided into single flights. A single flight is defined as the period between two consecutive stops, and a stop is defined as a time period without arm revolutions or with significantly slow revolutions.

Stop periods are determined after an experiment has finished as a pre-processing step before the analysis of captured data. The stop period durations are specific for each kind of insect. As the file with the recorded raw events is always available, the researcher can change the duration of this period and re-process the recorded data, gathering information about the insect's flight behaviour. The application presents a graphical interface (see Figure 14) where the user can set three parameters to define when a single flight starts and stops. The possibility of changing these parameters allows working with different insects. Moreover, the division in two separate applications, one for recording hardware events and the other for explaining them, enhances the flexibility and adaptability of this proposal. Details of the parameter follow:

- Minimum time to start a new single flight (MTNF). If the time between two arm half revolutions is longer than this parameter (MTNF), a new single flight is considered. That is, it is assumed the insect stopped and rested for a while, then started a new single flight.
- Minimum time of a half revolution (MTHR). If the time between two consecutive beam interrupts (*i.e.*, an arm half revolution) is shorter than this value, it is considered a backward movement or simply as an erroneous detection, because it means that the flying speed of the insect is too high. In this case, the second beam break is discarded and the next raw event is considered.
- Percentage of speed reduction (PSR). When the insect stops flying, there is still movement, but its speed gently decreases until stopping completely. The flight speed is more or less constant during a single flight, so a significant speed reduction may indicate that the insect has closed its wings and is moving only due to system inertia. When the average speed of the actual flight is reduced in the percentage indicated, the subsequent arm revolutions are discarded until a break in the flight or a significant increase of speed occurs. The meaning of the latter is that the insect has rested for a brief time, and so we consider the current single flight is still valid.

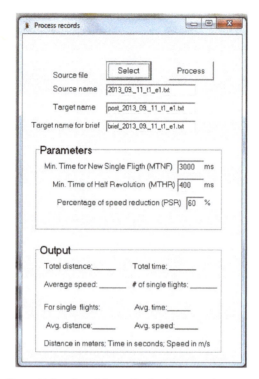

Figure 14. Interface of the application for processing events.

The graphical interface allows the user to select the file with recorded raw events as the source of the analysis, and it creates a new file adding the *post* prefix to the original name as the target file. In this file, the results of the whole experiment and single flights are stored using plain text. The information stored in the target file is described below:

- Source file name.
- All the data included in the source file related to experiment characteristics, like the age, sexual state, and sex of the insect, arm characteristics, and those previously described.
- Total distance flown by the insect as the cumulative sum of the distance of single flights.
- Total time flown by the insect as the cumulative sum of the duration of single flights.
- Average speed for the whole experiment.
- Number of single flights.
- Average distance for single flights.
- Average duration for single flights.
- Average speed for single flights.
- List of single flights, including for each flight, the following data: start and end time, number of revolutions (discarding those faster than the MTHR), elapsed time, distance flown, and average speed. This list is stored in plain text using a comma-separated values (csv) format [31,32] allowing the researcher to easily export this information to a statistical analysis application.

A second file is also created, adding the prefix *brief* to the source file name that contains the raw recorded events. In this file, only a brief summary of the results from the event analysis is stored in columns and the csv format. The purpose of this summary is to allow the automatic concatenation of lots of event-processed results to import all the information together into a statistical tool. The information is the same as in the full report except for the list of single flights. Each field in this file is stored in a column without headers. Concatenation may be easily done by means of batch scripts using operating system console commands.

4. Experimental Section

Experiments presented in this paper were devoted to test the proper mill operation studying different insects, and to collect insect flight data.

First of all, special calibration is not needed since there is neither analogue-digital conversion nor specific measurements capable of being calibrated during the mill operation. There are only two variables measured in this work. On the one hand, we measure the distance covered by the insect in the mill. This distance is derived from the radius of the arm (a fixed and known value measured with a precision ruler) and the number of revolutions completed by the insect. The second variable is the elapsed time between two consecutive interrupts of the infrared beam (a distance of a half revolution). To take the corresponding time measurements, we used the PC clock timing system that is precise enough (order of microseconds) to accurately measure the half revolutions elapsing time of the insect flight. Also, the proper functioning of the overall system has been manually verified by acting on the mill to generate all possible events, contrasting the information recorded by the system with the one observed by an operator with a stopwatch.

Several adults of three coleopteran borer species, *R. ferrugineus*, *S. laurasii*, and *M. galloprovincialis* were used.

A total of 163 *R. ferrugineus* unmated adults (86 males and 77 females) were used for the experiments, obtained from cocoons collected from infested *P. canariensis* palms in the town of Sueca, in eastern Spain (latitude N 39° 12′; longitude W 00° 18′), between January and December 2012. The cocoons were held in individual sterilized 100 ml plastic containers with perforated lids and maintained in a climatic chamber at 25 ± 2 °C and 65% ± 5% relative humidity. Adult emergence was checked once a day to determine their exact age and sex, after which the newly emerged weevils were returned to the containers. A piece of apple, replaced twice a week, was provided as a food source [33] until the insects were used in the tests.

S. laurasii adults were obtained from *Cupressocyparis leylandii* (A. B. Jacks. & Dallim.) Dallim. (Pinales: Cupressaceae) wood logs coming from gardens in the city of Valencia (latitude N 39° 28′; longitude W 00° 22′) in eastern Spain.

In the case of *M. galloprovincialis*, the adults were recovered from *Pinus halepensis* Mill. (Pinales: Pinaceae) wood logs coming from the 2013 fire in Cortes de Pallás (latitude N 39° 14′; longitude N 00° 56′) in the Valencia region of Spain. The wood logs for both species were kept in a climatic chamber under temperatures of 25 ± 2 °C and 65% ± 5% relative humidity. Adult emergence was checked once a day to determine their exact age and sex, and the emerged adults were kept in individualized containers. Table 1 shows the number of insects collected.

Table 1. Number of individuals collected for each insect species.

Species	Total	♂	♀
R. ferrugineus	163	86	77
S. laurasii	170	85	85
M. galloprovincialis	43	19	24

Each individual was weighed with a precision scale (Acculab; ALC-210.4, Bradford, PA, USA) and the length of its body was measured longitudinally with a digital calliper (Comecta Corp.; Barcelona, Spain). Table 2 shows the main parameters of the insects used in experiments.

Table 2. Main morphological characteristics, body weight and body length, for males and females of the different tested species.

Species	Avg Weight ± SE (g)	Min Weight (g)	Max Weight Max (g)	Avg. Length ± SE (mm)	Min Length (mm)	Max Length (mm)
R. ferrugineus ♂	1.02 ± 0.020	0.53	1.62	31.08 ± 0.25	24	38
R. ferrugineus ♀	1.19 ± 0.030	0.51	1.60	34.16 ± 0.29	26	39
R. ferrugineus ♂ + ♀	1.11 ± 0.020	0.51	1.60	32.83 ± 0.27	24	39
S. laurasii ♂	0.16 ± 0.003	0.10	0.24	17.05 ± 0.25	11.87	22.34
S. laurasii ♀	0.20 ± 0.060	0.11	0.33	18.75 ± 0.31	11.20	24.69
S. laurasii ♂ + ♀	0.18 ± 0.004	0.10	0.24	17.90 ± 0.21	11.20	22.34
M. galloprovincialis ♂	0.42 ± 0.030	0.23	0.65	22.00 ± 0.47	18	25
M. galloprovincialis ♀	0.34 ± 0.020	0.22	0.54	18.90 ± 0.27	18	24
M. galloprovincialis ♂ + ♀	0.38 ± 0.020	0.22	0.54	20.20 ± 0.28	18	24

For each weevil, a length of polyethylene foam (30x4x4 mm) is attached to its pronotum using cyanoacrylate glue (Super Glue-3, Henkel Ibérica; Barcelona, Spain) like shown in Figure 15. Then, the foam is fixed to two pins installed at the end of the arm. Five flight mills were simultaneously used in a climatic chamber, maintained at $25 \pm 2\,^{\circ}$ C, 65% $\pm$ 5% RH, and constantly lit by non-flickering 58 W fluorescent (Philips Ibérica; Madrid, Spain) and Grolux lamps (Osram Sylvania Inc.; Danvers, MA, USA).

Figure 15. Detail of the tethering of an *R. ferrugineus* adult to a piece of foam.

Because of the small size of *S. laurasii*, the arm for this insect was built with a carbon fibre rod 1 mm in diameter in order to reduce the mass the insect had to move, as described in Section 3.1. The arm structure, which includes the ball bearing, the washer, the arm, the two pieces of card stock, the foam, and pins used to attach the insects and the glue to keep all the parts joined, presented a total weight of 4.43 g. For the other two species, *R. ferrugineus* and *M. galloprovincialis*, the arm structure was built with a carbon fibre rod 2 mm in diameter, also described previously. The arm structure using the thicker rod presents a total weight of 6.90 g, including the same parts as the piece used for *S. laurasii*.

It is important to remark that changing the arm structure does not require any tools, since all parts remain joined together and the ball bearing rests freely on the vertical screw, stopped by the upper nut, and it can be easily removed and inserted.

Usually, these kinds of flying insects do not present tarsal reflex, so they do not start flying immediately after being left in the mill. This way, insects were left in the mill for 12 h, while the computer system monitored and recorded arm revolutions. After this period, computer records were observed to check that insects where able to fly when fixed to the arm. For *R. ferrugineus* and *M. galloprovincialis*, around 70% the individuals were able to fly (see Table 3). This percentage is similar to the data presented for other species [34]. For *S. laurasii* this percentage reduced to 22%. Further experiments would be needed to asses if this low number of flying individuals is because of the small size of the insect compared to the arm, or if it is an intrinsic property of this species.

Table 3. Number of tested insects and the number of individuals able to fly from each insect species.

Species	Total Tested	Total Flown	♂	♀
R. ferrugineus	163	110	58	52
S. laurasii	170	39	22	17
M. galloprovincialis	43	29	15	14

In all cases, there was no relationship between the insects that didn't fly and the mill they were attached to; several insects flew in all five mills. Recorded data was processed using the application described in Section 3.3.3. Different values for the minimum time to start a new single flight (MTNF), the minimum time of a half revolution (MTHR), and the percentage of speed reduction (PSR) were set for each kind of insect (see Table 4).

Table 4. Parameters used for processing recorded events.

Species	MTNF (ms)	MTHR (ms)	PSR (%)
R. ferrugineus	2000	400	70
S. laurasii	3500	400	70
M. galloprovincialis	3500	400	70

Tables 5–7 show the summary of the main flight parameters for *R. ferrugineus*, *S. laurasii*, and *M. galloprovincialis*, respectively. Parameters shown are number of single flights (NOF), total distance flown (TDF), longest single flight (LSF), flight duration (FD), average speed (AS), and maximum speed (MAXS).

Table 5. Summary of flight parameters for *R. ferrugineus* adults.

Parameters	Total (*n* = 110)			Males (*n* = 58)			Females (*n* = 52)		
	Mean ± SE	Max	Min	Mean ± SE	Max	Min	Mean ± SE	Max	Min
NOF	15.3 ± 2.8	233	1	20.1 ± 5	233	1	9.9 ± 1.5	59	1
TDF (m)	2780.2 ± 432.3	19,659.6	4	3309.2 ± 625	19,659.6	4	2190.1 ± 587.5	17,398.6	25.1
LSF (m)	1335.7 ± 229.9	11,244.9	4	1613.3 ± 317.3	9434.9	4	1026.1 ± 331.7	11,244.9	9
FD (min)	32.42 ± 5	260.7	0.1	39.1 ± 7.12	203.1	0.1	24.9 ± 6.9	260.6	0.4
AS (km/h)	3.7 ± 0.1	7.1	2.4	3.8 ± 0.1	7	2.4	4 ± 0.4	21	2.4
MAXS (km/h)	6.1 ± 0.1	9.3	2.9	6 ± 0.2	8.3	2.9	6.1 ± 0.2	9.3	3.4

Table 6. Summary of flight parameters for *S. laurasii* adults.

Parameters	Total (*n* = 39)			Males (*n* = 22)			Females (*n* = 17)		
	Mean ± SE	Max	Min	Mean ± SE	Max	Min	Mean ± SE	Max	Min
NOF	8.72 ± 1.48	36	1	7.77 ± 1.88	30	1	9.94 ± 2.41	36	2
TDF (m)	1126.81 ± 353.50	9468	8.25	1502.01 ± 582.75	9468	9.75	641.25 ± 279.15	4144.50	8.25
LSF (m)	454.75 ± 124.10	2841	3.75	586.19 ± 200.06	2841	3.75	284.65 ± 112.81	1543.50	4.50
FD (min)	33.46 ± 9.91	255.29	0.24	42.46 ± 16.03	255.29	0.24	21.82 ± 9.15	142.42	0.29
AS (km/h)	1.77 ± 0.06	2.77	1.10	1.88 ± 0.09	2.77	1.30	1.63 ± 0.08	2.26	1.10
MAXS (km/h)	2.87 ± 0.10	5.40	1.90	3.05 ± 0.17	5.40	2.06	2.64 ± 0.11	3.60	1.10

Table 7. Summary of flight parameters for *M. galloprovincialis* adults.

Parameters	Total (*n* = 29)			Males (*n* = 15)			Females (*n* = 14)		
	Mean ± SE	Max	Min	Mean ± SE	Max	Min	Mean ± SE	Max	Min
NOF	3.72 ± 0.75	19	1	2.95 ± 1.04	19	1	4.71 ± 1.10	15	1
TDF (m)	538.43 ± 114.40	2774.81	4.02	473.54 ± 170.63	2774.81	10.05	622.39 ± 154.82	1983.87	4.02
LSF (m)	316.04 ± 61.04	1629.11	4.02	269.93 ± 74.61	1112.54	6.03	375.69 ± 102.64	1629.11	4.02
FD (min)	9.46 ± 9.91	143.97	0.09	5.45 ± 1.91	32.22	0.23	14.66 ± 0.26	143.97	2.47
AS (km/h)	2.93 ± 0.16	5.91	2.33	2.71 ± 0.21	5.91	2.33	3.21 ± 0.43	5.91	3.55
MAXS (km/h)	4.82 ± 0.27	8.87	3.55	4.38 ± 0.36	8.87	3.95	5.39 ± 8.94	8.87	0.09

In order to validate the results obtained above, we have checked them with respect other studies about the *R. ferrugineus* (RPW) flight potential. In particular, we have found a couple of works in the literature: an outdoor study in [3], and recently in [15] an indoor study of RPW flight potential by means of a flight mill. We compared the results provided by [15] with the ones obtained with our flight mill. Unfortunately, the experimental setup of both works were not the same, so we have to take care when comparing their results. The main differences between both experimental works (apart of the flight mills) are the following:

(a) The RPW individuals used in our flight mill came from cocoons, so they have no previous flight experience. However, in [15] the RPW individuals have been captured from traps, being all of them adults with proved flight capacity.

(b) The age of individuals. In our experiments all the individuals are less than three weeks old, with and average age of 8.25 days. The age of individuals used in [15] is unknown, but it is expected to be higher than the ones used in our experiments.

(c) The individuals size. As with individuals age, there are also differences in the size and morphology of RPW individuals depending if they come from cocoons or traps. The average weight and length of the individuals used in our study was 1.11 grams, and 32.83 mm, respectively. Whereas in [15], the average weight and length was 1.32 grams and 26.36 mm. So, our insects are longer and lighter that the ones used in [15].

(d) In [15], each flight test lasted 24 h per individual, with an average temperature of $27 \pm 2\,^{\circ}$C. In our experiments the flight monitoring period was just the half (12 h) and the average temperature was $25 \pm 2\,^{\circ}$C.

In Figure 16, we show the summary of maximum distances obtained in the experiments of Hoddle's work [15] and the ones performed with our flight mill proposal. Taking into account the differences in the experimental setups, specially the differences in the flight test period and the individuals flight experience, we may observe that our tests show a significant high number (65%) of individuals that flew less than 1 Km when compared with the results found at [15]. However, the percentage of individuals flying from 1 to 20 Km is more or less equivalent in both studies. Our insects were not able to fly more than 20 Km in the 12 h flight test period.

The outdoor study described in [3] was able to estimate the flight potential of RPW individuals by means of mark, release and recapture techniques. The obtained results show that in average the individuals were able to fly between 1 and 7 Km in the lapse of 3 to 5 days.

Although there are differences between all the experiments we have analysed, we can conclude that (a) the results obtained in each of them are compatible; (b) their differences are mainly due to the experimental setups followed in each work; and (c) our flight mill design is able to measure the flight potential of RPW individuals.

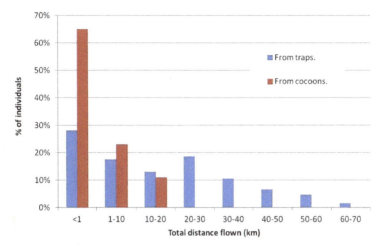

Figure 16. Total distance flown (in km) by individuals coming from traps [15] *versus* distance flown by individuals coming from cocoons (this work).

5. Conclusions

This work presents a detailed description of a flight mill and specialized software applications dedicated to collecting and analysing flight data in order to study the flight behaviour of different coleopteran species. Differences in size and weight of the species promote the design of a mill with a hardware structure easy to adapt to different species. Also, the software developed to process and analyse the collected data is able to adapt to differences among the species under study.

Regarding the mill hardware, the relationship between the insect masses and the mill arm that the insect must move during its flight is of utmost importance. The mill has to be stable and able to keep the insect flying a consistent path, so some mass and rigidity is needed, minimizing at the same time the mass of the arm to mimic the conditions of free flight as much as possible. To achieve both requirements, the moving parts of the mill have been reduced to a carbon fibre bar, a washer, and the outer ring of a miniature ball-bearing. Movement detection required the addition of two pieces of card stock, increasing the mass.

The moving parts are joined together in a unique piece than can be easily changed without the need for tools. This allows easily working with smaller or larger insects, adapting the mass of the arm to the physical characteristics of the insect. Two arms were designed and tested: one piece with a 1 mm fibre rod and a total weight of 4.43 g including all parts; the other one with a 2 mm fibre rod and a total weight of 6.90 g, including all parts.

The experiments conducted show that both the mill hardware and its software perform properly. We have also compared the results obtained with our flight mill with the ones from [3,15]. The comparative study shows that the results provided by our proposal are compatible with the ones shown in the other works.

Future work is directed to improve the hardware to reduce the mass of the moving parts, mainly affecting the washer and the card stock parts. Using lighter materials or another way to detect movement may result in a 20%–40% reduction in mass.

Acknowledgments: This research was partially funded by the Foundation of the Comunidad Valencia for Agroalimentary Research, Agroalimed, within the Project: Study of the flight behaviour and chromatic attraction in *Rhynchophorus ferrugineus* adults (Coleoptera: Curculionidae), and CICYT projects CTM2011-29691-C02-01 and TIN2011-28435-C03-01.

Author Contributions: Antonia Soto and Juan Antonio Ávalos conceived and designed the experiments; Juan Antonio Ávalos, Antonio Martí-Campoy, Francisco Rodríguez-Ballester and Manuel Pérez Malumbres designed and built hardware and software components; Juan Antonio Ávalos and Victoria Martínez-Blay performed the experiments; Juan Antonio Ávalos, Victoria Martínez-Blay, Antonia Soto and Antonio Martí-Campoy analyzed the data; Juan Antonio Ávalos, Antonio Martí-Campoy, Antonia Soto Sánchez, Francisco Rodríguez-Ballester and Manuel Pérez Malumbres wrote the paper.

Conflicts of Interest: The authors declare no conflict of interest.

References

1. Barranco, P.; De la Peña, J.; Cabello, T. Un nuevo curculiónido tropical para la fauna europea, *Rhynchophorus ferrugineus* (Olivier, 1790), (Coleoptera: Curculionidae). *Bol. Asoc. Esp. Entomol.* **1995**, *20*, 257–258.
2. Weissling, T.J.; Giblin-Davis, R.M.; Center, B.J.; Hiyakawa, T. Flight Behavior and Seasonal Trapping of *Rhynchophorus cruentatus* (Coleoptera: Curculionidae). *Ann. Entomol. Soc. Am.* **1994**, *87*, 641–647.
3. Abbas, M.; Hanounik, S.; Shahdad, A.; AI-Bagham, S. Aggregation pheromone traps, a major component of IPM strategy for the red palm weevil, *Rhynchophorus ferrugineus* in date palms (Coleoptera: Curculionidae). *J. Pest Sci.* **2006**, *79*, 69–73.
4. Chinchilla, C.M.; Oehlschlager, A.C.; Gonzalez, L.M. Management of red ring disease in oil palm through pheromone-based trapping of *Rhynchophorus palmarum* (L.). In *Porim International Palm Oil Congress "Update and Vision"*; Palm Oil Research Institute of Malaysia, Ministry of Primary Industries: Kajang, Malaysia, 1993; pp. 428–441.
5. Oehlschlager, A.C.; Chinchilla, C.M.; Gonzalez, L.M.; Jiron, L.F.; Mexzon, R.; Morgan, B. Development of a Pheromone-Based Trapping System for *Rhynchophorus palmarum* (Coleoptera: Curculionidae). *J. Econ. Entomol.* **1993**, *86*, 1381–1392.
6. Kloft, W.J.; Kloft, E.S.; Kanagaratnam, O.; Pinto, J.L.J.G. Studies on the use of radioisotopes for the control of the red palm weevil, *Rhynchophorus ferrugineus* F. by the sterile insect technique. *J. Coconut Res. Inst. Sri Lanka* **1986**, *4*, 11–17.
7. Hedin, J.; Ranius, T. Using radio telemetry to study dispersal of the beetle *Osmoderma eremita*, an inhabitant of tree hollows. *Comput. Electron. Agric.* **2002**, *35*, 171–180.
8. Rink, M.; Sinsch, U. Radio-telemetric monitoring of dispersing stag beetles: Implications for conservation. *J. Zool.* **2007**, *272*, 235–243.
9. McEwen, P.; Wyatt, T.D.; Reynolds, D.R.; Riley, J.R.; Armes, N.J.; Cooter, R.J.; Tucker, M.R.; Colvin, J.; Eigenbrode, S.D.; Bernays, E.A.; *et al.* In *Methods in Ecological and Agricultural Entomology*; Dent, D.R., Walton, M.; Eds.; Cab International: Wallingford, UK, 1997.
10. Riley, S.L.; Stinner, R.E. Recorder for Automatically Monitoring Tethered-insect Flight. *Ann. Entomol. Soc. Am.* **1985**, *78*, 626–628.
11. Sun, Y.; Fry, S.; Potasek, D.P.; Bell, D.; Nelson, B. Characterizing fruit fly flight behavior using a microforce sensor with a new comb-drive configuration. *Microelectromech. Syst. J.* **2005**, *14*, 4–11.
12. Cooter, R.J. *The Flight Potential of Insect Pests and Its Estimation in the Laboratory: Techniques, Limitations and Insights*; Central Association of Bee-Keepers: Upminster, UK, 1993.
13. Wales, P.J.; Barfield, C.S.; Leppla, N.C. Simultaneous monitoring of flight and oviposition of individual velvetbean caterpillar moths. *Physiol. Entomol.* **1985**, *10*, 467–472.
14. Schumacher, P.; Weyeneth, A.; Weber, D.C.; Dorn, S. Long flights in *Cydia pomonella* L. (Lepidoptera: Tortricidae) measured by a flight mill: Influence of sex, mated status and age. *Physiol. Entomol.* **1997**, *22*, 149–160.
15. Hoddle, M.S.; Hoddle, C.D.; Faleiro, J.R.; El-Shafie, H.A.F.; Jeske, D.R.; Sallam, A.A. How Far Can the Red Palm Weevil (Coleoptera: Curculionidae) Fly?: Computerized Flight Mill Studies With Field-Captured Weevils. *J. Econ. Entomol.* **2015**, *108*, 2599–2609.
16. Krogh, A.; Weis-Fogh, T. A roundabout for studying sustained flight of locusta. *J. Exp. Biol.* **1952**, *29*, 211–219.
17. Chambers, D.L.; O'Connell, T.B. A Flight Mill for Studies with the Mexican Fruit Fly. *Ann. Entomol. Soc. Am.* **1969**, *62*, 917–920.
18. Taylor, R.; Nault, L.R.; Styer, W.E.; Cheng, Z.B. Computer-Monitored, 16-Channel Flight Mill for Recording the Flight of Leafhoppers (Homoptera: Auchenorrhyncha). *Ann. Entomol. Soc. Am.* **1992**, *85*, 627–632.
19. Moriya, S. Automatic data acquisition systems for study of the flight ability of brown-winged green bug, *Plautia stali* Scott Heteroptera Pentatomidae. *Appl. Entomol. Zool.* **1987**, *22*, 19–24.

20. Hao, Y.N.; Miao, J.; Wu, Y.Q.; Gong, Z.J.; Jiang, Y.L.; Duan, Y.; Li, T.; Cheng, W.N.; Cut, J.X. Flight Performance of the Orange Wheat Blossom Midge (Diptera: Cecidomyiidae). *J. Econ. Entomol.* **2013**, *106*, 2043–2047.

21. Taylor, R.; Bauer, L.; Poland, T.; Windell, K. Flight Performance of Agrilus planipennis (Coleoptera: Buprestidae) on a Flight Mill and in Free Flight. *J. Insect Behav.* **2010**, *23*, 128–148.

22. Dingle, H. The relation between age and flight activity in the milkweed bug, Oncopeltus. *J. Exp. Biol.* **1965**, *42*, 269–283.

23. Gatehouse, A.G.; Hackett, D.S. A technique for studying flight behaviour of tethered Spodoptera exempta moths. *Physiol. Entomol.* **1980**, *5*, 215–222.

24. Dubois, G.; Vernon, P.; Brustel, H. A flight mill for large beetles such as *Osmoderma eremita* (Coleoptera: Cetoniidae). In *Saproxylic Beetles. Their Role And Diversity in European Woodland and Tree Habitats*; PENSOFT Publishers: Sofia, Bulgaria, 2009; pp. 219–224.

25. Chambers, D.; Sharp, J.; Ashley, T. Tethered insect flight: A system for automated data processing of behavioral events. *Behav. Res. Methods Instrum.* **1976**, *8*, 352–356.

26. Chen, H.; Kaufmann, C.; Scherm, H. Laboratory Evaluation of Flight Performance of the *Plum Curculio* (Coleoptera: Curculionidae). *J. Econ. Entomol.* **2006**, *99*, 2065–2071.

27. Abbott, D. *PCI Bus Demystified*; LLH Technology Publishing: Eagle Rock, VA, USA, 2000.

28. Hocking, B.; London, R.E.S. *The Intrinsic Range and Speed of Flight of Insects*; Transactions of the Royal Entomological Society of London, Royal Entomological Society: St Albans, UK, 1953.

29. Swart, B.; Cashman, M.; Gustavson, P.; Hollingworth, J. *Borland C++ Builder 6 Developer's Guide*; Sams: Indianapolis, IN, USA, 2002.

30. Hughes, J.; Dorn, S. Sexual differences in the flight performance of the oriental fruit moth, *Cydia molesta*. *Entomol. Exp. Appl.* **2002**, *103*, 171–182.

31. Shafranovich, Y. *Common Format and MIME Type for Comma-Separated Values (CSV) Files*; RFC 4180 (Informational), Updated by RFC 7111; Internet Engineering Task Force (IETF): Fremont, CA, USA, 2005.

32. Hausenblas, M.; Wilde, E.; Tennison, J. *URI Fragment Identifiers for the text/csv Media Type*; RFC 7111 (Informational); Internet Engineering Task Force (IETF): Fremont, CA, USA, 2014.

33. Llácer, E.; Santiago-Álvarez, C.; Jacas, J. Could sterile males be used to vector a microbiological control agent? The case of Rhynchophorus ferrugineus and Beauveria bassiana. *Bull. Entomol. Res.* **2013**, *103*, 241–250.

34. Jactel, H. Individual Variability Of The Flight Potential Of *Ips Sexdentatus* Boern. (Coleoptera: Scolytidae) in Relation to Day of Emergence, Sex, Size, and Lipid Content. *Can. Entomol.* **1993**, *125*, 919–930.

Article

Theoretical Design of a Depolarized Interferometric Fiber-Optic Gyroscope (IFOG) on SMF-28 Single-Mode Standard Optical Fiber Based on Closed-Loop Sinusoidal Phase Modulation with Serrodyne Feedback Phase Modulation Using Simulation Tools for Tactical and Industrial Grade Applications

Ramón José Pérez *, Ignacio Álvarez and José María Enguita

Department of Electrical Engineering, University of Oviedo, Ed. Torres Quevedo, Gijón Campus, Gijón 33204, Asturias, Spain; ialvarez@isa.uniovi.es (I.Á.); jmenguita@uniovi.es (J.M.E.)
* Correspondence: ramonjose.perez@lugo.uned.es; Tel.: +34-985-641-360

Academic Editor: Gonzalo Pajares Martinsanz
Received: 17 January 2016; Accepted: 20 April 2016; Published: 27 April 2016

Abstract: This article presents, by means of computational simulation tools, a full analysis and design of an Interferometric Fiber-Optic Gyroscope (IFOG) prototype based on a closed-loop configuration with sinusoidal bias phase- modulation. The complete design of the different blocks, optical and electronic, is presented, including some novelties as the sinusoidal bias phase-modulation and the use of an integrator to generate the serrodyne phase-modulation signal. The paper includes detailed calculation of most parameter values, and the plots of the resulting signals obtained from simulation tools. The design is focused in the use of a standard single-mode optical fiber, allowing a cost competitive implementation compared to commercial IFOG, at the expense of reduced sensitivity. The design contains an IFOG model that accomplishes tactical and industrial grade applications (sensitivity $\leqslant 0.055\,°/h$). This design presents two important properties: (1) an optical subsystem with advanced conception: depolarization of the optical wave by means of Lyot depolarizers, which allows to use a sensing coil made by standard optical fiber, instead by polarization maintaining fiber, which supposes consequent cost savings and (2) a novel and simple electronic design that incorporates a linear analog integrator with reset in feedback chain, this integrator generating a serrodyne voltage-wave to apply to Phase-Modulator (PM), so that it will be obtained the interferometric phase cancellation. This particular feedback design with sawtooth-wave generated signal for a closed-loop configuration with sinusoidal bias phase modulation has not been reported till now in the scientific literature and supposes a considerable simplification with regard to previous designs based on similar configurations. The sensing coil consists of an 8 cm average diameter spool that contains 300 m of standard single-mode optical-fiber (SMF-28 type) realized by quadrupolar winding. The working wavelength will be 1310 nm. The theoretical calculated values of threshold sensitivity and dynamic range for this prototype are $0.052\,°/h$ and 101.38 dB (from $\pm1.164 \times 10^{-5}\,°/s$ up to $\pm78.19\,°/s$), respectively. The Scale-Factor (SF) non-linearity for this model is 5.404% relative to full scale, this value being obtained from data simulation results.

Keywords: Interferometric Fiber-Optic Gyroscope (IFOG); closed-loop IFOG configuration; Integrated-Optical-Circuit (IOC); Phase Modulator (PM); Super-Luminiscent-Laser-Diode (SLD); Phase-Sensitive-Demodulation (PSD); serrodyne wave; Lyot depolarizer

Sensors **2016**, *16*, 604

1. Introduction

In all the electro-optical engineering areas, particularly in the design of high cost devices like Interferometric Fiber-Optic Gyroscopes (IFOGs), computational simulation resources can provide powerful and inestimable guidance. This stems from the rapidity, the reproducibility and the reliability of this kind of hardware to obtain the finished design of a preconceived model. Furthermore, it is possible to achieve substantial cost savings in components and time consuming model assembly in a laboratory's optical bank. Only after having obtained an ideal design, as much for the performance characteristics as for its adaptation to a specific application, it is suitable to initiate the laboratory manufacture stage for the previously designed prototype. In this article we show readers an aspect that is not usually found in the technical literature, namely how to realize the simulation of a classical IFOG system without having to make the real model in the laboratory. For this proposal three classical electro-optic simulation tools will be used: OptSim® (Synopsis™, Mountain View, CA, USA), MultiSim® (National Instruments™, Austin, TX, USA) and Matlab-Simulink® (MathWorks™, Natick, MA, USA). In the present decade the design trends in the IFOG field are focused on devices with very high performance (navigation-grade, sensitivity $\leqslant 0.001\ °/h$), mainly targeting aeronautics and spacecraft applications. Nevertheless, it is also possible to realize designs for certain applications that do not need such a high grade of performance (*i.e.*, tactical-grade, sensitivity $\leqslant \pm 0.01\ °/h$ or industrial-grade, sensitivity $\leqslant \pm 1\ °/h$). The latter mentioned will constitute the objective of the model presented. What follow next is a brief overview of the basis of IFOG performance.

The non-reciprocal phase shift between the two waves in counter-propagation (clockwise and counterclockwise) induced by rotation when both propagate across the sensing coil of optical-fiber, also known as the Sagnac effect, is usually given by the following expression (see, for instance, [1]):

$$\phi_S = \frac{2\pi LD}{\lambda c}\Omega \tag{1}$$

being L the total length (m) of the sensing coil, D its diameter (m), Ω the rotation rate (rad/s), and φ_S is the phase shift difference (rad), λ and c are the wavelength (m) and the speed of light (m/s) in free space, respectively, of the radiation emitted by the laser source. The proportionality factor that precedes the rotation-ratio is known as the scale-factor (SF) of the gyroscope, and it is a basic constructive constant that depends on geometric and optical parameters of the device. Taking the following initial values for the design: $L = 300$ m, $D = 0.08$ m and $\lambda = 1310$ nm, a value of 1.86 μrad/(°/h) is obtained for the SF. Detailed studies of the depolarization mechanism of optical counter-propagated waves within the fiber-optical sensing coil can be found in references [2–4]. The main advantage of the depolarization technique is that this approach allows using a single-mode optical-fiber for the sensing coil, with the consequent economic savings in the optical components costs of the gyroscope. This design is based on a conventional IFOG with sinusoidal phase modulation and a closed-loop feedback realized with classic analog electronic components, which provides a better stability and linearity of the gyroscope's SF, while using cost-competitive components.

The rest of the paper is organized as follows: the next section (Section 2) is focused on the design of the optical and electronic sub-systems of the model. Section 3 provides some important calculations and estimations of the performance of the design and Section 4 shows the simulation results (optical and electronic subsystems). Finally, Section 5 includes a discussion on simulation results and Section 6 collect the main conclusions of this paper.

2. Sensor Design

2.1. Design of the Optical System

The components of the optical system of this gyroscope are depicted in Figure 1. The light source is a 1310 nm superluminescent diode (SLD) with a low ripple Gaussian spectral profile. For this unit, the commercial reference SLD1024S of Thorlabs Inc. (Newton, NJ, USA) was used, with a DIL-14 pin assembly package, with FC/APC fiber pigtailing and realized in standard single-mode optical-fiber. This unit provides an adjustable optical power up to a 22 mW maximum level, although only 5 mW maximum level is needed for the present model. This unit uses an integrated thermistor to perform the temperature control, so that it is possible to obtain the stabilization of the power source in the spectral range. Accordingly with the temperature stabilization, the chip package must not exceed a maximum temperature of 65 °C. The directional optical coupler is four ports (2 × 2 configuration), with 50/50 output ratio, realized with the side-polished fiber-optic technique, and an insertion loss of 0.60 dB. The linear polarizer placed at the output of the directional input-output coupler is featured in polarization-maintaining fiber (PMF) with a 2.50 m length, insertion loss of 0.1 dB, and a polarization extinction ratio (PER) > 50 dB. The integrated optical circuit Integrated Optical Chip (IOC) performs the function of optical directional coupler at the input of the sensing fiber-optic coil (Y-Junction) and also the function of electro-optic phase modulator (PM). In a more advanced design, the linear fiber-optic polarizer can be replaced by an integrated approach, so that the former remains joined at the input of the IOC wave-guide [5]. This way, a bulk optic polarizer is avoided, which is an important contribution to the reduction of the whole space occupied by the optical system of the gyroscope.

The chosen PM is electro-optical class. Its electrodes remain parallel to the wave-guide channels obtained by diffusion of Ti on a lithium-niobate (LiNbO$_3$) substrate. The PM zone of the IOC includes two pairs of electrodes placed symmetrically with regard to the central axis of the integrated block. The output ports of the IOC remain connected, respectively, to the heads of the two Lyot depolarizers (both made on PM-fiber), with lengths L_1 and L_2, respectively. These Lyot depolarizers are realized in polarization-maintaining optical-fiber (PMF), connecting two segments appropriate lengths, so that the axes of birefringence of both form angles of 45°.

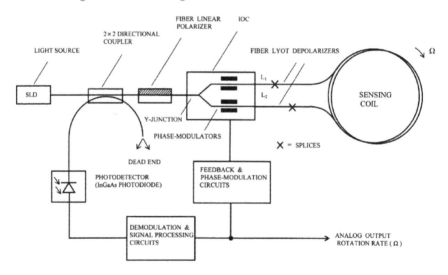

Figure 1. Electro-optical system configuration.

Taking into account the following values for PM fiber optic Lyot depolarizers: B = 1 × 10^{-4} = birefringence, λ = 1310 nm, then, coherence length L_c, beat length L_b and depolarization length L_D acquire the values collected in Table 1 (see their respective formulae). This table also shows the consequent calculated lengths L_1 and L_2 of both Lyot depolarizers (with their respective empirical formulae, as it can be seen).

Consequently, L_1 and L_2 Lyot depolarizer lengths add up to 26.20 cm and 52.40 cm, respectively. As it can be seen in Table 1, these calculations were realized taking into account a 26.20 μm value for coherence length L_c of the broadband light-source (emitting at 1310 nm wavelength) and 13.10 mm value for the beat length L_b of the optical fiber.

Table 1. L_c, L_b, L_D values and (L_1, L_2) Lyot depolarizer lengths.

Coherence Length L_c $L_C \gg 20\,\lambda$	Beat Length L_b $L_b = \frac{\lambda}{B} = \frac{\lambda}{|n_x - n_y|}$	Depolarization Length L_D $L_D \cong \frac{L_c L_b}{\lambda}$	Lyot Depolarizer Length L_1 $L_1 = L_D$	Lyot Depolarizer Length L_2 $L_2 = 2\,L_1$
26.20 [μm]	13.10 [mm]	26.20 [cm]	26.20 [cm]	52.40 [cm]

The two clockwise (CW) and counterclockwise (CCW) optical waves come from the sensing coil and join at the Y-Junction placed at the input of the IOC. The sensing coil consists of 300 m of optical standard single-mode fiber (commercial type SMF28), made by quadrupolar winding on a spool of 8 cm average-diameter, which provides 1194 turns. This optical fiber presents the following structural characteristics: Step refractive index, basis material = fused-silica, external coating = acrylate, core diameter = 8.2 μm, cladding diameter = 125 ± 0.7 μm, external coating diameter = 245 ± 5 μm, with the following optical parameters: n_{core} = 1.467, $n_{cladding}$ = 1.460, NA = 0.143, maximum attenuation = 0.35 dB/km at 1310 nm, h-parameter = 2×10^{-6} m^{-1}, dispersion coefficient ⩽ 18.0 ps/(nm × km) at 1550 nm, polarization dispersion coefficient ⩽ 0.2 ps/km$^{\frac{1}{2}}$, birefringence: B = 1.0×10^{-6}.

The chosen PM is electro-optical class. Its electrodes remain parallel to the wave-guide channels obtained by diffusion of Ti on a lithium-niobate (LiNbO$_3$) substrate. The PM zone of the IOC includes two pairs of electrodes placed symmetrically with regard to the central axis of the integrated block. The output ports of the IOC remain connected, respectively, to the heads of the two Lyot depolarizers (both made on PM-fiber), with lengths L_1 and L_2, respectively. These Lyot depolarizers are realized in polarization-maintaining optical-fiber (PMF), connecting two segments appropriate lengths, so that the axes of birefringence of both form angles of 45°.

2.2. Design of the Electronic System

In absence of rotation (Ω = 0 rad/s), the transit time of the two counter-propagated waves across the sensing coil is the same, being its value:

$$\tau = \frac{L}{(c/n_{core})} = \frac{n_{core}\,L}{c} \tag{2}$$

With the values of parameters adopted previously for the design of the model, and using 1194 turns wrapped on standard fiber-optic coil, the resultant value for the transit time is τ = 1.467 μs. On the other hand, the transit time value also determines the value of the modulation frequency f_m that must be applied to the phase modulator, given by the expression:

$$f_m = \frac{1}{2\tau} \tag{3}$$

resulting, for the present design, in a calculated value of 340.83 kHz. Equation (3) comes from the condition of maximum amplitude of the bias phase-difference modulation wave which is possible to formulate by the following expression:

$$\Delta\phi_{bias}(t) = 2\phi_0 \sin\left(\frac{2\pi f_m \tau}{2}\right) \cos\left[2\pi f_m \left(t - \frac{\tau}{2}\right)\right] \tag{4}$$

The condition of maximum amplitude needs the $2\pi f_m \tau = \pi$ relation to be satisfied (and then, Equation (3) is fulfilled). The block diagram of the electronic scheme for phase modulation and demodulation circuits is represented in Figure 2. A closed-loop configuration has been adopted with sinusoidal bias phase-modulation and serrodyne feedback phase modulation, taking as initial reference the state-of-the-art of demodulation circuits reported till now [6–11].

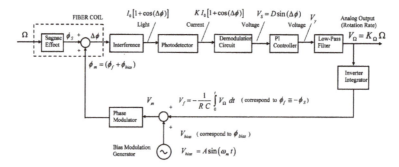

Figure 2. Analog closed-loop scheme for feedback phase-modulation configuration.

However, and this is the novelty, it has been changed the structure of feedback chain, adding now a new design of analog integrator which incorporates one FET transistor (2N4858, ON Semiconductor, Phoenix, AZ, USA) as depicted in Figure 3.

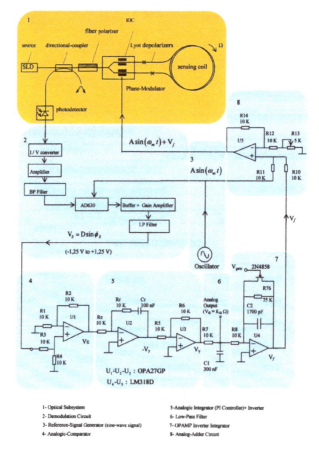

Figure 3. Primary block-diagram configuration of electro-optical and Phase-Sensitive-Demodulation (PSD) systems. The model is closed-loop configuration with sinusoidal bias phase-modulation and serrodyne feedback phase-modulation.

The function of this transistor is realizing periodically the shortcut of the capacitor voltage, therefore nulling instantaneously the voltage on feedback branch of integrator OPAMP. The time period for shorcut FET transistor is driving by the value of V_{gate} voltage, which, in turn, is controlled by one astable-based Flip-Flop circuit. Referring to Figure 3, block #7 generates a linear ramp voltage V_γ on its output and this ramp reset each time period driving by V_{gate} voltage. In this way, a resultant serrodyne-wave voltage is easily generated at the output of this integrator circuit, obtaining finally the same intended sawtooth voltage on feedback phase modulation chain as the reported on previous designs by literature [12–14]. Working as feedback phase modulation signal, the analog serrodyne-wave presents two important advantages with regard to the sinusoidal-one: (a) it is possible to generate the serrodyne wave easily by means of a simple integrator circuit (Miller integrator) with simple and low-cost electronic components and (b) the phase cancellation process inside the control loop becomes simpler and more efficient.

In accordance with the interference principle, the light intensity at the photodetector optical input presents the following form (for sinusoidal phase-modulation):

$$I_d(t) = \frac{I_0}{2}\left[1 + \cos(\Delta\phi)\right] = \frac{I_0}{2}\left\{1 + \left[J_0(\phi_m) + 2\sum_{n=1}^{\infty} J_{2n}(\phi_m)\cos(2n\omega_m t)\right]\cos\phi_S - 2\sum_{n=1}^{\infty} J_{2n-1}(\phi_m)\sin\left[(2n-1)\omega_m t\right]\sin\phi_S\right\} \quad (5)$$

being J_n the Bessel-function of the first kind of nth order. Here $\Delta\phi$ represents the effective phase-difference of the two counter-propagating optical waves on sensing coil. This value results from the combined action of the phase-modulation process ($\phi_m = \phi_{bias} + \phi_f$) and the Sagnac phase shift induced by the rotation-rate (ϕ_s). The output signal of the photodetector, in photocurrent form, is proportional to the light intensity at its optical input. This photocurrent signal is converted to voltage with a transimpedance amplifier that is placed at the entry of demodulation circuit. The demodulation circuit takes the task of extracting the information of the Sagnac rotation-induced phase shift (ϕ_s). The corresponding voltage signal at its output (V_S) scales as sine-function of the effective Sagnac phase-difference ϕ_s. The PI controller performs an integration of the V_S signal in time domain, so that a voltage signal (V_γ) is obtained, this signal growing almost linearly with the time. This latter signal is filtered by means of a low-pass-filter so that the corresponding output signal on voltage form (V_Ω) is a DC voltage value that is possible to consider to be almost proportional to the gyroscope rotation-rate Ω (when $\sin\phi_s \approx \phi_s$). Therefore, the analog output voltage signal V_Ω constitutes the measurement of the rotation rate of the system. The control system, as a whole, acts as the principle of phase-nulling. The phase-nulling process consists of generating a phase displacement ($\phi_m = \phi_{bias} + \phi$) in such a way that the phase-difference ϕ_f associated with the voltage output signal (V_f) is equal and with opposite sign with regard to the Sagnac phase-shift induced by the rotation rate, *i.e.*, $\phi_f = -\phi_s$. To achieve this, the feedback phase modulation circuit holds a sample of the output signal V_Ω. Note that this voltage signal is obtained at the end of the Low-Pass-Filter (LP Filter, Block 6 on Figure 3) and is proportional to rotation-rate Ω. An integration operation is needed for obtaining a linear ramp voltage to apply on phase-modulator. Then, integrates and inverts this signal by means of an operational integrator-inverter circuit, turning this signal into the following form:

$$V_f = -\frac{1}{RC}\int_0^t V_\Omega dt \quad (6)$$

This way, the time variation of the voltage signal V_f is a linear ramp, being its slope proportional to the rotation rate of the system (V_Ω). Figure 3 represents clearly the optical and electronic subsystems of the gyroscope, including the feedback phase-modulation and bias phase-modulation circuits for getting phase nulling process, both applied together to the PM. Referring now to Figure 3, then latter being the reference voltage for bias phase-modulation, see Figure 2), *i.e.*, $V_m = V_{bias} + V_f$. Therefore, the output signal of the phase modulator will be the sum of the phase-difference signals associated with the V_{bias} and V_f voltages, that is to say: $\phi_m = \phi_{bias} + \phi_f$. The error signal at the output of the comparator ($\Delta\phi$) tends to be nulled in average-time, due to the phase-cancellation (the average-time of the reference bias phase-modulation ϕ_{bias} is 0, so the following condition is fulfilled: $\Delta\phi = \phi_s + \phi_m$.

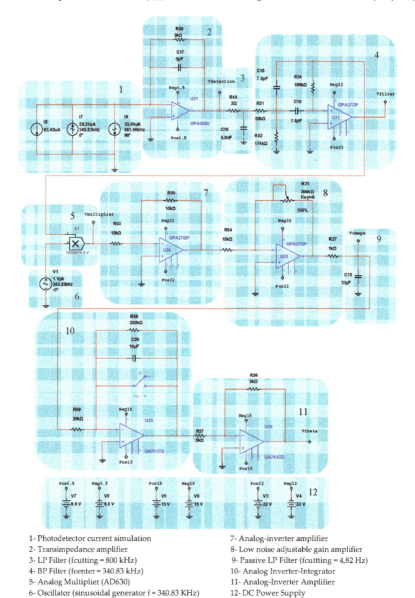

1- Photodetector current simulation
2- Transimpedance amplifier
3- LP Filter (fcutting = 800 kHz)
4- BP Filter (fcenter = 340.83 kHz)
5- Analog Multiplier (AD630)
6- Oscillator (sinusoidal generator f = 340.83 KHz)

7- Analog-inverter amplifier
8- Low noise adjustable gain amplifier
9- Passive LP Filter (fcuttting = 4,82 Hz)
10- Analog Inverter-Integrator
11- Analog-Inverter Amplifier
12- DC Power Supply

Figure 4. Detection and Phase-Sensitive-Demodulation (PSD) circuits.

The feedback phase-modulation circuit consists of an AC sine-wave signal generator that produces a voltage reference signal V_{bias} at 340.83 kHz for bias phase-modulation (block 3 of Figure 3), an analog comparator circuit (differential-operational-amplifier, block 4 of Figure 3) that generates an error voltage signal V_ε, an analog Proportional-Integral (PI) controller followed by one inverter-amplifier (block 5 of Figure 3), and a LPF that yields a DC V_Ω voltage signal proportional to the rotation-rate (block 6 on Figure 3). The inverter-amplifier on block 5 produces the inversion of the $-V_\gamma$ signal, obtaining the V_γ voltage signal. The DC V_Ω output voltage after passive the LP Filter on block 6 is integrated by the Integrator circuit on block 7 and, then converted into the V_f feedback voltage signal, as calculated from Equation (6), consisting on constant frequency and variable amplitude serrodyne-wave which is applied to one of the two inputs of an analog adder featured with a non-inverter operational-amplifier (the other input is connected to AC signal generator, block 8 on Figure 3). Therefore, the voltage output signal of this analog adder is the V_m voltage signal that realizes the sum of the V_{bias} and V_f voltage signals, as described previously.

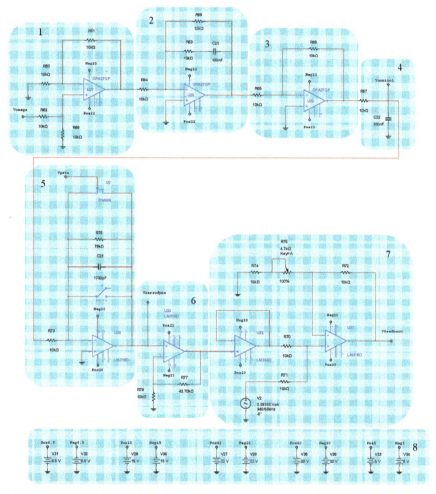

1- Analog comparator   5- Analog integrator with periodic discharge (reset) capacitor circuit (JFET 2N4858 transistor)
2- Analog PI controller   6- Non-inverter analog amplifier
3- Analog inverter   7- Analog and adjustable gain adder
4- Passive LP Filter   8- DC Power supply

Figure 5. Analog controller circuit (includes blocks #1, #2, #3 and #4) and serrodyne feedback phase-modulation circuit (includes blocks # 5, #6 and #7).

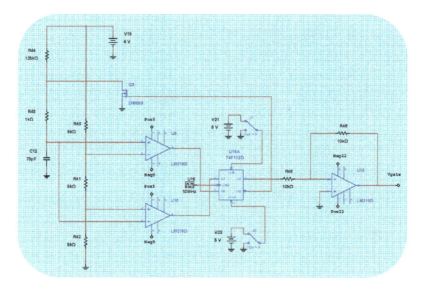

Figure 6. V_{gate} Voltage Signal generator (Astable Pulse Generator + J-K Flip-Flop + Analog Inverter).

Figure 4 represents the detail block-diagram of electronic scheme for detection and Phase-Sensitive-Demodulation (PSD) circuits. It consists basically of twelve functional blocks: (1) Photodetector simulated output current; (2) Transimpedance amplifier (current to voltage converter); (3) LP Filter (f_c = 800 kHz); (4) Band-Pass-Filter (BP Filter, f_{center} = 340.83 kHz); (5) Analog multiplier (AD630); (6) Sinusoidal Oscillator (f = 340.83 kHz); (7) Analog inverter amplifier; (8) Low-noise adjustable-gain amplifier; (9) LP Filter (f_c = 4.82 Hz) [15]; (10) Analog integrator filter (for rotation-angle determination); (11) Inverter OPAMP; the output voltage V_{theta} of this inverter allows obtaining the draft experienced by the system from a certain time (initialization time); block (12) DC Power Supplies. Figure 5 represents in detail the analog PI controller and feedback phase modulation circuits. Figure 6 represents the V_{gate} Voltage Signal generator circuit. This circuit consists, as it can be seen, on one sequence of an Astable Pulse Generator plus J-K Flip-Flop plus an Analog Inverter.

3. Calculations and Estimations

This design has been simulated using Matlab-Simulink®, MultiSim™ and OptSim™. The parameters of the model were chosen as: fiber coil length L = 300 m, fiber coil diameter D = 80 mm, number of turns in the coil N = 1194, light source wavelength λ = 1310 nm, average-power at the optical detector input P_d = 145.61 µW, and responsivity of the InGaAs photodetector R = 0.68678 µA/µW (note that the original version of the OptSim™ software only allows implementing APD-type photodetectors on optical circuit design, consequently an APD-PIN equivalent current-conversion will be necessary for connecting the simulation results to IFOG prototype designed in this article which possesses a PIN photodetector).

The open-loop scale factor K_0 can be calculated (being $c \approx 3 \times 10^8$ m/s the speed of light in vacuum) as:

$$K_0 = \frac{2\pi LD}{\lambda c} \tag{7}$$

The beat length of the optical fiber, L_b, can be calculated from its optical birefringence (B) as:

$$L_b = \frac{\lambda}{B} = \frac{\lambda}{|n_x - n_y|} \tag{8}$$

where n_x and n_y are the refractive indexes of the two orthogonally polarized modes along the x and y directions. For this model, the following performance parameters have been analysed: sensitivity threshold [16], dynamic range, and scale factor (SF) [17]. The values calculated (using the formulae) and estimated (by the results of the simulations) for such parameters are shown in Table 2. In this table the third column shows the value calculated directly by the formula and the fourth shows estimated results from the optical and electronic simulations.

Table 2. Performance parameters of the IFOG prototype (analog closed loop configuration).

Parameter	Calculation Formula	Calculated Value	Estimated Value	Unit
Sensitivity Threshold	$\Delta\Omega = \frac{2}{K_0}\sqrt{\frac{e}{P_d R t}}$	0.05,193,796	0.05,193,820	$[°/h]$
Dynamic Range	$20\log\left(\frac{\Omega_{max}}{\Omega_{min}}\right)$	101.38	101.38	[dB]
	$\Omega_{max} = \frac{\lambda c}{12LD}$	± 78.185	± 78.185	$[°/s]$
	$\Omega_{min} \approx \frac{\sqrt{hL_b}}{LD}$	$\pm 1.164 \times 10^{-5}$	$\pm 1.164 \times 10^{-5}$	$[°/s]$
Scale Factor	$SF = \frac{2\pi LD}{\lambda c}$	0.3837	0.3664	$\left[\frac{rad}{\left(\frac{rad}{s}\right)}\right]$

The sensitivity threshold considers the SNR at photodetector optical input provided by the optical simulation, and the dynamic range and scale factor are determined by the sine function non-linearity (assuming the maximum value $\phi_s = \pm\pi/6$). In the formulae, h is the h-parameter of the optical-fiber and t is the average integration time.

4. Simulation Results

Three different kinds of computer simulations have been realized. First, the control system simulation has been realized using Simulink™ for determining the 2% settling-time t_s of the complete electro-optic system. Second, an optical system simulation has been realized using OptSim™ for obtaining the optical interference signal at the PIN photodetector optical input and its main and representative values: Average optical power and Signal-to-Noise-Ratio (SNR). Third and finally, the electronic circuit simulation made with MultiSim™, to obtain the V_Ω DC voltage as image of the rotation rate of the system, and then, for obtaining the output graph-response of gyroscope unit.

Figure 7 represents the electro-optical system of designed IFOG-model. It is depicted as a parametrized block-diagram with its corresponding trasfer functions. The transfer function for each block is obtained taking into account the optoelectronic parameters relative to each IFOG component. The normalized transfer function ($TF_{closed-loop}$) of the whole closed-loop system is shown in a label in Figure 8. The step-response curve of the closed-loop IFOG system (obtained with Simulink®) is also shown in Figure 8. A settling time t_s (2%) of 1.39 ms is obtained. This value can be used to estimate a value for the initialization time of the final gyroscope unit. Optical subsystem simulation results (realized by means of the OptSim™ software) are presented in Figures 9–14. Figure 9 presents the optical schematic circuit of the designed model for obtaining its optical performance.

Figure 10 presents the sinusoidal electrical signal provided by the AC signal generator and applied to the PM as bias phase-modulation signal. Figure 11 presents the power spectral density as a function of frequency obtained at the photodetector optical input (central frequency is 288.844 THz).

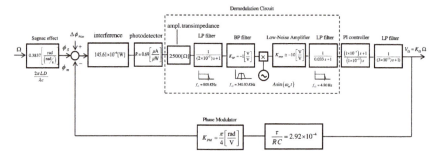

Figure 7. Parametrized block-diagram of the gyroscopic model system (sinusoidal bias phase-modulation and serrodyne feedback phase-modulation. Initial parameters: P_d = 145.61 µW, L = 300 m, D = 0.08 m, λ = 1310 nm and R = 0.69 µA/µW.

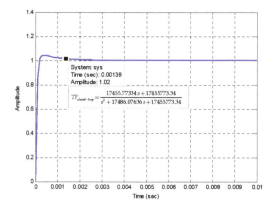

Figure 8. Time-response curve of designed IFOG model (closed-loop system). The input signal applied to system is a unit-step time function. The t_s (2%) settling time obtained is 1.39 ms, as it can be seen on the label.

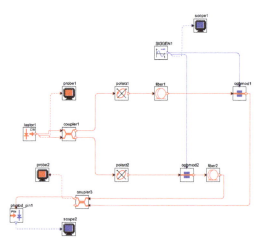

Figure 9. Optical circuit setup of the designed IFOG gyroscope for computer simulation (OptSim™).

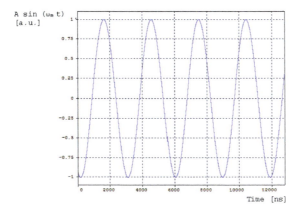

Figure 10. Bias-sinusoidal voltage signal provided by the AC signal generator and applied to Phase-Modulator (PZT, scope 1 in Figure 9). Sinusoidal curve parameters are: $f_m = \left(\frac{\omega_m}{2\pi}\right) = 340.83kHz$, $A = 1 [a.u.]$.

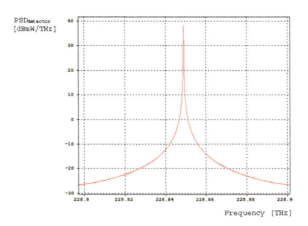

Figure 11. Power-Spectral-Density (PSD) curve obtained at the photodetector optical-input (probe 2 in Figure 9).

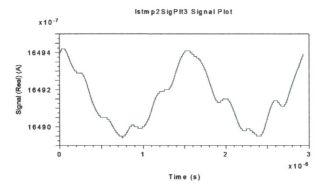

Figure 12. Interferometric current signal obtained at APD-equivalent-photodetector electrical output (after electrical BP Filter with f_{center} = 340.83 kHz) when $\Omega = \pm 10\,°/s$ is applied to system.

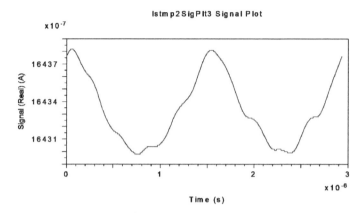

Figure 13. Interferometric current signal obtained at APD-equivalent-photodetector electrical output (after electrical BP Filter with f_{center} = 340.83 kHz) when Ω = ±20 °/s is applied to system.

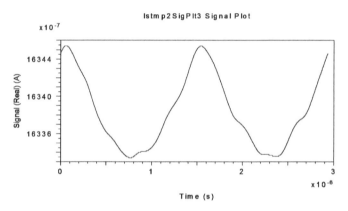

Figure 14. Interferometric current signal obtained at APD-equivalent-photodetector electrical output (after electrical BP Filter with f_{center} = 340.83 kHz) when Ω= ±30 °/s is applied to system.

Considering 210 µW as average optical power providing by light source, 145.61 µW were obtained at photodetector optical input, which means a power loss of −9.837808 dBm. Equation (9) allows the calculation of Photon-Shot-Noise photocurrent at photodetector (I_{sn}), taking into account 100 µA for photocurrent average value at its electrical output:

$$I_{sn} = \sqrt{\frac{e^2 q \lambda}{hc} P_{max-detector} \Delta f} \tag{9}$$

In this equation, the following values are assumed: e = 1.6 × 10^{-19} C, q = 0.65 (*quantum efficiency of the photodetector*), λ = 1310 nm, h = 6.626 × 10^{-34} Js (Planck constant), Δf = 1 Hz, and $P_{max\text{-}detector}$ = 100 × 10^{-6} W. Then, calculated value for I_{sn} is 3.312008 × 10^{-12} [A]. Note that the lower the Photon-Shot-Noise photocurrent value is, the lower the threshold sensitivity is and, therefore, the bigger the accuracy of the IFOG-sensor is. On the other hand, it is necessary to say that for a low level of optical-power coupled into the photodetector, the main optical noise source of FOG-sensor is quantum Photon-Shot-Noise (excess RIN can be neglected). This way, in accordance with Photon-Shot-Noise photocurrent above calculated, the threshold sensitivity of gyro-sensor Ω_{lim} (that is to say, the minimum rotation-rate which the gyro-sensor is able to measure) can be calculated by Equation (10):

$$\Omega_{lim} \cong \left(\frac{hc^2}{\pi eqLDP_{max}} \right) I_{sn} \tag{10}$$

In this equation the following values are taken into account: $D = 300$ m (*fiber coil length*), $P_{max-detector} = 100 \times 10^{-6}$ W and the rest are the same as those in Equation (9). Then, calculated value for Ω_{lim} is 0.05193796 [°/h], as collected in Table 2.

Figures 12–14 represent the electrical interferometric signal (APD equivalent photo-current after electrical BP filtering, $f_{center} = 340.83$ kHz) detected by an APD equivalent photodetector when $\Omega = \pm 10$ °/s, $\Omega = \pm 20$ °/s and $\Omega = \pm 30$ °/s, respectively, are applied to the system.

This is because the block-mode simulation only offers measurements realized by an APD equivalent photodetector as optical output of the system. The average mean values of APD photo-currents are, respectively, 1649.20, 1643.30 and 1633.80 µA which corresponds to 99.873, 99.515 and 98.940 µA for the PIN-equivalent photodiode. Note that in this interval the average current decreases almost linearly as rotation-rate increases linearly. These curves agree with theoretical interferometric curves as calculated on the optical input photodetector.

The results of electronic circuit simulation (realized by the MultiSim™ software) collect the waveform voltages on the following test-point voltages: $V_{detection}$, V_{filter}, $V_{multiplier}$, V_{theta}, $V_{serrodyne}$ and V_{gate} (referring to Figures 4–6). All these values are obtained on electronic circuits when $\Omega = +30$ °/s rotation-rate is applied to system and are gathered in Figures 15–20. Figure 15 shows the detected output voltage after the transimpedance amplifier ($V_{detection}$, see Figure 4). Figure 16 represents the output voltage after the BP Filter (V_{filter}, see Figure 4).

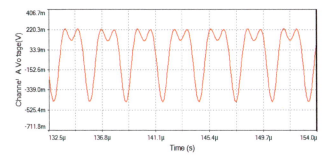

Figure 15. $V_{detection}$ voltage signal (after the transimpedance amplifier) for $\Omega = +30$ °/s.

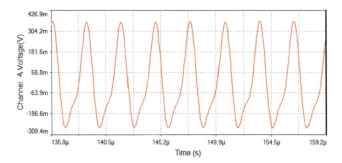

Figure 16. V_{filter} output voltage after the BP Filter for $\Omega = +30$ °/s.

Figure 17 represents output voltage after the analog Multiplier ($V_{\text{multiplier}}$, Figure 4). Figure 18 represents output voltage after the Angle analog integrator (V_{theta}, Figure 4). Figure 19 represents output voltage after the Analog Integrator ($V_{\text{serrodyne}}$: A sawtooth-voltage with constant frequency and variable amplitude, this amplitude depending on V_{Ω} voltage value). Finally, Figure 20 represents V_{gate} generated by the pulse generator circuit and applied to the gate of the J2N4858 FET transistor (see the circuit in Figure 6).

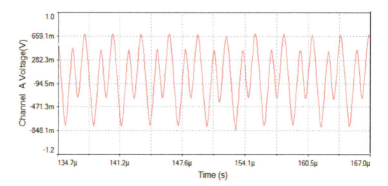

Figure 17. $V_{\text{multiplier}}$ output voltage after the Analog Multiplier for $\Omega = +30\,^{\circ}/\text{s}$.

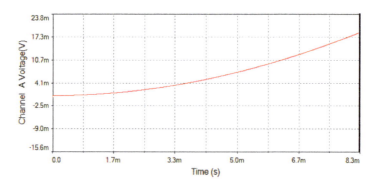

Figure 18. V_{theta} output voltage after the Angle analog integrator for $\Omega = +30\,^{\circ}/\text{s}$.

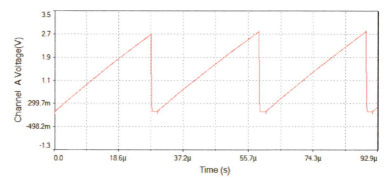

Figure 19. $V_{\text{serrodyne}}$ after the Analog Integrator (feedback voltage signal to the Phase-Modulator) for $\Omega = +30\,^{\circ}/\text{s}$.

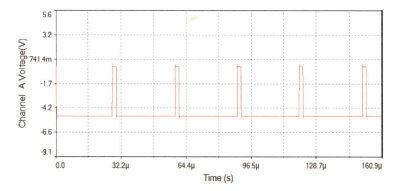

Figure 20. V_{gate} voltage generated by the pulse generator circuit (fixed frequency f = 32.59 kHz).

The expansion of Equation (5) with only the contribution of first two time-component harmonics allows obtaining an approximate value for detected $I_d(t)$ photo-current. The result of this approximation is Equation (11):

$$I_d(t) \cong \frac{I_0}{2}\left[1 + J_0(\phi_m)\cos\phi_S\right] + I_0 J_2(\phi_m)\cos(2\omega_m t)\cos\phi_S - I_0 J_1(\phi_m)\sin(\omega_m t)\sin\phi_S \quad (11)$$

being I_0 the maximum value of detected photo-current and ϕ_m the amplitude of differential phase-modulation. Assuming the value $\phi_m = 1.80$, this value corresponding to the maximum value of $J_1(\phi_m)$ function, the following Bessel functions calculations are obtained: $J_0(1.80) \cong 0.33999$, $J_1(1.80) \cong 0.58150$ and $J_2(1.80) \cong 0.30611$. Then, taking into account 100 µA as the DC average detected photodetector-current and, after some numerical adjusts, Equation (11) yields the following analytical value:

$$I_d(t) \cong 74.63\left[1 + 0.34\cos\phi_S\right] + 45.69\cos(2\omega_m t)\cos\phi_S - 86.79\sin(\omega_m t)\sin\phi_S \ [\mu A] \quad (12)$$

This analytical expression allows one to calculate for each rotation-rate Ω value (*i.e.*, ϕ_s, the Sagnac phase-shift) the DC term and the 1st and 2nd harmonics terms. This terms can later be introduced as current DC and AC generators on the MultiSim™ circuit simulation program (block 1 on Figure 4). By this means, V_Ω can be measured on the simulated circuit (see Figure 4), so that a table with V_Ω value *versus* Ω [°/s] value can be made. Table 3 lists the correlation data obtained from demodulation circuit for the measured output-voltage signal V_Ω [mV] *versus* input rotation rate Ω [°/s] of the system. Figure 21 shows the graphic representation between both variables corresponding to mentioned data table. After appropriate calculations and taking into account the theoretical value of the Scale Factor (SF) of the gyroscope that appeared on Table 2, a linear function can be obtained for the best fitting of the output response-curve. This linear function is obtained by the least square fitting method. Table 3 also includes the values $(V_\Omega)_{\text{lin}}$ [mV] of this linear fitting, the module of the differential values $\Delta(V_\Omega)$ [mV], and the module $|\Delta(V_\Omega)/(V_\Omega)_{\text{lin}}|$ of the ratio values.

Table 3. Output data and linear fitting for output response curve of designed IFOG prototype.

Ω [°/s]	0	± 10	± 20	± 30	± 40	± 50	± 60	± 70	± 80	± 90	± 100	± 110
V_Ω [mV]	0	± 305	± 613	± 931	± 1280	± 1609	± 1970	± 2337	± 2708	± 3085	± 3450	± 3800
$(V_\Omega)_{\text{lin}}$ [mV]	0	± 338.7	± 677.5	± 1016	± 1355	± 1694	± 2032	± 2371	± 2710	± 3049	± 3387	± 3726
$|\Delta(V_\Omega)|$ [mV]	0	33.73	64.46	85.19	74.92	84.65	62.38	34.11	1.834	36.44	62.71	73.98
$\dfrac{|\Delta(V_\Omega)/}{(V_\Omega)_{\text{lin}}|\%}$	0	9.959	9.514	8.385	5.529	4.997	3.070	1.439	0.068	1.195	1.851	1.986

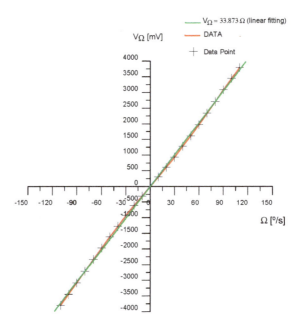

Figure 21. Output response curve V_Ω [mV] *versus* Ω [°/s] (in red colour) and best linear fit (least square fitting method) $(V_\Omega)_{lin}$ [mV] *versus* Ω [°/s] (in green colour), of the gyroscopic sensor prototype.

The $\Delta(V\Omega)$ [mV] value is defined as:

$$\Delta V_\Omega = V_\Omega - (V_\Omega)_{lin} \ [mV] \tag{13}$$

from correlation values of both curves (output data curve and linear fitting curve), it can be determined the non-linearity percentage coefficient of the SF, defined as the percentage of the standard deviation, which can be calculated by the following expression:

$$SF - NonLinearity(\%) = \sqrt{\frac{1}{N}\sum_{i=1}^{N}\left|\frac{(\Delta V_\Omega)_i}{(V_{\Omega lin})_i} \times 100\right|^2} \tag{14}$$

so that in our case, with $N = 23$ and taking the values obtained from Table 2, this expression yields a value of 5.404%.

5. Discussion of Simulation Results

The results obtained for the performance parameters of IFOG model designed in this article (threshold sensitivity = 0.052 °/h, dynamic range = ±78.19 °/s, Scale Factor non-linearity = 5.404%) are sufficient for industrial grade gyroscopic applications, such as stabilization and positioning of mobile platforms or inertial-navigation systems for terrestrial robots and automotive vehicles [18–20].

The effects of the different types of optical noise which take place are not critical in the specific design of this sensor, since its operation works in a medium level of optical-power and the SNR is relatively high at the photodetector's optical-input (SNR > 100 dB). The most important type of optical-noise for this sensor is Photon-Shot-Noise on the photodetector, with a 3.31 pA noise-equivalent-current value, this value being much less than 100 µA that is the average photocurrent value for photodetector electrical output signal in zero rotation-rate conditions. This type of noise is inevitable since it owes to intrinsic quantum-mechanical phenomenon in photoconductivity (electron-hole production by photon shoot). Photon-Shot-Noise scales as $1/\sqrt{P_{\text{max}-\text{detector}}}$, so it diminishes as optical power on photodetector increases (or what is the same, the SNR increases).

The Relative Intensity Noise (RIN) is an important issue in this design, since it works at a medium-level of average optical-power coupled to photodetector optical-input (145.61 µW average optical-power value). This type of noise stems from two causes: (1) the two interfering optical waves do not come to the photodetector with the same optical power level, due to polarization crosstalk between the two orthogonal polarizations states along the entire length of the sensing fiber-coil (due to fiber birefringence); and (2) the light source is low-coherence (broadband source), thereby producing several beat wavelengths, which add at the photodetector optical-input, causing a variation in relative intensity on every point of photodetector's response-curve. This noise can be minimized by reducing the optical power emitted by the light source, but a very large reduction in optical power also lowers the SNR at the photodetector, so that to maintain it at a high-level, the optical power emitted by light source cannot be reduced greatly. An alternative way to effectively reduce the effect of excess RIN noise is to operate the gyroscope at a bias point close to a black fringe, that is, a phase bias close to π instead of $\pi/2$. This way, the sensitivity is proportional to the slope of the raised cosine response curve $\sin \phi_m$ (being ϕ_m the phase-bias), while the excess RIN noise is proportional to the actual power on bias (then, $1 + \cos \phi_m$) which is the response-curve of IFOG. If the choice is $\phi_m = 0.9\,\pi$, for example, sensitivity is reduced by a factor $\sin(0.9\,\pi)/\sin(1.80) = 0.317$ while excess RIN noise experiences a reduction five times higher, since $[1 + \cos(0.9\,\pi)/[1 + \cos(1.80)] \cong 0.063$. Furthermore, it results an improvement for theoretical SNR due to Photon-Shot-Noise. As a result, such an excess RIN reduction technique allows in practice to get a total noise very close to the theoretical Photon-Shot-Noise, as calculated previously for our considered IFOG model. Estimations of excess RIN before and after this correction are showed next in Table 4 The formulae for computing Photon-Shot-Noise, Excess-RIN and Full-Noise are shown in Equations (A1)–(A6). All of them are collected below (see the Appendix for calculations) at the end of this section.

Table 4. Photon-Shot-Noise and excess RIN noise before ($\phi_m = 1.80$) and after ($\phi_m = 0.9\,\pi$) correction.

Noise Source	Before Correction ($\phi_m = 1.80$)	After Correction ($\phi_m = 0.9\,\pi$)
Photon-Shot-Noise	$\Delta\Omega = \Omega_{\text{lim}} \cong 0.052 \, [°/h]$	$\Delta\Omega = \Omega_{\text{lim}} \cong 0.043 \, [°/h]$
Excess RIN	$\Delta\Omega = \Omega_{\text{lim}} \cong 0.235 \, [°/h]$	$\Delta\Omega = \Omega_{\text{lim}} \cong 0.015 \, [°/h]$
Full Noise = Photon-Shot-Noise+ Excess RIN	$\Delta\Omega = \Omega_{\text{lim}} \cong 0.239 \, [°/h]$	$\Delta\Omega = \Omega_{\text{lim}} \cong 0.050 \, [°/h]$

The noise associated with the fiber non-linear Kerr effect is based on the electro-optical phenomenon which consists in changes experienced by refractive index of the optical-fiber caused when it is excited by an optical wave that varies in amplitude. This occurs by the fluctuation of the optical power level of light source. In the case of the gyroscopic system, this optical power variation coupled to the fiber-coil causes changes on its refractive index, which results in a phase change in the optical wave propagated along the length of the optical fiber-coil. This change can be evaluated as a phase-equivalent-noise, and could be diminished efficiently using a low coherence light source (broadband source). Another important aspect is providing the light source with a thermal stabilization system to achieve a constant level of optical power emission.

The thermal Shupe effect is due to local temperature gradients along the fiber coil length. These temperature gradients induce phase changes in the optical waves traveling through the fiber. This effect can be minimized performing an appropriate winding of fiber-coil, so that a uniform temperature distribution is achieved throughout its entire length. The quadrupolar winding (number of turns in each layer of coil equal to an integer multiple of four) fulfills this condition. Other minor optical noise sources with less effect on the optical signal detected by photodetector are due to backscattering and reflections phenomena along the length of the sensing fiber-coil. A serious disadvantage for this model design is that the results of optical simulation do not allow realizing the evaluation of the main sources of optical noise. Only an average optical power and SNR values at the photodetector optical input can be obtained.

Regarding the electrical noise generated by the electronic circuits, the most important is white noise (thermal-noise or Johnson noise), which spreads equally over all the frequencies. An appropriate way for overcoming this noise source is performing a selective filtering at the frequency of the desired signal and fitting later the gain of the amplification stages to increase the electrical SNR at the output. In the case of the designed IFOG circuits, a strict design of LP Filter and BP Filter is necessary after photodetector-amplifier. It is crucial for the good performance of demodulation circuit and, therefore, the good linearity of the output response-curve of the designed gyroscope.

6. Conclusions

An IFOG prototype was theoretically designed by means of optical and electronic simulation tools. The conventional IFOG design with sinusoidal phase modulation is based on an open-loop configuration. The main innovation of IFOG designed here is the use of a simple closed-loop configuration with sinusoidal bias phase modulation. Its electro-optical system is realized by means of cost competitive optical and electronic components. Furthermore, the proposed design also allows reaching substantial progress in the stability and linearity of the Scale Factor (SF), dynamic range and threshold sensitivity of the gyroscope, compared to previous models proposed with the same fiber-optic coil length (L = 300 m). The cost advantage in the optical subsystem is obtained by means of optical wave depolarization by using two Lyot depolarizers, both realized in optical fiber. This allows using a sensing coil made of optical standard fiber, instead of a special polarization maintaining fiber, which is much more expensive. On the other hand, the electronic subsystems (detection, demodulation, bias and feedback phase-modulation) are based on conventional analog electronics, using classical components which are high precision and cost competitive, so that it also contributes to achieving a reasonable cost, and at the same time optimizing quality/price ratio of the final device. On the other hand, an interesting observation is that if the entire volume occupied by the device does not suppose a major restriction (this condition is fulfilled in certain applications), it is possible to get an additional saving costs by means of a particular optical subsystem design. This design can be based on a suitable selection of bulk optical components: The Integrated Optical Circuit (IOC) can be replaced with two 2 × 2 fiber optical couplers (SMF fiber), a fiber polarizer and a fiber-based electro-optic phase modulator (PZT), because until today the IOC is not a standard manufacturing item. In the same way, a SLD source light can be replaced by an alternative broadband source like an Erbium-Doped-Fiber-Amplifier (EDFA), and for optical-wave's depolarization a new solution based on bulk-optics can be adopted, as crystal Lyot depolarizers.

Supplementary Materials: Supplementary material are available online at http://www.mdpi.com/1424-8220/16/5/604/s1.

Acknowledgments: We are grateful to Oviedo University-Spain, specifically to Publication Department for the support that this institution has given us for full access to scientific articles published in international high impact factor magazines; also we are grateful to Computer Science Department for to have provided us the required access to software-resources.

Author Contributions: Ramón José Pérez made the designs, simulations and calculations; Ignacio Álvarez and José María Enguita supervised the work providing insights and design ideas and helped in the interpretation of the results; all the authors collaborated in the preparation of the manuscript.

Conflicts of Interest: The authors declare no conflict of interest.

Appendix A. Calculations

Contribution due only to photon-shot-noise (threshold sensitivity) before correction:

$$\Delta\Omega = \Omega_{\lim} \cong \left(\frac{hc^2}{\pi eqLDP_{\max}}\right)I_{sn} = \left[\frac{6.624\times10^{-34}\times(3\times10^8)^2}{\pi\times(1.6\times10^{-19})\times0.65\times300\times0.08\times(100\times10^{-6})}\right]\times3.312008\times10^{-12} = \tag{A1}$$

$$\approx 2.5180234\times10^{-7}\,[\text{rad/s}] = 2.5180234\times10^{-7}\times\left(\frac{180^\circ}{\pi}\right)\times\left(\frac{3600\,\text{s}}{1\,\text{h}}\right)\approx0.052\left(\tfrac{\circ}{h}\right)$$

Contribution due only to excess Relative-Intensity-Noise (RIN) before correction:

$$\Delta\Omega = \frac{2}{K_0}\sqrt{\frac{\lambda^2}{4c\Gamma t}} = \frac{1}{K_0}\frac{\lambda}{\sqrt{c\Gamma t}} \cong \frac{1}{0.3837}\frac{1310\times10^{-9}}{\sqrt{(3\times10^8)\times(30\times10^{-9})\times1}} \cong 1.1387\times10^{-6}\left[\frac{\text{rad}}{\text{s}}\right]\approx0.2349\left[\frac{\circ}{h}\right] \tag{A2}$$

Full contribution due to photon-shot-noise + excess RIN before correction:

$$\Delta\Omega = \frac{2}{K_0}\sqrt{\frac{e}{P_dRt}+\frac{\lambda^2}{4c\Gamma t}} = \frac{2}{0.3837}\sqrt{\frac{1.60\times10^{-19}}{(145.61\times10^{-6})\times0.68678\times1}+\frac{(1310\times10^{-9})^2}{4\times(3\times10^8)\times(30\times10^{-9})\times1}} = \tag{A3}$$

$$\cong 1.1567\times10^{-6}\left[\frac{\text{rad}}{\text{s}}\right]\approx0.2386\left[\frac{\circ}{h}\right]$$

Contribution due only to Photon-Shot-Noise (threshold sensitivity) after correction:

$$\Delta\Omega = \frac{2}{K_0}\sqrt{\frac{e}{P_dRt}} = \frac{2}{0.3837}\sqrt{\frac{1.60\times10^{-19}}{(145.61\times10^{-6})\times0.68678\times1}} \approx 0.2084941\times10^{-6}\left[\frac{\text{rad}}{\text{s}}\right]=0.043\left[\frac{\circ}{h}\right] \tag{A4}$$

Contribution due only to excess RIN after correction:

$$\Delta\Omega = \frac{2}{K_0}\sqrt{\frac{\lambda^2}{4c\Gamma t}}\times0.063 \cong 1.1387\times10^{-6}\left[\frac{\text{rad}}{\text{s}}\right]\times0.063\approx0.2349\times0,063\left[\frac{\circ}{h}\right]=0.015\left[\frac{\circ}{h}\right] \tag{A5}$$

Full contribution due to Photon-Shot-Noise + excess RIN after correction:

$$\Delta\Omega = \frac{2}{K_0}\sqrt{\frac{e}{P_dRt}+\frac{\lambda^2}{4c\Gamma t}} \cong \frac{2}{0.3837}\sqrt{\frac{1.60\times10^{-19}}{(145.61\times10^{-6})\times0.68678\times1}+\frac{(1310\times10^{-9})^2\times0.063^2}{4\times(3\times10^8)\times(30\times10^{-9})\times1}} = \tag{A6}$$

$$\cong 0.24267\times10^{-6}\left[\frac{\text{rad}}{\text{s}}\right]\approx0.0501\left[\frac{\circ}{h}\right]$$

References

1. Ashley, P.R.; Temmen, M.G.; Sanghadasa, M. Applications of SLDs in fiber optical gyroscopes. In Proceedings of the Test and Measurement Applications of Optoelectronic Devices 104, San Jose, CA, USA, 18 April 2002; pp. 104–115.
2. Burns, W.K.; Kersey, A.D. Fiber-optic Gyroscopes with Depolarized Light. *J. Light. Technol.* **1992**, *10*, 992–998. [CrossRef]
3. Szafraniec, B.; Sanders, G.A. Theory of polarization evolution in interferometric fiber-optic depolarized gyros. *J. Light. Technol.* **1999**, *17*, 579–590. [CrossRef]
4. Kintner, E.C. Polarization control in optical-fiber gyroscopes. *Opt. Lett.* **1981**, *6*, 154–156. [CrossRef] [PubMed]
5. Lefèvre, H.C.; Vatoux, S.; Papuchon, M.; Puech, C. Integrated optics: A practical solution for the fiber-optic gyroscope. In Proceedings of the Fiber Optic Gyros: 10th Anniversary Conference 101, Cambridge, MA, USA, 11 March 1987; pp. 101–112.
6. Kim, B.Y.; Lefèvre, H.C.; Bergh, R.A.; Shaw, H.J. Response of Fiber Gyros To Signals Introduced at the Second Harmonic of the Bias Modulation Frequency. In Proceedings of the Single Mode Optical Fibers 86, San Diego, CA, USA, 8 November 1983. [CrossRef]

7. Kim, B.Y.; Shaw, H.J. Gated phase-modulation feedback approach to fiber-optic gyroscopes. *Opt. Lett.* **1984**, *9*, 263–265. [CrossRef] [PubMed]

8. Kim, B.Y.; Shaw, H.J. Gated phase-modulation approach to fiber-optic gyroscope with linearized scale factor. *Opt. Lett.* **1984**, *9*, 375–377. [CrossRef] [PubMed]

9. Kim, B.Y.; Shaw, H.J. Phase reading, all-fiber-optic gyroscope. *Opt. Lett.* **1984**, *9*, 378–380. [CrossRef] [PubMed]

10. Böhm, K.; Petermann, K. Signal Processing Schemes for The Fiber-Optic Gyroscope. In Proceedings of the Fiber Optic Gyros: 10th Anniversary Conference 101, Cambridge, MA, USA, 11 March 1987. [CrossRef]

11. Moeller, R.P.; Burns, W.K.; Frigo, N.J. Open-loop output and scale-factor stability in a fiber-optic-gyroscope. *J. Light. Technol.* **1989**, *7*, 262–269. [CrossRef]

12. Ebberg, A.; Schiffner, G. Closed-loop fiber-optic gyroscope with a sawtooth phase-modulated feedback. *Opt. Lett.* **1985**, *10*, 300–302. [CrossRef] [PubMed]

13. Kay, C.J. Serrodyne modulator in a fibre-optic gyroscope. *IEEE Proc. J. Optoelectron.* **1985**, *132*, 259–264. [CrossRef]

14. Yahalom, R.; Moslehi, B.; Oblea, L.; Sotoudeh, V.; Ha, J.C. Low-cost, compact fiber-optic gyroscope for super-stable line-of-sight stabilization. In Proceedings of the IEEE/ION Position Location and Navigation Symposium (PLANS), Indian Wells, CA, USA, 4–6 May 2010; pp. 180–186.

15. Çelikel, O.; San, S.E. Establishment of all digital closed-loop interferometric fiber-optic-gyroscope and Scale factor comparison for open-loop and all digital closed-loop configurations. *IEEE J. Sens.* **2009**, *9*, 176–186. [CrossRef]

16. Sandoval-Romero, G.E.; Nikolaev, V.A. Límite de detección de un giroscopio de fibra óptica usando una fuente de radiación superluminiscente. *Rev. Mex. Fís.* **2002**, *49*, 155–165.

17. Medjadba, H.; Simohamed, L.M. Low-cost technique for improving open-loop fiber optic gyroscope scale factor linearity. In Proceedings of the International Conference on Information and Communication Technologies, Damascus, Syria, 24–28 April 2006; pp. 2057–2060.

18. Bennett, S.; Emge, S.R.; Dyott, R.B. Fiber Optic Gyros for Robotics. Available online: http://www-personal. acfr.usyd.edu.au/nebot/sensors/Fiber%20Optic%20Gyro/fog_robots.pdf (accessed on 15 October 2014).

19. Emge, S.; Bennet, S.M.; Dyot, R.B.; Brunner, J.; Allen, D.E. Reduced minimum configuration fiber optic gyro for land navigation applications. In Proceedings of the Fiber Optic Gyros: 20th Anniversary Conference, Denver, CO, USA, 4 August 1996. [CrossRef]

20. Bennett, S.M.; Emge, S.; Dyott, R.B. Fiber optic gyroscopes for vehicular use. In Proceedings of the IEEE Conference on Intelligent Transportation System (ITSC'97), Boston, MA, USA, 9–12 November 1997; pp. 1053–1057.

Article

Damage Detection Based on Power Dissipation Measured with PZT Sensors through the Combination of Electro-Mechanical Impedances and Guided Waves

Enrique Sevillano *,†, Rui Sun † and Ricardo Perera †

Department of Structural Mechanics, Technical University of Madrid, C/ José Gutiérrez Abascal, 2,
Madrid 28006, Spain; rui.sun@alumnos.upm.es (R.S.); ricardo.perera@upm.es (R.P.)
* Correspondence: enrique.sevillano@upm.es; Tel.: +34-913-363-278; Fax: +34-913-363-004
† These authors contributed equally to this work.

Academic Editor: Gonzalo Pajares Martinsanz
Received: 21 March 2016; Accepted: 29 April 2016; Published: 5 May 2016

Abstract: The use of piezoelectric ceramic transducers (such as Lead-Zirconate-Titanate—PZT) has become more and more widespread for Structural Health Monitoring (SHM) applications. Among all the techniques that are based on this smart sensing solution, guided waves and electro-mechanical impedance techniques have found wider acceptance, and so more studies and experimental works can be found containing these applications. However, even though these two techniques can be considered as complementary to each other, little work can be found focused on the combination of them in order to define a new and integrated damage detection procedure. In this work, this combination of techniques has been studied by proposing a new integrated damage indicator based on Electro-Mechanical Power Dissipation (EMPD). The applicability of this proposed technique has been tested through different experimental tests, with both lab-scale and real-scale structures.

Keywords: structural health monitoring; PZT sensors; electro-mechanical impedance; guided wave; electro-mechanical power dissipation; non-destructive testing

---

1. Introduction

A vast number of aerospace, civil, and mechanical infrastructures usually work under very demanding conditions, continuously subjected to both static and dynamic loads and frequently in severe environments, which can lead to either a gradual deterioration of the structure or its sudden failure. For that reason, Structural Health Monitoring (SHM) technologies have received increasing attention in recent years. Many authors have focused their research activities towards the development of these damage detection methods. In general terms, all these methods are based on the definition of different strategies that allow the structural health to be assessed with the purpose of obtaining a quantification and, if possible, the location of any damage present in the structure [1]. In order to achieve this goal, the output responses of the structure to a given input excitation are analyzed before and after damage, and therefore the definition of a baseline of the structure is first required. After collecting all the corresponding measurements, each of the damaged state measurements is compared with the defined baseline, so that every modification of the expected response of the structure can be translated into a damage indication.

Many different SHM techniques have been used over the years, based on either global or local monitoring methods [2]. Although all global approaches are usually successfully applied and well accepted [3–5], they are based on the lowest modes of vibration, and so are not appropriate for monitoring reduced sensing regions, which is necessary to develop an accurate and efficient early detection methodology. In this sense, a whole set of local methods has been proposed in the literature [6], many of them sharing the same drawbacks pointed out by Balla *et al.* [2], and for that reason all efforts have been placed lately on the development of "smart structures" [7,8]. Among all the proposed technologies and smart sensing solutions, transducers based on piezoelectric ceramic materials (PZT sensors in particular) are the most promising ones [9]. These materials develop an electric response over their surfaces when a certain mechanical stress is directly applied on them, according to the direct piezoelectric effect; but they are also capable of developing the reverse phenomenon, producing mechanical stresses when an electric field crosses through the piezoelectric material, according to the inverse piezoelectric effect [2]. This capability enables the material to be used both as a sensor and as an actuator simultaneously [10]. By using these smart sensors, different local damage detection techniques have been proposed, although the two most widely developed ones are those based on electro-mechanical impedance measurements and those based on guided waves.

Damage detection based on electro-mechanical impedance measurements was first developed by Liang *et al.* [11]. The electrical impedance of the PZT can be directly related to the mechanical impedance of the host structural component where the PZT transducers are attached. Since the structural mechanical impedance will be affected by the presence of structural damage, any change in the electrical impedance spectra might be an indication of a change in structural integrity [12–14]. On the other hand, guided waves were used for non-destructive testing (NDT) purposes for the first time by Woriton in [15]. Typically, an array of sensors is attached to the structure and used to inject a pulse of guided-wave energy. The same array is then used to record the reflected signals, and the position of these signals in the time domain can be easily related to the position of features (e.g., flanges, welds) in the structure. Any signal that cannot be related to a known feature is assumed to be a defect. As the number of features becomes higher, their associated signals merge together obscuring the reflected signals from damage. A good review of different applications of this technique can be found in [16].

Hence, these two techniques have been widely proven to have a great potential for SHM purposes. However, few new approaches have been proposed so far [17,18], at least to the knowledge of the authors, that combine the advantages of each of them. Furthermore, more complicated structures need to be tested. In that sense, a new approach is proposed in this paper in order to combine impedance signatures with guided waves to evaluate the presence of different typologies of damage in different structures. A new damage indicator, Electro-Mechanical Power Dissipation (EMPD), is firstly introduced in this paper, and then tested using two controlled lab-scale experiments for it as well as a real-scale complex structure.

2. Electro-Mechanical Power Dissipation as a New Damage Indicator

2.1. Basis of Piezoelectric Sensing for SHM

Lead-zirconate-titanate (PZT) materials have been subject to increasing interest in recent years due to their light weight and variety of shapes and sizes [10,19–21], besides their wide range of applications in many research fields. In particular, they provide very promising sensing and monitoring solutions, and so these smart materials have been widely applied for Structural Health Monitoring purposes. With these smart sensors, the electrical impedance can be measured at high frequency ranges so that the wavelength of the excited motion is small and sensitive enough to detect local damage [13,22]. This constitutes the basis of the so-called Electro-Mechanical Impedance Method (EMI), whose core idea is that the presence of damage in the structure under study will affect its mechanical properties and, in turn, the electro-mechanical properties of the PZT patch, which can be directly measured by means of an impedance analyzer. The coupled relationship between the electrical and mechanical impedances was first introduced by Liang *et al.* [11] as follows:

$$Y(\omega) = \frac{I_0}{V_i} = j\omega a \left[\bar{\varepsilon}_{33}^T - \frac{Z_s(\omega)}{Z_s(\omega) + Z_a(\omega)} d_{3x}^2 \hat{Y}_{xx}^E \right] \tag{1}$$

where $Y(\omega)$ is the electrical admittance (inverse of impedance), V_i is the input voltage to the PZT actuator; I_0 the output current from the PZT; a, $\bar{\varepsilon}_{33}^T$, d_{3x}^2, and $\hat{Y}_{xx}^E$ are the geometry constant, complex dielectric constant, piezoelectric coupling constant, and complex Young's modulus of the PZT in a state without stresses, respectively; $Z_s(\omega)$ and $Z_a(\omega)$ are the impedances of the structure and the PZT actuator, respectively. As has been proved (Park *et al.* [23]), the real part is less sensitive to ambient temperature change, compared to the imaginary part [18]. Furthermore, the imaginary part is mostly related with the capacitance of the piezoelectric transducer. This makes it harder to extract the mechanical data from the imaginary component, making it less suitable for damage identification purposes. Because of this, the real part of the impedance is usually used for the EMI method, and so this has been done in this paper.

However, piezoelectric materials offer wider possibilities rather than those related to the EMI method. Among all of them, guided elastic waves are increasingly used in SHM applications [24]. These waves are called guided waves because the propagation of the elastic wave is confined by the boundaries of the structure itself [25–27], as will be the case of the experiments carried out in this work. Because of the confined nature of these waves, it has been demonstrated that they can travel relatively long distances through the material, achieving a long sensing range for some damage detection applications [18,28,29]. Hence, this technique allows a wider sensing region than the EMI method. However, in most of applications found in the literature, the guided waves technique is often limited to simple structures [10,18,26], basically lab-scale structures made of aluminium and steel, or some composite structures in the most complex cases. In this work, more complex structures will be studied in order to explore the capabilities of this technique.

More details about this technique can be found in [24,27].

2.2. Definition of Electro-Mechanical Power Dissipation

One of the most interesting features of PZT sensors is that they constitute the key element of two of the most studied and successful NDT techniques nowadays: the EMI method and guided waves. However, even though these two methodologies have demonstrated great performance, few efforts have been made so far, at least to the knowledge of the authors, in the development of an integrated method that can combine both impedance signatures and guided waves in order to take advantage of the whole potential of these smart sensors. One of the few studies addressing this challenge is found in [18], where An *et al.* proposed an integrated damage diagnosis procedure based on those two different signals. However, in that study, impedance signatures and guided waves were treated separately so that two different damage indicators were obtained, both of them being weighted and accordingly combined afterwards in order to achieve a unique and integrated damage indicator. In this sense, the work presented in this paper aims to achieve a more integrated combination of the physical properties of both kinds of signals into a single one. By doing this, the necessity of weighting different damage indicators would be avoided, and thus the damage identification might be done with less *a priori* information.

To perform an advantageous combination of the information yielded by impedance signatures and guided waves, a reinterpretation of both should be carried out. On the one hand, the impedance curves used in this work will provide the evolution of the real part of the coupled electro-mechanical impedance of both the sensor and the host structure within a frequency range (in this case, from 10 to 100 kHz) and in a small area around each PZT sensor. On the other hand, with the guided waves, a signal in time domain, with a particular frequency and maximum amplitude, will be sent from one transmitter sensor to a receiver sensor, measuring the distortion between the voltage sent and received. In a simplistic way, an analogy can be established between the system PZT-host structure and a simple electrical system by means of including the impedance and voltage data measured in the traditional Ohm's Law, and assuming as negligible all the possible mathematical or physical inaccuracies of this approach. By doing this, and by assuming that no information about the current is available for any of the experiments carried out in this work (only voltage and electro-mechanical impedances are measured), the plane definition of power can be rewritten as follows

$$P = V \cdot I = \frac{V_m^2}{R} = \frac{V_m\,(V_e,t)^2}{R} \tag{2}$$

where $V_m\,(V_e,t)$ is the instantaneous voltage of the guided wave measured at the second sensor, which depends on the time as well as on the voltage of excitation V_e, both real values at a particular frequency ω_{gw}; R represents just the real part of the ideal impedance (resistance). The dependence between V_m and V_e is just considered in this work regarding the amplitude and shape of the guided waves. Hence, if any of these parameters were changed for the voltage of excitation, it would have a direct impact on the corresponding parameters of the measured voltage, which would happen within a given time frame.

From Equation (2), an instantaneous Electro-Mechanical Power Dissipation (EMPD) is defined from the impedance as follows

$$\text{EMPD}\,(t) = Re\left(\text{ifft}\left(\frac{\text{fft}\,(V_m\,(V_e,t))^2}{Re\left(\overline{Z}\,(\omega)|_{\omega_{gw}} \right)} \right) \right) \tag{3}$$

where $\overline{Z}\,(\omega)$ is the average of the electro-mechanical impedances measured at the frequency ω_{gw} obtained from the PZT transmitter and receiver sensors; Re indicates that only the real part of the electro-mechanical impedance is considered, fft denotes the Fast Fourier Transform, and ifft the Inverse Fast Fourier Transform. As the Inverse Fast Fourier Transform provides a set of complex values, just the real part of the EMPD will be used in this damage detection procedure.

At this point, it is worth clarifying that several assumptions have been made in order to define this time-dependent variable (EMPD):

i   The guided waves used in this work are sent and received at a particular frequency, while the impedance signatures are obtained by performing a sweep in the frequency domain. This implies that, after expressing both signals in the frequency domain, the electro-mechanical impedance will show a different amplitude at each frequency. However, after applying the corresponding filter, just one significant non-zero amplitude will be found for the measured voltage in the frequency domain. For this reason, just the corresponding value at this frequency is taken from the electro-mechanical signal.

ii   As it is not the purpose of this paper to provide a rigorous mathematical explanation about the analogy between the EMPD and the one measured on a real electric circuit, the authors have not considered any phase dependency between the voltage and the current, which has been possible due to the fact that absolute magnitudes have been used in Equations (2) and (3).

A similar approach, although in a very different application, can be found at [30], where power dissipation is determined by measuring all the changes in electric impedances that a dither piezo presents to an oscillator. Although the theory behind the approach presented in [30] is certainly more complex, it may be of interest to those readers wishing for a more detailed and rigorous theoretical explanation.

2.3. Damage Detection Procedure

The overall procedure of the proposed integrated damage detection technique is illustrated in Figure 1. This process starts by measuring, separately, the impedance signatures and the guided waves for a baseline stage of the structure, if possible, in a healthy state. Thanks to the combination of both signals at a particular frequency, it is then possible to obtain EMPD in the time domain for different damage conditions or stages. All possible deviations in the EMPD signal, when compared with the baseline stage, are finally interpreted as a damage increment through the definition of a damage indicator. The detailed procedure of this proposed techniques can be summarized as follows:

(a) Collection of baseline impedance signals and guided waves. In the case of the electro-mechanical impedance, one signature is obtained per PZT sensor embedded in the host structure. In order to enhance the accuracy of the process, several measurements can be taken in order to work in the subsequent stages with the average of all those measurements. Once the impedance signatures are stored, guided waves will be measured, obtaining two different signals between every two sensors: one sent from the first sensor and measured at the second one, and vice versa.

(b) All experimental procedures imply the presence of environmental noise, and given that these sets of measurements are taken at high frequencies, the following step is to clean each signal from undesired noises. The influence of the noise on the electro-mechanical impedance is already minimized thanks to the impedance analyzer used, which employs a measuring technique that calculates the average between a number of neighbouring sample points, but a bandpass filter (which depends on the particular application) is needed for the guided wave signals.

(c) To compute EMPD, the Fast Fourier Transform (FFT) is applied to all guided wave signals in order to obtain their corresponding values in the frequency domain. As each of these signals is sent and measured at a particular frequency, a single point would be expected in the frequency domain. However, there is still some negligible noise present after the filtering process, which is going to reveal some more frequencies after the FFT is applied. The corresponding value of the electro-mechanical impedance at that frequency must be selected, so that the Electro-Mechanical Power can be defined using an extension of Equation (2). As only one impedance signature is available per sensor, the value considered in this case has been the average of the impedances measured at the corresponding two consecutive sensors used to measure each guided wave.

(d) The Inverse Fast Fourier Transform (IFFT) is applied in order to obtain the expression of the EMPD in time domain. The signal obtained in this step will be used as a baseline in the subsequent iteration, so only the appearance of new damage with respect to the baseline stage should be reflected by the damage index.

(e) Repeat steps (a) to (d) for each of the studied damage cases of each experiment, so that the corresponding EMPD signatures can be obtained for each health condition of the structure.

(f) Once the EMPD is obtained for each of the damage scenarios, the following step is to define an appropriate damage indicator that allows the analysis of the health condition of the corresponding structure. The root mean square deviation (RMSD) is the most commonly used indicator to assess damage [20,31,32], and is computed from the difference in the EMPD value at each timepoint as

$$RMSD\,(\%) = \sqrt{\frac{\sum_{i=1}^{n}[EMPD_0\,(t_i) - EMPD_1\,(t_i)]^2}{\sum_{i=1}^{n}EMPD_0\,(t_i)^2}} \cdot 100 \qquad (4)$$

where $EMPD_0\,(t_i)$ is the EM power dissipation of the PZT measured at a previous stage, which might agree with the healthy condition of the structure; $EMPD_1\,(t_i)$ is the corresponding value at a subsequent stage, which might agree with a post-damage stage, at the *ith* timepoint; n is the number of timepoints. For the RMSD index, the larger the difference between the baseline reading and the subsequent reading, the greater the value of the index denoting changes in structural dynamic properties which can be due to damage.

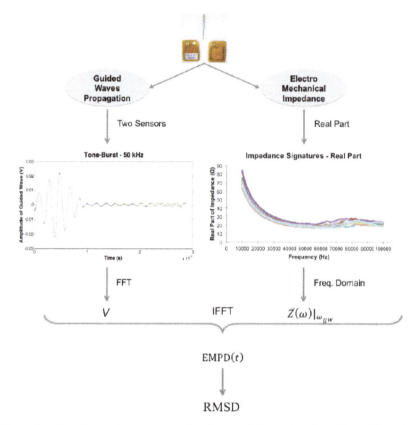

Figure 1. Overall procedure to extract the Damage Indicator based on Electro-Mechanical Power Dissipation.

3. Experimental Study for Lab-Scale Structures

In order to check the proposed methodology, some previous experimental tests were carried out. These included two lab-scale structures such as a simple bolt-jointed aluminium beam and an FRP-strengthened concrete specimen. These two tests were considered as the basis for a third experimental study, which consisted in the identification of debonding on a full-scale reinforced concrete beam strengthened with an FRP strip.

At this point, it is important to remark on the influence of the environmental conditions on the measurements taken to build the EMPD signals. Regarding the impedance signatures, some slight variations were observed between consecutive measurements, as many authors have reported in the literature referenced in this work. In order to prevent this variability affecting the final damage predictions, averaging has been made in all the experiments performed in this work, following the recommendations found in the literature. This minimizes the influence of these variations or deviations in the whole damage prediction procedure. Exactly the same problem was observed when measuring guided waves at each sensor. Once again, averaging was the adopted solution, as explained below at each of the experiments. However, in this case, some outliers were also found for some measured signals. As the number of outliers found were small enough (no more than five times in the whole procedure), and as each signal was checked *in situ* after measured, the authors decided to neglect these signals and just take new measurements.

On the other hand, temperature changes might affect measured characteristics, as some authors have addressed in the literature [18,33,34]. However, most of these studies including temperature effects allow large temperature range variations, e.g., between −20 °C and 40 °C [18], which is not the intention of the work presented in this paper. Each of the lab-scale experiments were performed in approximately the same conditions, which were in a room maintained at 24°C during the whole experiment. The authors estimate that temperature variations no larger than 2 °C could be found in this room. Regarding the experiment real-scale specimen, the same conditions were maintained, this time with a temperature of 27 °C and with the same estimated temperature deviations. For this reason, the authors did not consider temperature effects as a source of uncertainty for the results presented in this section.

3.1. Bolt Loosening Detection in an Aluminium Lap Joint

The first test was carried out on a lab-scale bolt jointed aluminium specimen consisting of two beams, the first one with dimensions of $45 \times 5 \times 0.5$ cm^3 and the second one with dimensions of $27 \times 5 \times 0.5$ cm^3, connected by four bolts with a diameter of 10 mm (Figure 2). Two identical P-876 DuraAct Patch Transducers [35] were bonded to both sides of the lap joint at a distance of 20 cm between them by using an epoxy adhesive. The purpose of this test was to evaluate the ability of the proposed approach to detect the damage due to loose bolts by inducing the following four stages or damage scenarios in the analyzed specimen: (a) No damage (D0), for which five different and independent measures were taken and used as raw data (D1 to D5); (b) Damage 1 (D1): Loose bolt #3 by one half-cycle; (c) Damage 2 (D2): Loose bolt #3 by another half-cycle; (d) Damage 3 (D3): Loose bolt #2 by one-half-cycle. It is important to remark here that all the necessary measurements were taken one after another for each subsequent damage condition. In this way, for the D0 condition, both impedances and guided waves were initially measured. Once these measurements were collected, damage D1 was induced and the new stage was measured and so on.

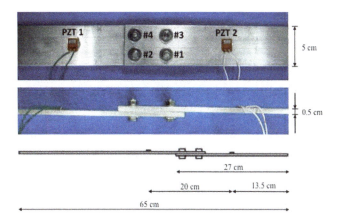

Figure 2. Bolt-jointed aluminium beam with two PZT transducers.

The two PZT sensors were individually connected to two different channels of the 3499B multiplexor, which was used to make a multiple connection between the HP 4192A Impedance Analyzer and the PZT sensors. Both, the multiplexor and the impedance analyzer were from Agilent Technologies. To measure the impedances at each sensor, a sinusoidal sweep voltage with a 1 volt amplitude was applied to the PZT sensors over a frequency range of between 10 kHz and 100 kHz. Five impedance signatures were measured for each damage case, so the raw data were obtained by averaging the responses of five measurements. Figure 3 shows the overall experimental setup for the impedance measurements.

Figure 3. Experimental setup for impedance measurements on the bolt-jointed aluminium beam.

Once all the impedance data were collected at each damage step, each sensor was afterwards and alternatively connected to the guided waves system, which consists on the following components: a waveform generator that sends the selected waveform to the structure in the form of an input excitation on the first sensor, an oscilloscope to register the response of the structure on the second sensor, and a conventional laptop controlling both of them (Figure 4). Both the waveform generator and the oscilloscope were acquired from Agilent Technologies. In this test, the same 10-cycle tone-burst was sent in all cases at 50 kHz with a peak amplitude of 10 volts, and five time signals were measured and averaged. Finally, a bandpass filter with 1 kHz and 500 kHz low and high cutoff frequencies was used to improve the quality of the guided signals.

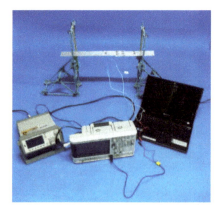

Figure 4. Experimental setup for guided waves measurements on the bolt-jointed aluminium beam.

The frequency used for the guided waves, which is the same as that used to generate the EMPD signatures, has been selected based on previous results, both published by the authors of this paper and found in the literature as well. Usually, guided waves work much better with high frequencies, even higher than those used by the EMI method [18], but it has also been shown in the literature that the amplitude of the impedance signatures decreases with the frequency. Thus, a trade-off must be found so that both guided waves and impedance signatures can be equally significant in the calculation of the EMPD, and so 50 kHz was set in order to test the methodology proposed in this work.

Figures 5 and 6 show, respectively, the EMPD obtained when the guided tone-burst is sent from sensor 1 to sensor 2, and vice versa. As can be immediately verified from these figures, there is a visible variation in the EMPD every time new damage is induced in the bolt-jointed aluminium beam, which indicates that the EMPD is sensitive to the presence of, at least, this kind of damage in this sort of lab-scale structures. Furthermore, this variation in the EMPD, which appears to be affecting the amplitude of the measured signal, is not exactly the same in both figures, since it slightly depends on the wave direction. This conclusion might not be too obvious to extract from these figures, but it is quite easy to check by a simple comparison between the corresponding arrays of output data at each damage level. An example of this is shown in Figure 7 for damage D0 and the wave travelling in both directions. Nevertheless, further information might be extracted if Equation (4) is used to translate these data into a damage indicator.

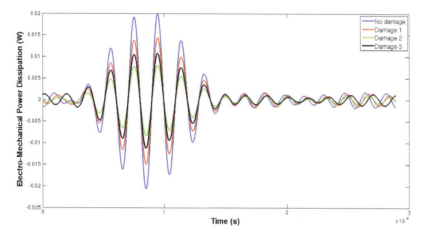

Figure 5. EMPD at 50 kHz. Tone-burst sent from sensor 1 to sensor 2.

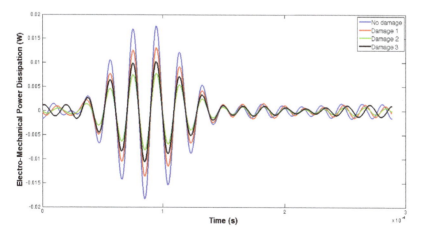

Figure 6. EMPD at 50 kHz. Tone-burst sent from sensor 2 to sensor 1.

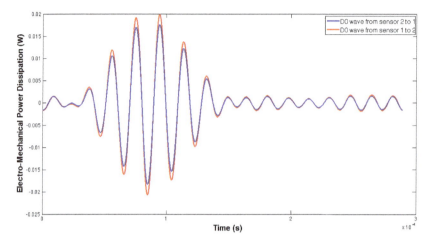

Figure 7. Differences in the EMPD in the aluminium lap-joint for damage D0 when the wave travels opposite directions.

Figure 8 shows the computed values of the RMSD corresponding to the two previous figures, one value per damage stage and wave direction. As commented above, this figure shows a clear sensitivity to the appearance of damage in all cases. Nevertheless, a clear difference cannot be found between single and multiple damage scenarios. In the two single damage scenarios, D1 and D2, the RMSD is higher when the tone-burst travels towards sensor 2 (1.78% higher for D1 and 3.7% higher for D2, where damage is more severe), and this sensor turns out to be the one closer to the induced damage (bolt #3). From this statement it is tempting to extract the general conclusion that the variation in the EMPD will be higher when the damage is closer to the sensor in which the tone-burst is received, and this might work in that way for all single damage scenarios, but it is certainly not true when addressing the case of a multiple damage scenario (D3). In this case, even though there is new damage induced close to sensor 1 (bolt #2), the RMSD calculated in the other direction is still higher, which clearly contradicts the conclusions from the single damage scenario. However, this might make sense given that the damage severity at bolt #3 is higher than the new damage at bolt #2. Also, considering the proximity between the two bolts makes the process more difficult.

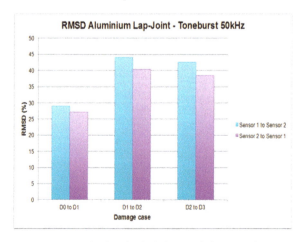

Figure 8. RMDS values for the bolt-jointed aluminium beam.

By means of this comparison, it can be stated that the EMPD is more sensitive to the appearance of new damages (D2 and D3) than the traditional EMI, given that the RMSD values are significantly lower after the first damage scenario when using the EMI method instead of the EMPD, even though additional damage was induced in the specimen. Therefore, it is possible to obtain a better estimation of the severity of the total damage in the specimen by using the EMPD signatures, making them more sensitive to the presence of damage. However, attending to the results after damages D1 and D2, it has to be pointed out that the estimation of damage localization is not really improved with respect to the EMI method, particularly in the case of damage D2, since the differences in the RMSD values between sensors are larger for the EMI method. On the other hand, both methods give an incorrect localization for damage D3, which is the most severe and complex in the specimen.

These damage predictions can be now compared to the ones obtained from the traditional EMI method, which are shown in Figure 9.

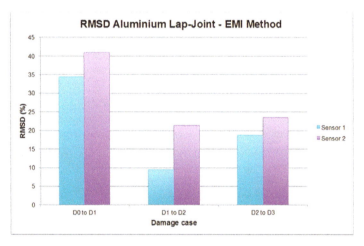

Figure 9. RMDS values for the bolt-jointed aluminium beam corresponding to the EMI method.

Thus, with this first experiment, the sensitivity of the EMPD to the presence of damage, in both single and multiple scenarios, has been demonstrated to be higher than with the EMI method. Furthermore, the use of the EMPD seems to be promising when localizing damage in single damage scenarios, although accurate damage localization is apparently not possible when multiple damages are induced in the structure. Actually, at least in this experiment, the localization does not improve the results obtained with the EMI method. In order to clarify this point and to test the real potential of the proposed methodology, more experiments have been carried out in this work.

3.2. Induced Debonding on an FRP-Strengthened Concrete Specimen

The second test was carried out over a concrete specimen with dimensions $31.3 \times 9.5 \times 7.5$ cm^3 which was strengthened with an external FRP strip, with dimensions $29.5 \times 5 \times 0.18$ cm^3, bonded by using an epoxy resin adhesive. Three identical P-876 DuraAct Patch Transducers from Piceramics [35] with the thickness of 0.5 mm were symmetrically bonded along the length of the specimen using the same epoxy adhesive as before (Figure 10). The overall experimental setups for the measurements are the same as the ones shown in Figures 3 and 4 and the same measurement conditions were imposed on both the impedance signatures and the guided waves.

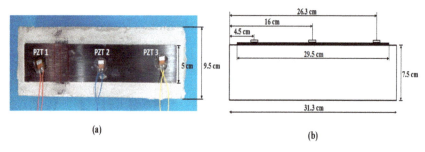

(a)   (b)

Figure 10. FRP-strengthened concrete specimen with three PZT transducers.

The purpose of this final test was to simulate, over a lab-scale concrete specimen, the debonding failure mode that usually comes up in real structures strengthened with FRP strips. In order to detect that failure through the application of the proposed method, the following four stages or damage scenarios were induced (all by means of holes drilled in the resin adhesive) in the analyzed specimen: (a) No damage (D0); (b) Damage 1 (D1): a 5 mm debonding located 7.25 cm away from the left end of the specimen; (c) Damage 2 (D2): amplification of the debonding practised in the previous stage towards the PZT number 2, up to a total debonding length of 1 cm; (d) Damage 3 (D3): repetition of the previous step up to a total debonding length of 2.5 cm (Figures 11 and 12).

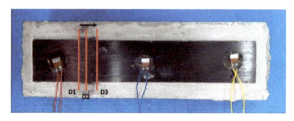

Figure 11. Damage scenario for an FRP-strengthened concrete specimen (I).

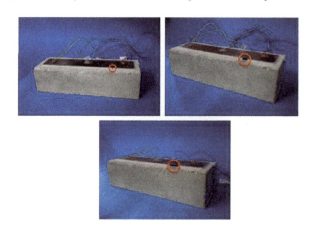

Figure 12. Damage scenario for an FRP-strengthened concrete specimen (II).

In this case, both the structural properties and boundaries, as well as the nature of the damage, are totally different from those used in the previous example, so different EMPD results are expected. These differences can actually be found in Figures 13 and 14 where not only the shape of the EMPD is different from the previous example, but also the EMPD values depending on the wave direction, which is much clearer than in the previous example (Figure 15).

As in the previous example, these differences encountered in the EMPD can be evaluated though the RMSD index in order to obtain conclusions about damage presence in the concrete specimen. These values are, again, calculated once per damage stage and wave direction, as shown in Figure 16.

In this test, a continuous damage scenario is studied. It cannot be considered either a single or multiple damage scenario in the same sense as in the previous example, since it is not possible to specify the location of each damage at a particular position. From the initial location of damage (case D1) defined at the concrete-FRP interface, an extension towards sensor 2 was practised for the other two damage cases, D2 and D3, with the purpose of increasing the severity and extension of the damage. As seen in Figure 16, the RMSD for D1 is higher when the guided wave travels towards sensor 1 (49.84% *vs.* 18.87%), which means, according to the conclusions extracted in the previous example, that the damage D1 is close to this sensor, as it actually is (Figure 11). Furthermore, from the noticeable difference in the RMSD values for this case, given that if the guided wave travels in the opposite direction the value is 30.97% lower, it is reasonable to say that the distance between the damage position and sensor 1 is much lower than that between the damage and sensor 2, which turns out to be true.

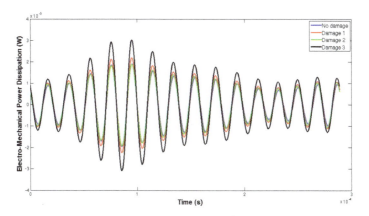

Figure 13. EMPD at 50 kHz. Tone-burst sent from sensor 1 to sensor 2.

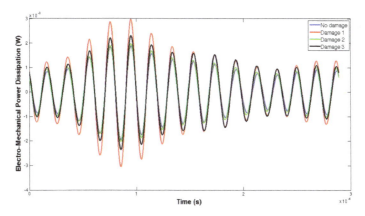

Figure 14. EMPD at 50 kHz. Tone-burst sent from sensor 2 to sensor 1.

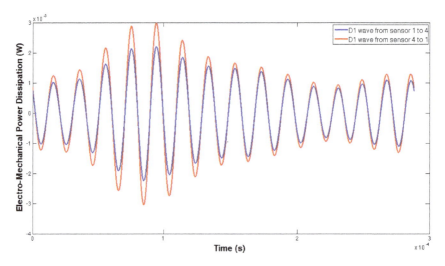

Figure 15. Differences in the EMPD in the concrete specimen for damage D1 when the wave travels opposite directions.

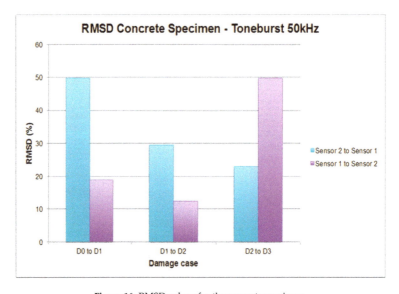

Figure 16. RMSD values for the concrete specimen.

For the D2 damage case, whose severity is higher than in the D1 scenario, again the RMSD is higher when the guided wave travels towards sensor 1, with a difference of 17.08% this time. According to this, it is again concluded that the damage is located closer to sensor 1. However, the RMSD value is quite smaller than that corresponding to D1, when a similar value might be expected. This reduction, from the experience of the authors, is quite normal given that the first hole made in the adhesive implies bigger alterations of the mechanical properties of the specimen than those induced by the second one. In a real-scale specimen, a hole of those characteristics would be more than sufficient to initiate the failure mechanism due to debonding propagation. Furthermore, it should not be forgotten that the RMSD values are computed between two consecutive stages. Therefore, this parameter should only reflect any change between one stage and the consecutive one, so consistently higher RMSD values cannot be expected for the same sensor at different damage stages, since damage is being extended along the specimen in the opposite direction (from sensor 1 to sensor 2 in this case).

Finally, in the D3 damage case, the RMSD values are much higher when the EMPD is calculated from the measurements at sensor 2, which, according to the reasoning proposed in this paper, would erroneously mean that this damage is much closer to sensor 2 this time, given that the maximum extent of this hole reached the midpoint between the two sensors. Nevertheless, considering just the RMSD values calculated when the guided wave travels towards sensor 1, it is quite clear that the value corresponding to each of the damage cases is lower each time, which suggests that the new debonding between the concrete and the FRP is further each time from this sensor, as can be verified from Figure 12. Taking into account again the expression defined in Equation (4) for the RMSD indicator, the values computed between D2 and D3 stages should only capture the new debonding practiced at the interface from the D2 stage.

By making a comparison, as in the previous example, with the RMSD obtained by applying the traditional EMI method, shown in Figure 17, it is even clearer in this case that the proposed methodology is quite more sensitive to damage than the traditional EMI method. Furthermore, it might be true that the proposed methodology fails in the localization of damage D3, but the values of the RMSD shown in Figure 16 are more representative of the real damage present in the concrete specimen after inducing D3, which is actually the biggest damage induced in this example (see Figure 12). In comparison to the damage found by the EMPD, the values obtained with the traditional EMI can be considered negligible, so even the conclusions about damage location cannot be taken into account. In terms of RMSD differences between sensors 1 and 2 at each damage stage, it is clear as well that damage conclusions are much more evident by using the EMPD.

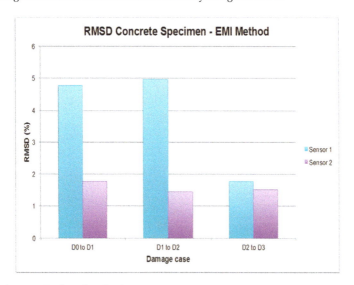

Figure 17. RMSD values for the concrete specimen corresponding to the EMI method.

Again, this methodology shows promising results when dealing with concentrated damages (D1), even despite the complexity of both the structural system and the nature of the damage. However, the location of the damage is not equally good when the extent of damage increases (in the case of the debonding, although accurate results have been presented for D2), or when it is affecting different sensing regions along the beam (bolt-jointed beam), although those are scenarios that should not be reached in any real structure if a good early detection system is properly developed and deployed. Nevertheless, the sensitivity of the EMPD to damage appearance is clearly demonstrated, even for more complex specimens.

4. Debonding Assessment in a Real-Scale RC Beam Externally Strengthened with FRP

A third experimental study was carried out to test the feasibility of the proposed approach when applied to a real-scale complex structure, such as that corresponding to a reinforced concrete beam externally strengthened with an FRP strip. The failure mode of this kind of structure is critical because it is usually due to the sudden and brittle debonding of the FRP reinforcement originating from an intermediate flexural crack [36–38]. Detection of debonding in its initial stage is essential to prevent future failure, which might be catastrophic.

4.1. Test Description

The geometric dimensions and the reinforcement layout in the sections are illustrated in Figure 18. The identified material properties are: (a) for concrete, the elastic moduli, compressive strength, mass density, and Poisson coefficient were taken to be $E = 24,858$ MPa, $f_c = 24.64$ MPa, $\rho = 2350$ kg/m^3, and $\upsilon = 0.2$, respectively; (b) for steel reinforcement, the elastic moduli, elastic limit, mass density, and Poisson coefficient were taken to be $E = 210,000$ MPa, $f_y = 510$ MPa, $\rho = 7850$ kg/m^3 and $\upsilon = 0.3$, respectively; (c) for FRP external reinforcement, the elastic moduli and Poisson coefficient were taken to be $E = 150,000$ MPa and $\upsilon = 0.35$, respectively, with a thickness of $e_{FRP} = 1.4$ mm; (d) for adhesive, the shear moduli was assumed to be $G = 4300$ MPa and the thickness $e_{ad} = 3.45$ mm.

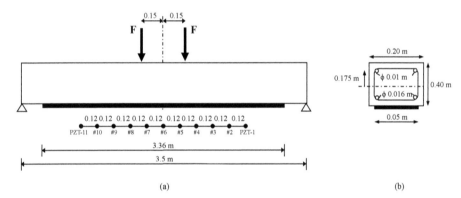

Figure 18. (a) Geometry, loading scheme, and sensor location map for the RC beam; (b) Cross-section of the beam.

In the test programme performed, the strengthened beam was subjected to a series of four-point increasing static load tests with the purpose of gradually introducing the cracks into the specimen. After each static test, the impedance was measured by using 11 identical P-876 DuraAct Patch Transducers [35] of 0.5 mm thickness which were externally bonded with an epoxy adhesive along the FRP strip with a constant spacing of 12 cm (Figure 18). For this, both the impedance signatures and the guided waves were measured under the same conditions explained for the two previous examples. After that, four different loading stages were considered by applying the static loads of 26 (stage D1), 40 (stage D2), 65 (stage D3), and 100 kN (stage D4). Before measuring both the impedances and guided waves, the beam was previously unloaded after each loading stage. The experimental set-up is shown in Figure 19.

Figure 19. Experimental setup for the reinforced concrete beam externally strengthened with an FRP strip.

4.2. Results and Discussion

The selected guided wave was sent between every two consecutive working sensors, and the EMPD values obtained from all those measurements are shown in Figures 20–27. As can be easily verified, the EMPD presents different shapes depending on the sensing region, on the damage scenario, as well as on the direction of motion of the guided waves, as in the two previous examples, which comes to confirm the high sensitivity of this new indicator to the presence of damage, even when dealing with structural systems of such complexity as the one addressed in this section. Furthermore, the differences in the EMPD depending on the direction of the tone-burst is clearer in this case than in the previous two examples, for almost all damage cases.

Each of the graphics shown in Figure 20 demonstrates the presence of the series of damage stages generated in the RC strengthened beam, and it is even possible to discriminate which one could have a higher impact on each sensing region. However, this is just a qualitative analysis of a really complex damage scenario, as shown in Figure 28. Thus, a better understanding will be achieved through the analysis of the RMSD values, as in the previous examples. These values are presented in Figure 29.

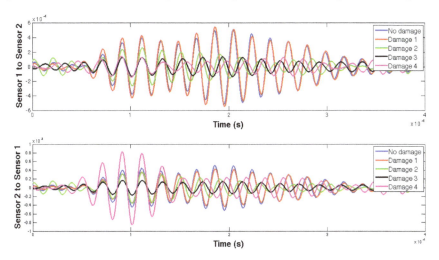

Figure 20. EMPD at 50 kHz. Tone-burst sent and measured between sensors 1 and 2.

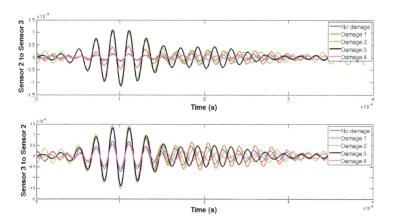

Figure 21. EMPD at 50 kHz. Tone-burst sent and measured between sensors 2 and 3.

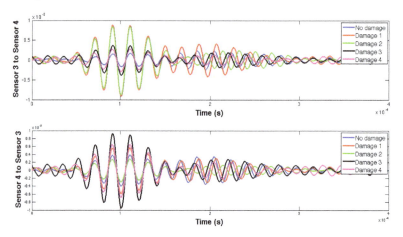

Figure 22. EMPD at 50 kHz. Tone-burst sent and measured between sensors 3 and 4.

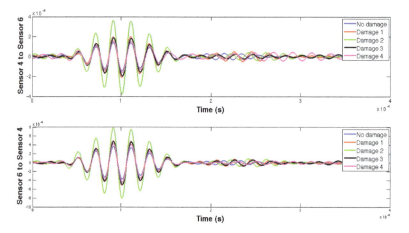

Figure 23. EMPD at 50 kHz. Tone-burst sent and measured between sensors 4 and 6.

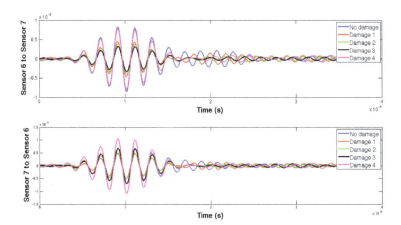

Figure 24. EMPD at 50 kHz. Tone-burst sent and measured between sensors 6 and 7.

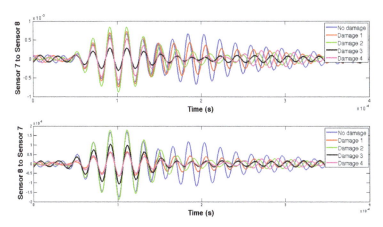

Figure 25. EMPD at 50 kHz. Tone-burst sent and measured between sensors 7 and 8.

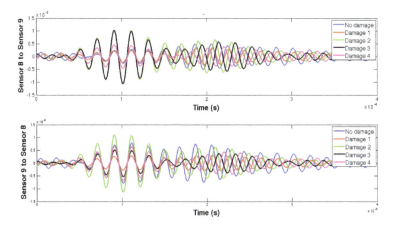

Figure 26. EMPD at 50 kHz. Tone-burst sent and measured between sensors 8 and 9.

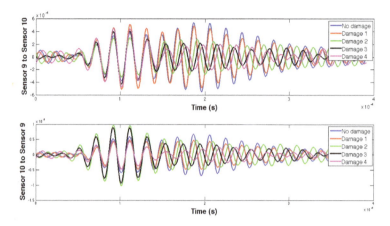

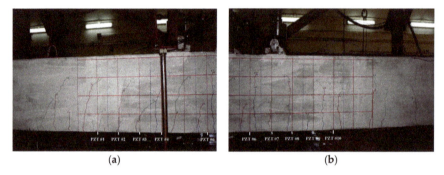

Figure 27. EMPD at 50 kHz. Tone-burst sent and measured between sensors 9 and 10.

Figure 28. Cracking map of the beam after the last loading stage. (**a**) Left side of the beam; (**b**) right side of the beam.

From Figure 29, it is quite obvious that every loading step generates a set of damages that influence the calculated value of the EMPD. In this case, just by looking at the RMSD obtained values, the first conclusion is that the EMPD is quite sensitive to all damage present in the structure, regardless of its nature (debonding or flexural cracks). This conclusion can be stated due the fact that for every damage stage, almost all measurements give RMSD values close to or higher than 50%. Furthermore, the most remarkable conclusion corresponds to the first loading step, after which no visible flexural cracks were appreciated on the concrete surface (Figure 28), at least apparently. However, for this damage stage, all measurements clearly detect damage with high RMSD values, which means that there were minor damages generated in the vicinity of the interface between the concrete and the FRP. Hence, thanks to the high sensitivity of this integrated damage detection procedure, it is possible to detect damages at a very early stage of the loading procedure, even in spite of the complexity of the structural system under study. This, however, was not possible to determine so clearly by means of the traditional EMI, whose corresponding RMSD values are shown in Figure 30. In this last figure, all damage indications are much lower than those shown in Figure 29 for all sensors and damage cases, so with this test on a real-scale structure, it is fully demonstrated that the EMPD shows higher potential than the traditional EMI on its own, at least in terms of sensitivity to damage, minor damage in particular. Thanks to this, even small and non-visible incipient flexural cracks can be detected on the structure.

Sensors **2016**, *16*, 639

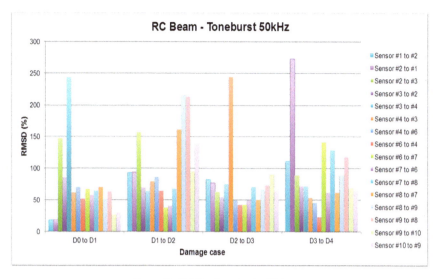

Figure 29. RMSD values for the RC beam.

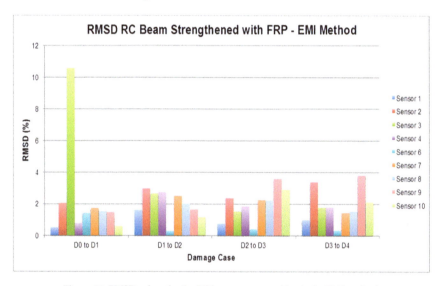

Figure 30. RMSD values for the RC beam corresponding to the EMI method.

On the other hand, in Figure 29 there are several RMSD values that reach values around 150% and above, clearly far from the rest of the results. This, in the opinion of the authors, could suggest the presence of two different types of damage: while smaller RMSD values would indicate the presence of flexural cracks, these last values would indicate physical damages in the vicinity of the interface between FRP and concrete, which means an origin for debonding appearance. Some of these high values are obtained around sensor 3 for damages D1, D2, and D3 (D1: 147% when the tone-burst is sent from sensor 2 to sensor 3 and 243% from sensor 3 to 4; D2: 156% from sensor 2 to 4; D3: 244% from sensor 4 to 3), suggesting thus the potential presence of debonding in this region of the beam, which comes to be true as can be verified from Figure 31, where the debonding damages after the last loading step until failure are shown. However, this conclusion is apparently not confirmed by any sensor for the second debonding present in the sensing region between sensors 6 and 7. Therefore, further analysis is needed in this region, for which Figure 32 is proposed.

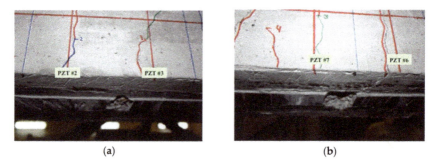

(a)   (b)

Figure 31. Damages with concrete cover separation between sensors 2 and 3 (**a**) and sensors 6 and 7 (**b**) for the RC beam.

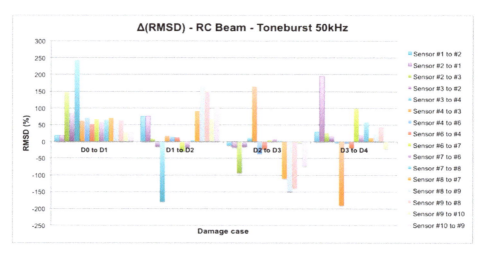

Figure 32. RMSD increments between loading steps for the RC beam corresponding to the EMPD method.

In Figure 31, the RMSD increments from one damage case to the following one is shown, so that a positive value means higher damage severity in the subsequent state, while a negative value means that a lower damage has been induced in the subsequent state. In this sense, just the positive values are important for our purpose, since the interest of this analysis is to find sensing regions likely to suffer a failure mechanism initiated by debonding appearance, which will happen when damage indications grow significantly during the last loading steps. Some high indications are found around sensors 8 and 9 for damage D2, in both Figures 29 and 31 corresponding to two different flexural cracks larger than 15 cm appearing in this region after damage D2. However, none of these indications increase again in the fourth loading step, which suggests that no debonding damage was initiated in this region. The contrary of this happens then in the sensing region between sensors 6 and 7. As can be appreciated in Figure 31, there is an indication slightly higher than 50% after the first loading step (D1), and then the damage is not significantly increased after D2 and D3, but suddenly there is an increase of almost 100% after the last measured loading step, D4, when the EMPD is calculated from measurements taken at sensor 7, which suggests that in this region the interface between concrete and FRP is likely to suffer debonding as well (Figure 32). Although it is true that this debonding is not so obvious as the one detected between sensors 2 and 3, with this methodology it is still possible to analyze each of the sensing regions with a higher level of detail and more accuracy than with the traditional EMI, which is not even capable of offering a clear distinction between stages of damage (Figure 30).

Finally, there is an indication higher than 250%, from sensor 2 to sensor 1 for damage stage D4, which is not related to any debonding either. In this case, as can be seen in Figure 28, a new flexural crack bigger than 10 cm appeared right in the position of sensor 1, which is probably the reason for this unusual RMSD value almost at the end of the loading procedure. Once again, the methodology proposed in this paper enhances the performance of the traditional EMI method, given that this damage is not even noticeable in Figure 30, although there is certainly an important crack generated right in the position of sensor 1, as indicated above.

5. Conclusions

A new damage indicator has been proposed in this paper based on Electro-Mechanical Power Dissipation (EMPD), as a result of combining impedance signatures and guided waves into an integrated damage detection procedure. This proposed methodology supposes a new approach in the SHM field, and it has been successfully tested on two different lab-scale structures. The first of these structures was selected mainly in order to prove the sensitivity of the EMPD to the presence of damage, using for that purpose a well-known structure that has been widely used before in the literature. With the second specimen, the goal pursued and achieved in this paper is double. Firstly, thanks to the proposed methodology, guided waves have been used to test more complex structures than what the authors have found in the literature, even though a good number of references suggest that they are not suitable to be used in complex structures. Secondly, the studied concrete specimen is subjected to a more complex series of damage scenarios, involving different heterogeneous materials and complicated behaviours at the interfaces between these materials, and it has been proven how guided waves, through the application of this integrated procedure, can successfully contribute to positive damage identification even for instances of damage as complicated as those studied for this specimen.

Furthermore, some conclusions can also be extracted about damage location, just by paying attention to the direction of motion of the guided wave. In that sense, good results have been presented for single damage scenarios in both lab-scale structures, although the results needs to be improved in the case of multiple damage scenarios.

Finally, the EMPD has also been demonstrated to be sensitive to damage in a real and complex structural system such as the RC beam externally strengthened with FRP. In this case, extremely high sensitivity has been demonstrated even to minor damage at a very early stage of the loading procedure, which means that early detection can be successfully performed.

Hence, this new integrated method implements a promising combination of impedance signatures and guided waves that has been demonstrated to be sensitive to different damages, and which is suitable to work even for real-scale complex structural systems.

Acknowledgments: The authors acknowledge the support for the work reported in this paper from the Spanish Ministry of Economy (project BIA2013-46944-C2-1-P). The writers are also grateful to Ángel Arteaga, Ana de Diego, and Daniel Cisneros from the Eduardo Torroja Institute for Construction Science (CSIC) for their help in making possible the experimental work reported in this paper in Section 4.

Author Contributions: E.S. and R.P. conceived and designed the experiments; E.S. performed the experiments; E.S. and R.S. analyzed the data; E.S. and R.P. wrote the paper.

Conflicts of Interest: The authors declare no conflict of interest.

Abbreviations

SHM	Structural Health Monitoring
NDT	Non-Destructive Testing
PZT	Lead-Zirconate-Titanate
EMI	Electro-Mechanical Impedance
EMPD	Electro-Mechanical Power Dissipation
FFT	Fast Fourier Transform
IFFT	Inverse Fast Fourier Transform

References

1. Sohn, H.; Farrar, C.R.; Hemez, F.M.; Shunk, D.D.; Stinemates, D.W.; Nadler, B.R.; Czarnecki, J.J. *A Review of Structural Health Monitoring Literature: 1996–2001*; Los Alamos National Laboratory Report LA-13976-MS; Los Alamos National Labs: Los Alamos, NM, USA, 2004.
2. Bhalla, S.; Soh, C.K. High frequency piezoelectric signatures for diagnosis of seismic/blast induced structural damages. *NDT&E Int.* **2004**, *37*, 23–33.
3. Chang, P.C.; Flatau, A.; Liu, S.C. Review Paper: Health Monitoring of Civil Infrastructure. *Struct. Health Monit.* **2003**, *2*, 257–267. [CrossRef]
4. Fang, S.E.; Perera, R.; De Roeck, G. Damage identification of a reinforced concrete frame by finite element model updating using damage parameterization. *J. Sound Vibr.* **2008**, *313*, 544–559. [CrossRef]
5. Perera, R.; Marín, R.; Ruiz, A. Static-dynamic multiscale structural damage identification in a multi-objective framework. *J. Sound Vibr.* **2013**, *332*, 1484–1500. [CrossRef]
6. Chang, P.C.; Liu, S.C. Recent Research in Nondestructive Evaluation of Civil Infrastructures. *J. Mater. Civil Eng.* **2003**, *15*, 298–304. [CrossRef]
7. Leo, D.J. *Engineering Analysis of Smart Material Systems*; John Wiley & Sons: New Jersey, NJ, USA, 2007.
8. Laflamme, S.; Cao, L.; Chatzi, E.; Ubertini, F. Damage Detection and Localization from Dense Network of Strain Sensors. *Shock Vibr.* **2016**, *2016*, 2562949. [CrossRef]
9. Teng, J.G.; Chen, J.F.; Smith, S.T.; Lam, L. *CFRP Strengthened RC Structures*, 1st ed.; John Wiley and Sons: West Sussex, UK, 2002.
10. Giurgiutiu, V. *Structural Health Monitoring with Piezoelectric Wafer Active Sensors*, 1st ed.; Elsevier Inc.: San Diego, CA, USA, 2008.
11. Liang, C.; Sun, F.P.; Rogers, C.A. Coupled electro-mechanical analysis of adaptive material systems determination of the actuator power consumption and system energy transfer. *J. Intell. Mater. Syst. Struct.* **1994**, *5*, 12–20. [CrossRef]
12. Yang, Y.; Divsholli, B.S. Sub-Frequency Interval Approach in Electromechanical Impedance Technique for Concrete Structure Health Monitoring. *Sensors* **2010**, *10*, 11644–11661. [CrossRef] [PubMed]
13. Sevillano, E.; Sun, R.; Gil, A.; Perera, R. Interfacial crack-induced debonding identification in FRP-strengthenined RC beams from PZT signatures using hierarchical clustering analysis. *Compos. Part B Eng.* **2015**, *87*, 322–335. [CrossRef]
14. Min, J.P.S.; Yun, C.B.; Lee, C.G.; Lee, C. Impedance-based structural health monitoring incorporating neural network technique for identification of damage type and severity. *Eng. Struct.* **2012**, *39*, 210–220. [CrossRef]
15. Worlton, D.C. Experimental confirmation of Lamb waves at megacycle frequencies. *J. Appl. Phys.* **1961**, *32*, 967–971. [CrossRef]
16. Raghavan, A. Guided-Wave Structural Health Monitoring. Ph.D Thesis, The University of Michigan, Ann Arbor, MI, USA, 2007.
17. Providakis, C.P.; Angeli, G.M.; Favvata, M.J.; Papadopoulos, N.A.; Chalioris, C.E.; Karayannis, C.G. Detection of Concrete Reinforcement Damage Using Piezoelectric Materials—Analytical and Experimental Study, World Academy of Science, Engineering and Technology. *Int. J. Civil Environ. Struct. Constr. Archit. Eng.* **2014**, *8*, 197–205.
18. An, Y.K.; Sohn, H. Integrated impedance and guided wave based damage detection. *Mech. Syst. Sign. Process.* **2012**, *28*, 50–62. [CrossRef]
19. Saafi, M.; Sayyah, T. Health monitoring of concrete structures strengthened with advanced composite materials using piezoelectric transducers. *Compos. Part B Eng.* **2001**, *32*, 333–342. [CrossRef]

20. Giurgiutiu, V.; Reynolds, A.; Rogers, C.A. Experimental Investigation of E/M Impedance Health Monitoring for Spot-Welded Structural Joints. *J. Intell. Mater. Syst. Struct.* **1999**, *10*, 802–812. [CrossRef]
21. Giurgiutiu, V.; Harries, K.; Petrou, M.; Bost, J.; Quattlebaum, J.B. Disbond detection with piezoelectriz wafer active sensors in RC structures strengthened with CFRP composite overlays. *Earthq. Eng. Vibr.* **2003**, *2*, 213–223. [CrossRef]
22. Wang, D.S.; Yu, L.P.; Zhu, H.P. Strength monitoring of concrete based on embedded PZT transducer and the resonant frequency. In Proceedings of the Symposium on Piezoelectricity, Acoustic Waves and Device Applications (SPAWDA), Xiamen, China, 10–13 December 2010; pp. 202–205.
23. Park, G.; Farrar, C.R.; Rutherford, A.C.; Robertson, A.C. Piezoelectric active sensor self-diagnosis using electric admittance measurements. *J. Vibr. Acoust.* **2006**, *128*, 469–476. [CrossRef]
24. Raghavan, A.C. Review of Guided Wave Structural Health Monitoring. *Shock Vibr. Digest* **2007**, *39*, 91–114. [CrossRef]
25. Viktorov, I.A. *Rayleigh and Lamb Waves*, 1st ed.; Plenum Press: New York, USA, 1967.
26. Rose, J.L. *Ultrasonic Waves in Solid Media*; Cambridge University Press: Cambridge, UK, 1999.
27. Ostachowicz, W.; Kudela, P.; Krawczuk, M.; Zak, A. *Guided Waves in Structures for SHM: The Time-Domain Spectral Element Method*, 1st ed.; Wiley: West Sussex, UK, 2012.
28. Alleyne, D.; Cawley, P. The interaction of Lamb waves with defects. *IEEE Trans. Ultrason. Ferroelectr. Freq. Control* **1992**, *39*, 381–397. [CrossRef] [PubMed]
29. Cawley, P.; Alleyne, D. The use of Lamb wave for the long range inspection of large structures. *Ultrasonics* **1996**, *34*, 287–290. [CrossRef]
30. Hsu, J.; WP, M.L.; Bascom, S.D. Nanometer distance regulation using electromechanical power dissipation. U.S. Patent No. 5,886,532, 23 March 1999.
31. Perais, D.M.; Tarazaga, P.A.; Inman, D.J. A study on the correlation between PZT and MFC resonance peaks and adequate damage detection frequency intervals using the impedance method. In Proceedings of the International Conference on Noise & Vibration Engineering (ISMA), Leuven, Belgium, 18–20 September 2006; pp. 18–20.
32. Saravanan, T.J.; Balamonica, K.; Priya, C.B.; Gopalakrishnan, N.; Murthy, S.G.N. Non-Destructive Piezo electric based monitoring of strength gain in concrete using smart aggregate. In Proceedings of the International Symposium Non-Destructive Testing in Civil Engineering (NDT-CE), Berlin, Germany, 15–17 September 2015.
33. Krishnamurthy, K.; Lalande, F.; Rogers, C.A. Effects of temperature on the electrical impedance of piezoelectric sensors. *Proc. SPIE 2717* **1996**, *2003*, 451–463.
34. Annamdas, V.G.M.; Yang, Y.; Soh, C.K. Influence of loading on the electromechanical admittance of piezoceramic transducers. *Smart Mater. Struct.* **2007**, *16*, 1888–1897. [CrossRef]
35. PI Piezo Technology, P-876 DuraAct Patch Transducer Whitepaper. 2014. Available online: http://piceramic.com/product-detail-page/p-876-101790.html (accessed on 1 February 2014).
36. Pesic, N.; Pilakoutas, K. Concrete beams with externally bonded flexural FRP-reinforcement: Analytical investigation of debonding failure. *Compos. Part B Eng.* **2003**, *34*, 327–338. [CrossRef]
37. Liu, S.T.; Oehlers, D.J.; Seracino, R. Study of Intermediate Crack Debonding in Adhesively Plated Beams. *J. Compos. Constr.* **2007**, *11*, 175–183. [CrossRef]
38. Sun, R.; Sevillano, E.; Perera, R. A discrete spectral model for intermediate crack debonding in FRP-strengthened RC beams. *Compos. Part B* **2015**, *69*, 562–575. [CrossRef]

Article

A Continuous Liquid-Level Sensor for Fuel Tanks Based on Surface Plasmon Resonance

Antonio M. Pozo [1], Francisco Pérez-Ocón [1,*] and Ovidio Rabaza [2]

[1] Department of Optics, Faculty of Science, Edificio Mecenas, Campus Universitario de Fuentenueva, University of Granada, 18071 Granada, Spain; ampmolin@ugr.es
[2] Department of Civil Engineering, University of Granada, 18071 Granada, Spain; ovidio@ugr.es
* Correspondence: fperez@ugr.es; Tel.: +34-958-241-000; Fax: +34-958-248-533

Academic Editor: Gonzalo Pajares Martinsanz
Received: 18 March 2016; Accepted: 13 May 2016; Published: 19 May 2016

Abstract: A standard problem in large tanks at oil refineries and petrol stations is that water and fuel usually occupy the same tank. This is undesirable and causes problems such as corrosion in the tanks. Normally, the water level in tanks is unknown, with the problems that this entails. We propose herein a method based on surface plasmon resonance (SPR) to detect in real time the interfaces in a tank which can simultaneously contain water, gasoline (or diesel) and air. The plasmonic sensor is composed of a hemispherical glass prism, a magnesium fluoride layer, and a gold layer. We have optimized the structural parameters of the sensor from the theoretical modeling of the reflectance curve. The sensor detects water-fuel and fuel-air interfaces and measures the level of each liquid in real time. This sensor is recommended for inflammable liquids because inside the tank there are no electrical or electronic signals which could cause explosions. The sensor proposed has a sensitivity of between 1.2 and 3.5 RIU^{-1} and a resolution of between 5.7×10^{-4} and 16.5×10^{-4} RIU.

Keywords: fuel tanks; fuel level; air-fuel-water interfaces; plasmonic sensor; surface plasmon resonance

---

1. Introduction

The presence of water inside fuel tanks currently poses a problem. The water can originate from condensation and, in regions prone to floods, there is also a higher risk of water infiltrating in the tanks. Also, the water combines with sulfur and other chemical components of the fuel to corrode the inside of the tank [1]. Being denser than the fuel, the water lies at the bottom of the tank, and when the water surpasses the maximum permitted level, it must be removed from the tank.

Currently, it is therefore essential in the fuel industry to have methods to detect the presence of water in fuel tanks and furthermore measure the water level in real time, but fuel tanks in refineries and petrol stations normally use rudimentary methods—for instance, a stick with a special paste (water-finding paste) is inserted into the tank (if the tank is not very deep). When in contact with water, this paste changes color, indicating the presence but not the level of the water in the tank [2].

Different types of sensors have been proposed to measure the level of fuel in tanks, such as sensors based on ultrasonic Lamb waves [3], capacitive sensors [4], pressure sensors [5], and sensors based on optical fiber [6,7]. All these sensors work when a single type of liquid occupies the tank. However, when the tank contains fuel and water, other methods have been proposed to detect the water content in the tank and to measure the levels of the water and fuel in the tank. These methods are based on acoustics [8], microwave reflection [1], reflectometry [9,10], electrode arrays [11], magnetic floats [12], pressure sensors [13,14], and capacitance sensors [15]. All these methods have advantages and disadvantages [16,17].

In the present paper, we have designed an optical sensor that indicates, in real time, the level of air, the level of water, and the level of the gasoline or diesel in the tank. Our sensor is based on surface plasmon resonance (SPR) and overcomes many of the limitations of the sensors proposed to date. It is a safe and rapid-response device that can be used in inflammable or explosive environments such as fuel tanks, as opposed to capacitance sensors or those based on electrodes. Our sensor does not contain movable parts, as opposed to sensors based on floats or pressure sensors, which can become obstructed and are susceptible to mechanical damage. Our device has no problems of friction, heating, or hysteresis, nor does it have the drawback of susceptibility to acoustic or electromagnetic interference.

2. Design of the Plasmonic Sensor

The operational principle is based on surface plasmon resonance. Noble metals have a dense assembly of negatively charged free electrons in an equally charged positive-ion background. If an external optical field is applied at one point in the metal, the local density of free electrons at that place in the metal changes due to the force of the field applied. A metal-dielectric interface supports surface plasma oscillations, which are charge-density oscillations (free electrons) along the metal-dielectric interface. Surface plasmon (SP) is the quantum of these oscillations. The SPs are associated with a longitudinal electric field (TM-polarized or p-polarized) that has its maximum at the metal-dielectric interface itself and decays exponentially both on the metal as well as on the dielectric medium.

For SPs to be excited, the condition of resonance must be fulfilled, according to which the wave-vector of the excitation light along the metal-dielectric interface should be equal to that of SPs. One way of achieving this is to excite SPs by an evanescent wave, using a configuration based on a prism. In this case, the resonance condition is given by the following expression [18]:

$$\sin\theta = \sqrt{\frac{\varepsilon_m \varepsilon_s}{\varepsilon_p (\varepsilon_m + \varepsilon_s)}} \tag{1}$$

where ε_m, ε_s, ε_p are the dielectric constants of the metal, the medium and the hemispherical prism, respectively, and θ is the angle of incidence respect to the normal on the prism base. Under resonance conditions, the energy of the incident light is transferred to the SPs, resulting in a sharp dip in the intensity of the light reflected at the interface of the prism base and the medium in contact. This occurs at an angle greater than the critical angle.

Figure 1 shows the plasmonic sensor that we propose, based on the Kretschmann configuration [19]. It is formed by a hemispherical prism of SF 10 glass ($n = 1.7231$) [20]. At the base of the prism is a layer of magnesium fluoride (MgF$_2$, $n = 1.38$ [21]) and then a layer of gold ($n = 0.12517 + 3.3326i$) [22]. The gold layer is the one in contact with the medium (air, water, gasoline or diesel); we used gold as the outer layer because of its chemical stability. A p-polarized laser beam with a wavelength of 632.8 nm strikes the prism with a normal incidence to the hemispherical prism surface by an optical fiber with its endface adhered to the prism surface. The angle of incidence with respect to the normal in the MgF$_2$ is θ (>critical angle). Finally, the light reflected at the prisma-MgF$_2$ interface is received by another optical fiber and conducted to an optical power meter.

When the light strikes the hemispherical prism-MgF$_2$ interface at an angle greater than the critical angle, an evanescent wave is generated and is propagated along the hemispherical prism-MgF$_2$ interface. This evanescent wave could excite two surface plasmon polaritons (SPPs) depending on the thickness of the layers.

The sensor that we propose is based on intensity interrogation. This has a great advantage: as opposed to sensors based on angular interrogation, our sensor does not require moving parts. In sensors based on angular interrogation, when the medium changes (and therefore the refraction index), the resonance angle also changes. Therefore, in these types of sensors, when the medium changes, it is necessary to adjust the incidence angle of the laser beam and move the detector to locate the corresponding angle to the minimum in the reflectance curve. In the case of our sensor, this operates by the modulation in the reflected intensity. The theoretical modeling of SPR reflectance was carried out by using transfer-matrix method to solve the Fresnel equations for the multilayer stack [23,24] with WinSpall software package. We designed the sensor in such a way that when the sensor is in contact with air, a low reflectance value is registered, when the sensor is in contact with water, a mean reflectance value is registered, and when it is in contact with gasoline or diesel, high reflectance values are registered. For this to happen, the sensor has to work at a fixed angle of the laser beam and the reflected beam at 65.5°.

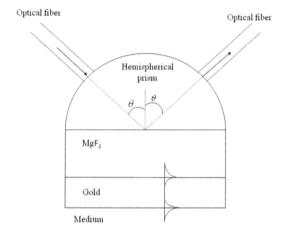

Figure 1. Scheme of the plasmonic sensor. On the left part of the hemispherical prism, the optical fiber transports the incident radiation and on the right, another optical fiber collects the reflected radiation. Also, the surface plasmon polaritons (SPP) are shown propagating through the MgF_2-gold and gold-medium interfaces.

To have the conditions mentioned above (low reflectance in air, mean reflectance in water, and high reflectance in gasoline/diesel), the thicknesses found for the layers of metal and dielectric were 48 nm for the gold layer and 190 for the MgF_2 layer. As will be seen in Section 4, with this design, the sensor can reliably distinguish air, water, and different types of gasoline or diesel starting from the reflectance value measured with a detector situated at the end of the optical fiber.

3. Gauge Construction

The device is composed of two modules. The first part is the module in which the laser beam reaches the sensor (see Figure 2). The encoded radiation from a laser of wavelength of 632.8 nm is incident on a transparent material to which a voltage is applied.

In the first stage, the light from the laser is injected in one only optical fiber. With the electro-optical prism we get the same signal from this optical fiber injected into all the necessary optical fiber (see Figures 3 and 4) so that the laser beam exiting the electro-optic prism is sequentially redirected to each optical fiber which transports the light to each plasmonic sensor (Figure 4). An array of plasmonic sensors is located inside the tank. The entrance optical fiber of each plasmonic sensor is glued to a hemispherical miniprism. The entry of light in each hemispherical miniprism is perpendicular (Figure 1). The angle between normal to the base of each hemispherical miniprism and the direction of the light within each hemispherical miniprism has to be 65.5°, as commented in the previous section (Figure 1).

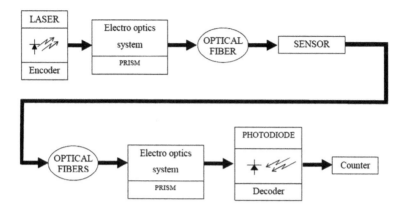

Figure 2. Diagram of the emission-detection system of the device.

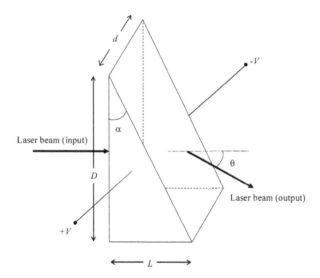

Figure 3. In the electro-optic prism, the deflection angle θ is controlled by the application of a voltage.

In the second stage (see Figure 2), the laser beam exiting each sensor of the array is guided through each optical fiber. The process of the light path is similar to the path of the illumination system. Now we have a bundle of optical fibers, as many as minisensors (hemispherical prisms) and the light from each optical fiber has to be sequentially injected into only one (Figure 4 with light in opposite direction). The path of the light is the opposite of the stage before, but now the end of the path is the photodiode instead of the sensor.

By means of an electro-optical prism, we inject the light from each optical fiber into only one and, from there, into the photodiode. Both electro-optical prisms work with the same clock signal to synchronize the light signal in the entrance and exit stages. After the photodiode, there is a counter to identify each sensor of the array. The first light emission corresponds to the first sensor, the second emission to the second sensor, and so on. As we know the position of each sensor of the array, we can determine the height of each liquid or level of the air in the tank.

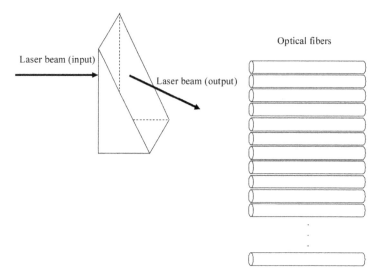

Figure 4. Scheme of the laser beam entering sequentially all the optical fibers. Output endface of each optical fiber is glued to a plasmonic sensor.

The optical fiber is used to conduct light from the laser to the hemispherical prism and from there to the photodetector. The essential reason is that the optical fiber is not attacked by any of the liquids used, and it transports light instead of electrical signals to avoid any possible explosion in the fuel tanks. The working principle of the electro-optical prism is as follows:

The laser beam can be deflected dynamically by using a prism with an electrically controlled refractive index. The angle of deflection introduced by a prism of small apex angle α and refractive index n is $\theta \approx (n-1)\,\alpha$ (see Figure 3). An incremental change of the refractive index Δn caused by an applied electric field E corresponds to an incremental change of the deflection angle [25],

$$\Delta\theta = \alpha\Delta n = \frac{1}{2}\alpha r n^3 E = \frac{1}{2}\alpha r n^3 \frac{V}{d} \qquad (2)$$

where r is the Pockels coefficient or the linear electro-optic coefficient, n is the refractive index of the material, E the applied electric field, V the applied voltage to the material and d the prism width. Depending on the maximum deflection angle required, two or more prisms can be cascaded to increase this angle.

An optical beam of width D and wavelength λ, has an angular divergence:

$$\delta\theta \approx \frac{\lambda}{D} \qquad (3)$$

To minimize that angle, the beam should be as wide as possible, ideally covering the entire width of the prism itself. For a given maximum voltage V corresponding to a scanned angle $\Delta\theta$, the number of independent spots (resolution) is given by:

$$N \approx \frac{|\Delta\theta|}{\delta\theta} = \frac{\frac{1}{2}\alpha r n^3 \frac{V}{d}}{\frac{\lambda}{D}} \qquad (4)$$

taking into account that:

$$\alpha \approx \frac{L}{D} \qquad (5)$$

and the half-wave voltage (the applied voltage necessary to get a phase retardation π) is:

$$V_\pi = \left(\frac{d}{L}\right)\left(\frac{\lambda}{rn^3}\right) \tag{6}$$

substituting Equations (4) and (5) in Equation (6), we get:

$$N \approx \frac{V}{2V_\pi} \tag{7}$$

Therefore, the voltage that we have to apply to the prism is given by [25]:

$$V \approx 2NV_\pi \tag{8}$$

where N is the number of optical fibers to illuminate (see Figure 4).

The second electro-optical prism works in the same way, but the light comes from the multiplicity of optical fibers to the prism (Figure 4 with the light in the opposite direction) and from it to a single optical fiber and from there to the photodiode.

If the tank is open and the ambient illumination changes, radiation from the exterior could enter the optical fibers. If the illumination conditions change, the radiation in the optical fibers changes; in this case, we could have a variable beam in the optical fiber and therefore a different signal in the output of the photodiode for the same entry signal of the laser beam [26]. For this reason, the laser beam is encoded and the photodiode has to decode the signal. If external radiation enters the optical fibers, the photodiode disregards it, so that the only radiation considered in the photodiode is the beam exiting of the optical fibers. The laser beams of all optical fibers are incident on a single photodiode.

The final signal (height of each liquid or air level) is analyzed by a computer from which the sensor is controlled. These levels are selected by the operator. The alarms can be sounds or visual keys for the blind or deaf, to warn of a dangerous situation. Furthermore, the data can be sent by Internet in real time to a remote point so that the tank can be controlled at all times regardless of its physical location.

The weakest part of the device might appear to be the electro-optical prisms, but these are parts of optical communications and are in fact not weak. All the parts of the optical device, except one part of the optical fiber are outside the tank and it can be sealed in a hermetic box to avoid being broken.

The optical fiber is glued to the minisensors but currently this special glue is extremely strong, so that there are no problems in the sense that the optical fiber could come unglued. The minisensors in the tanks could be housed inside a cylinder (in the tank), for instance, to protect them when the fuel is poured into the tanks. This cylinder has to be open at the same level of the tank.

The coating process is as follows: the base of the hemispherical prism is firstly cleaned with a solution consisting of ethanol and diethyl ether at a 1:1 ratio, rinsed with deionized water, and then dried with nitrogen. The substrate is sequentially coated with a 190 nm MgF_2 layer and a 48 nm gold layer to construct the sensor chip. The gold layer is coated using magnetron sputtering with the layer thickness measured by a quartz crystal oscillator thickness monitor. The substrate-heating and bias-voltage techniques are used in the coating process to improve the uniformity and firmness of the gold layer thickness. The MgF_2 layer is deposited using evaporation coating with the layer thickness measured by a step profiler. The MgF_2 crystals are used as the evaporation materials and the weight is controlled for by the specific layer thickness in the coating process [27].

4. Results and Discussion

Figure 5 shows the reflectance registered by a photodiode as a function of the angle of incidence of the light in the MgF_2 layer. Figure 5 shows the reflectance curves when the sensor is in contact with air ($n = 1$), with water ($n = 1.33$), and with gasoline or diesel. For the refractive index of gasoline and diesel, a range was taken of between 1.40 and 1.48, which corresponds to the values for different types of gasoline and diesel [28–33]. As commented in Section 2, the sensor that we propose (see Figure 1) is based on the intensity interrogation method, which uses a dielectric MgF_2 layer to excite two plasmons. These surface plasmon resonances can be clearly seen as two minimums in the reflectance when the sensor is in contact with air. In this case, the reflectance dip at 37.7° is associated with the air/gold SPP, while the dip at 65.5° with the MgF_2/gold one. Our sensor works at a fixed angle of 65.5°, the angle of incidence in the MgF_2 layer. When the medium in contact with the sensor changes, the resonance characteristics change also, altering the shape of the reflectance curves.

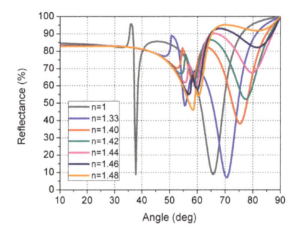

Figure 5. Reflectance curves as a function of the angle of incidence of the light in the prism. These represent the curves when the sensor is in contact with air ($n = 1$), water ($n = 1.33$), and gasoline or diesel fuel ($n = 1.40 - 1.48$). The thickness of the gold layer is 48 nm, and the thickness of dielectric layer of the MgF_2 is 190 nm.

The values of the refractive indices of the gasolines and diesels measured in our laboratories are 1.43 and 1.42 for Efitec 95 Neotech and Efitec 98 Neotech gasoline, respectively, and 1.46 for the two types of diesel (e+ Neotech diesel and e + 10 Neotech diesel). The four fuels are from the REPSOL company. The measurements were made with an Phywe 62,409.00 optical refractometer (PHYWE Systeme GmbH & Co. KG, Göttingen, Germany.). We have also measured the refractive-index values for 589.3 nm and 632.8 nm by minimum-deviation methods using an spectrogoniometer, and the results varied only in the third decimal of the refractive index, as reported by other authors [34,35]. As can be seen, the experimental measurements of our gasolines and diesels are within the range of values for which we have made the calculations.

As reflected in Figure 6 (detail from Figure 5 for an angle range of between 60° and 70°, where the angle of interest is shown), when the sensor is in contact with air a reflectance value of 9.0% is registered; when it is in contact with water, the value is 49.1%; and when it is in contact with gasoline or diesel the values are 80.4%, 86.4%, 90.0%, 91.5%, and 91.0% for refractive-index values of 1.40, 1.42, 1.44, 1.46, and 1.48, respectively. It is important to note that our sensor provides reflectance values above 80% for gasoline and diesel. In this way, the sensor in the tank continuously distinguishes between air, water, and gasoline or diesel, since when the reflectance measured with the photodiode is lower than 10%, the sensor that is in this position would be indicating that in this position corresponds to air. If a given sensor in the tank provided a reflectance of around 49.1%, this would indicate that there would be water. Finally, if the sensor provided a reflectance higher than 80%, this would indicate gasoline or diesel.

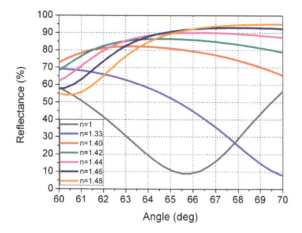

Figure 6. Detail of Figure 6. Reflectance curves from Figure 5 for incidence angles of between 60° and 70°.

We have also made the calculations for other thicknesses (28 and 68 nm). As can be seen in Figure 7, when the sensor is in contact with water, the reflectance is very close to the reflectance when the sensor is in contact with gasoline or diesel. This could cause the sensor to fail to distinguish reliably between water and gasoline or diesel. A similar failure occurs for gold layer thicknesses greater than 48 nm. In this case, (Figure 8), if we consider the angle of 75°, the reflectance curves for air, water, and gasoline/diesel are very close together. Therefore, we designed the sensor in such a way that the reflectance for air is low (9.0%), medium for water (49.1%), and high for gasoline/diesel (>80.4%). We achieved this for a gold thickness of 48nm. In this way, the sensor can easily distinguish water from gasoline/diesel.

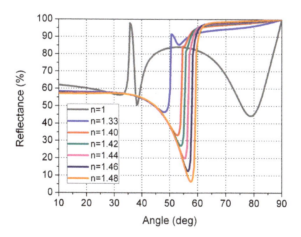

Figure 7. Reflectance curves as a function of the angle of incidence of the light in the prism for a gold layer of 28 nm. The thickness of the dielectric layer of the MgF_2 is 190 nm.

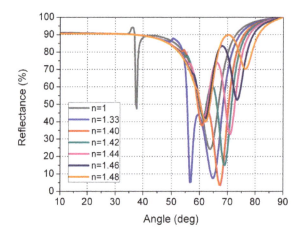

Figure 8. Reflectance curves as a function of the angle of incidence of the light in the prism for a gold layer of 68 nm. The thickness of the dielectric layer of the MgF$_2$ is 190 nm.

Tanks in petrol stations usually have a capacity of 20,000–50,000 liters stored in a cylinder. The inner diameter of the base of the cylinder is approximately 2.5 m. The cylindrical tanks are placed parallel to the ground to avoid deep excavation, so that the height of the sensor would be 2.5 m.

The critical height of water in petrol station tanks is approximately 15 cm. If a hemispherical prism is located every 2 cm, in this case, the accuracy of the measurement of the interface position would be high but we would have to use 125 hemispherical prisms; however, if we locate one every 5 cm, we would have less accuracy but could use only 50 hemispherical prisms. Even with an accuracy of 5 cm, we would know whether the height of water is the adequate or not. In practice, we must reach a compromise between accuracy and cost.

One drawback of the sensors based on wavelength interrogation is that they require a spectrometer to analyze the light spectrum at the exit of the sensor [36,37]. In the case of our sensor, we would need 50 spectrometers or 125 spectrometers (depending on the accuracy of the measurement of the interface position mentioned above) to analyze the light spectrum leaving each optical fiber. Such a large quantity of spectrometers would enormously increase the complexity, size, and cost of the sensor.

In addition, spectrometers need a certain amount of time to measure a spectrum. For example, Coelho *et al.* [37] used an optical spectrum analyzer with a resolution of 0.2 nm and a sampling rate of 1 spectrum *per* min, and a BraggMETER FS2200SA interrogator (eTester, Verona, VA, USA)) with a resolution of 1 pm at a sampling rate of 1 spectrum *per* s. Considering that our sensor works by sequentially redirecting the laser beam exiting the electro-optic prism to each optical fiber, our sensor would work more slowly if it were based on wavelength interrogation due to the time spent by the spectrometers for analyzing the output spectrum of each fiber.

Also, on some occasions, it is necessary to apply a special treatment of the signal in sensors based on wavelength interrogation, as for example applying fast Fourier transform smoothing filters to the spectral data to reduce the noise due to the low power level of the light that reaches the spectrometer [37]. This type of signal processing or similar ones would also slow down and complicate the functioning of our sensor.

In short, we have designed our sensor based on intensity interrogation because if it were based on wavelength interrogation the sensor would be not only more complex, but bigger, slower, and more expensive too.

The sensitivity of our sensor can be calculated from the change in reflectance *per* unit of change of refractive index. With respect to gasolines and diesels, we have considered the mean value of the refractive index (1.44) and the mean reflectance value (0.879), since the reflectance for gasolines and diesels does not vary linearly with the refractive index. Therefore, to calculate the sensitivity, we have considered 1, 1.33, and 1.44 for the refractive index and 0.09, 0.491, and 0.879 for the reflectance. Considering two consecutive values, we get a sensitivity of 1.2 and 3.5 RIU^{-1}, where RIU indicates the units of the refractive index.

The sensor resolution depends upon the accuracy with which the monitored SPR parameter can be determined by the specific sensing device and as such is limited by sensor-system noise [38]. To calculate the resolution of the sensor, we divide the accuracy of the photodetector by its sensitivity. Considering an accuracy of 0.2% in the signal registered by the photodetector [38,39], we get a resolution of 16.5×10^{-4} and 5.7×10^{-4}. Below, we compare our sensor, in terms of sensitivity and resolution, with other proposed in the literature.

Lin *et al.* [40] report sensitivity values of 3.8 and 4.16 RIU^{-1} for a tapered optical-fiber sensor based on localized surface plasmon resonance with an operating range of 0.07 RIU. The resolutions are 3.7×10^{-5} and 3.2×10^{-5} RIU, considering a standard deviation of noise of the sensor output of 1.42×10^{-4} and 1.35×10^{-4}, while in our case we have considered an accuracy of 0.2%. The great operating range of our sensor (0.48 RIU) makes the sensitivity somewhat lower. On the other hand, the resolution is also different, since it depends on the accuracy of the optical signal registered by the photodetection system used.

Chen *et al.* [41] have proposed an optical fiber biosensor based on silver nanoparticles, with a sensitivity of 160%/RIU and an operating range of 0.07 RIU. This value is within the range of our sensor (120%/RIU and 350%/RIU). To calculate the resolution, the authors considered a value for the standard deviation of noise of the sensor output of 0.75%, and they got a resolution of 4.68×10^{-3} RIU. Considering the same value of 0.75%, we get resolutions of 2.1×10^{-3} and 6.2×10^{-3} RIU for our sensor.

Wang *et al.* [42] report a resolution of 2×10^{-4} RIU for an optical fiber sensor based on Kretschmann's configuration. The operating range of their sensor is 0.025 RIU and they consider an accuracy of 0.1% in the signal registered. If we use 0.1% for the calculation of the sensitivity of our sensor, we get resolutions of 2.8×10^{-4} and 8.2×10^{-4} RIU.

We can also compare the sensitivity of our sensor with the sensitivity of the sensors based on the wavelength interrogation. For example, Liu *et al.* [36] present a surface plasmon resonance sensor based on silver-coated hollow-fiber structure for the detection of liquids with high refractive index. The operating range of this sensor is 0.06 RIU and is based on wavelength interrogation. This sensor has a resolution of between 0.8×10^{-4} and 5.1×10^{-4} RIU.

Coelho *et al.* [37] have proposed two hybrid sensors based on a fiber Bragg grating for monitoring organic solvents in high-refractive-index edible oils. With the configuration in intensity interrogation, they get a resolution of 2.1×10^{-4} RIU for one of them and 2.3×10^{-4} RIU for the other in transmission mode and 1.4×10^{-4} RIU in the reflection mode. The first of the sensors is also interrogated in wavelength, providing a resolution of 5.8×10^{-4} RIU in transmission mode.

As indicated, our sensor has a sensitivity and a resolution similar to those of other sensors based on intensity interrogation, and even those based on wavelength interrogation. In general, sensors based on wavelength or angular interrogation have greater sensitivity and resolution than those based on intensity interrogation [38]. However, our sensor does not require a high resolution, given that the aim of the sensor proposed is simply to distinguish between air, water, and gasolines/diesels. Also, taking into account that we do not need to distinguish between different types of gasoline/diesel (refractive index between 1.40 and 1.48), a resolution of 10^{-2} RIU would be sufficient to distinguish between media with indexes of 1, 1.33, and 1.40–1.48. Finally, with respect to the sensors based on angular interrogation, the advantage of our sensor is that it does not require moving parts, as mentioned in Section 2.

5. Conclusions

Here, we present an optical sensor for fuel tanks based on surface plasmon resonance. The sensor detects not only water-fuel and fuel-air interfaces, but can also measure the level of every liquid contained in the tank in real time. A major advantage is that this sensor can measure the height of the fuel (gasoline or diesel) without distinguishing between them, whereas other sensors have to be specific for gasoline or diesel.

This sensor serves for inflammable liquids such as gasoline, diesel, other crude-oil derivatives, acetones, alcohols, *etc.* because there are no electrical or electronic signals inside the tank but only light radiation. This makes it impossible to generate electrical sparks that could cause explosions. The laser and the photodetector, which work with electronic circuits, remain outside the tank.

As we are encoding light emitted by the laser, the influence of any external illumination that could enter the tank is avoided. Therefore, the device measures the height of several liquid levels in any type of tank, whether opaque, translucent, or transparent.

In addition, the device continuously measures the levels of all kinds of liquids that do not attack the optical fiber, hemispherical prism, or the plasmonic structures used.

With respect to the light-emission system and the photodetection system, our device uses only one laser and only one photodiode. Only one laser illuminates all the optical fibers and only one photodetector receives the radiation from all the optical fibers.

By knowing the tank geometry and the spacing of the sensors, we can track the volume (in real time) of the fuel and water in the tank, and therefore we can also calculate the flow rate, *i.e.*, how much liquid enters and leaves the tank per time unit.

The overall system is equipped with a visual and audible alarm that can be regulated to any liquid height. Also, the gauge is controlled by a PC that can store the data, print them or send them in real time by Internet worldwide. Finally, it is not necessary to carry out periodic calibrations.

Author Contributions: A. M. Pozo and F. Pérez-Ocón conceived of and designed the sensor; A. M. Pozo performed the calculations; O. Rabaza looked for the appropriate materials of the sensor. A. M. Pozo, F. Pérez-Ocón and O. Rabaza wrote the paper.

Conflicts of Interest: The authors declare no conflict of interest.

Abbreviations

The following abbreviations are used in this manuscript:

SPP   Surface plasmon polaritons
SPR   Surface plasmon resonance
TM   Transversal magnetic

References

1. Khalid, K.; Grozescu, L.V.; Tiong, L.K.; Sim, L.T.; Mohd, R. Water detection in fuel tanks using the microwave reflection technique. *Meas. Sci. Technol.* **2003**, *14*, 1905–1911. [CrossRef]
2. Thorogood, A.J. Water-Finding Probe. UK Patent Application GB2259366 A, 10 March 1993.
3. Sakharov, V.E.; Kuznetsov, S.A.; Zaitsev, B.D.; Kuznetsova, I.E.; Joshi, S.G. Liquid level sensor using ultrasonic Lamb waves. *Ultrasonics* **2003**, *41*, 319–322. [CrossRef]
4. Shim, J. Liquid level measurement system using capacitive sensor and optical sensor. *J. Korean Soc. Mar. Eng.* **2013**, *37*, 778–783. [CrossRef]
5. Niu, Z.; Zhao, Y.; Tian, B.; Guo, F. The novel measurement method of liquid level and density in airtight container. *Rev. Sci. Instrum.* **2012**, *83*. [CrossRef] [PubMed]
6. Vázquez, C.; Gonzalo, A.B.; Vargas, S.; Montalvo, J. Multi-sensor system using plastic optical fibers for intrinsically safe level measurements. *Sens. Actuators A Phys.* **2004**, *116*, 22–32. [CrossRef]
7. Antunes, P.; Dias, J.; Paixao, T.; Mesquita, E.; Varum, H.; André, P. Liquid level gauge based in plastic optical fiber. *Measurement* **2015**, *66*, 238–243. [CrossRef]

8.	Bardyshev, V.I. Acoustic and Combined Methods for Measuring the Levels of Two-Layer Liquids. *Acoustor Phys.* **2002**, *48*, 518–523. [CrossRef]
9.	Di Sante, R. Time domain reflectometry-based liquid level sensor. *Rev. Sci. Instrum.* **2005**, *76*. [CrossRef]
10.	Bruvik, E.M.; Hjertaker, B.T.; Folgerø, K.; Meyer, S.K. Monitoring oil-water mixture separation by time domain reflectometry. *Meas. Sci. Technol.* **2012**, *23*. [CrossRef]
11.	Casanella, R.; Casas, O.; Pallàs-Areny, R. Oil-Water Interface Level Sensor Based on an Electrode Array. In Proceedings of the IEEE Instrumentation and Measurement Technology Conference, Sorrento, Italy, 24–27 April 2006; pp. 710–713.
12.	Chen, X.G.; Gu, D.W.; Qu, Y. Real-time estimation of oil quantity in crude oil tanks. *IEEE Proc. Sci. Meas. Technol.* **2006**, *153*, 108–112. [CrossRef]
13.	Skeie, N.O.; Mylvaganam, S.; Lie, B. Using multi sensor data fusion for level estimation in a separator. In Proceedings of the 16th European Symposium on Computer Aided Process Engineering and 9th International Symposium on Process Systems Engineering, Garmisch-Partenkirchen, Germany, 9–13 July 2006; Volume 21, pp. 1383–1388.
14.	Arvoh, B.K.; Skeie, N.O.; Halstensen, M. Estimation of gas/liquid and oil/water interface levels in an oil/water/gas separator based on pressure measurements and regression modelling. *Sep. Purif. Technol.* **2013**, *107*, 204–210. [CrossRef]
15.	Lu, G.R.; Hu, H.; Chen, S.Y. A Simple Method for Detecting Oil-Water Interface Level and Oil Level. *IEEJ Trans. Electr. Electr.* **2010**, *5*, 498–500. [CrossRef]
16.	Jin, B.; Liu, X.; Bai, Q.; Wang, D.; Wang, Y. Design and Implementation of an Intrinsically Safe Liquid-Level Sensor Using Coaxial Cable. *Sensors* **2015**, *15*, 12613–12634. [CrossRef] [PubMed]
17.	Meribout, A.; Naamany, A.A.; Busaidi, K.A. Interface Layers Detection in Oil Field Tanks: A Critical Review. In *Expert Systems for Human, Materials and Automation*; Vizureanu, P., Ed.; INTECH Open Access: Rijeka, Croatia, 2011; pp. 181–208.
18.	Gupta, B.D.; Srivastava, S.K.; Verma, R. Physics of Plasmons. In *Fiber Optic Sensors Based on Plasmonics*; World Scientific Publishing Co. Pte. Ltd.: Singapore, 2015; pp. 21–53.
19.	Kretschmann, E.; Raether, H. Radiative decay of non radiative surface plasmons excited by light. *Z. Naturf. Part A Astrophys. Phys. Phys. Chem.* **1968**, *A23*, 2135–2136. [CrossRef]
20.	SCHOTT Optical Glass 2014. Description of Properties. Available online: http://www.schott.com/advanced_optics/english/products/optical-materials/optical-glass/optical-glass/index.html (accessed on 17 May 2016).
21.	Dodge, M.J. Refractive properties of magnesium fluoride. *Appl. Opt.* **1984**, *23*, 1980–1985. [CrossRef] [PubMed]
22.	Babar, S.; Weaver, J.H. Optical constants of Cu, Ag, and Au revisited. *Appl. Opt.* **2015**, *54*, 477–481. [CrossRef]
23.	Heavens, O.S. Thin Films Optics. In *Optical Properties of Thin Solid Films*; Dover Books on Physics: New York, NY, USA, 1965; pp. 45–96.
24.	Ohta, K.; Ishida, H. Matrix formalism for calculation of electric field intensity of light in stratified multilayered films. *Appl. Opt.* **1990**, *29*, 1952–1959. [CrossRef] [PubMed]
25.	Saleh, B.E.A.; Teich, M.C. Electro-Optics. In *Fundamentals of Photonics*, 2nd ed.; John Wiley & Sons, Inc.: New York, NY, USA, 2007; pp. 799–831.
26.	Pérez-Ocón, F.; Rubiño, M.; Abril, J.M.; Casanova, P.; Martínez, J.A. Fiber-optic liquid-level continuous gauge. *Sens. Actuators A Phys.* **2006**, *125*, 124–132. [CrossRef]
27.	Zhang, P.; Liu, L.; He, Y.; Zhou, Y.; Ji, Y.; Ma, H. Noninvasive and Real-Time Plasmon Waveguide Resonance Thermometry. *Sensors* **2015**, *15*, 8481–8498. [CrossRef] [PubMed]
28.	Dombrovsky, L.A.; Sazhin, S.S.; Mikhalovsky, S.V.; Wood, R.; Heikal, M.R. Spectral properties of diesel fuel droplets. *Fuel* **2003**, *82*, 15–22. [CrossRef]
29.	Kawano, M.S.; Cardoso, T.K.M.; Possetti, G.R.C.; Kamikawachi, R.C.; Fabris, J.L.; Muller, M. Sensing biodiesel and biodiesel-petrodiesel blends. In Proceedings of the 22nd International Conference on Optical Fiber Sensors, 84215X, Beijing, China, 15–19 October 2012; pp. 1–3.
30.	Possetti, G.R.C.; Muller, M.; Fabris, J.L. Refractometric optical fiber sensor for measurement of ethanol concentration in ethanol-gasoline blend. In Proceedings of the International Microwave and Optoelectronics Conference (IMOC), SBMO/IEEE MTT-S, Curitiba, Brazil, 3–6 November 2009; pp. 606–616.

31. Roy, S. Fiber optic sensor for determining adulteration of petrol and diesel by kerosene. *Sens. Actuators B Chem.* **1999**, *55*, 212–216. [CrossRef]
32. Li, D.C.; Zhang, X.L.; Zhu, R.; Wu, P.; Yu, H.X.; Xu, K.X. A Method to Detect the Mixed Petrol Interface by Refractive Index Measurement with a Fiber-Optic SPR Sensor. *IEEE Sens. J.* **2014**, *14*, 3701–3707. [CrossRef]
33. Sadrolhosseini, A.R.; Moksin, M.M.; Yunus, W.M.M.; Talib, Z.A. Application of Surface Plasmon Resonance Sensor in Detection of Water in Palm-Oil-Based Biodiesel and Biodiesel Blend. *Sens. Mater.* **2011**, *23*, 315–324.
34. Sadrolhosseini, A.R.; Moksin, M.M.; Yunus, W.M.M.; Talib, Z.A. Surface Plasmon Resonance Characterization of Virgin Coconut Oil Biodiesel: Detection of Iron Corrosion using Polypyrrole Chitosan Sensing Layer. *Sens. Mater.* **2012**, *24*, 221–232.
35. Sadrolhosseini, A.R.; Moksin, M.M.; Yunus, W.M.M.; Mohammadi, A.; Talib, Z.A. Optical Characterization of Palm Oil Biodiesel Blend. *J. Mater. Sci. Eng.* **2011**, *5*, 550–554.
36. Liu, B.H.; Jiang, Y.X.; Zhu, X.S.; Tang, X.L.; Shi, Y.W. Hollow fiber surface plasmon resonance sensor for the detection of liquid with high refractive index. *Opt. Express* **2013**, *21*, 32349–32357. [CrossRef] [PubMed]
37. Coelho, L.; Viegas, D.; Santos, J.L.; de Almeida, J.M.M.M. Optical sensor based on hybrid FBG/titanium dioxide coated LPFG for monitoring organic solvents in edible oils. *Talanta* **2016**, *148*, 170–176. [CrossRef] [PubMed]
38. Homola, J.; Yee, S.S.; Gauglitz, G. Surface plasmon resonance sensors: Review. *Sens. Actuators B Chem.* **1999**, *54*, 3–15. [CrossRef]
39. Ronot-Trioli, C.; Trouillet, A.; Veillas, C.; Gagnaire, H. Monochromatic excitation of surface plasmon resonance in an optical fibre refractive-index sensor. *Sens. Actuators A Phys.* **1996**, *54*, 589–593. [CrossRef]
40. Lin, H.Y.; Huang, C.H.; Cheng, G.L.; Chen, N.K.; Chui, C.H. Tapered optical fiber sensor based on localized surface plasmon resonance. *Opt. Express* **2012**, *20*, 21693–21701. [CrossRef] [PubMed]
41. Chen, J.P.; Shi, S.; Su, R.X.; Qi, W.; Huang, R.L.; Wang, M.F.; Wang, L.B.; He, Z.M. Optimization and Application of Reflective LSPR Optical Fiber Biosensors Based on Silver Nanoparticles. *Sensors* **2015**, *15*, 12205–12217. [CrossRef] [PubMed]
42. Wang, S.F.; Chiu, M.H.; Chang, R.S. New idea for a D-type optical fiber sensor based on Kretschmann's configuration. *Opt. Eng.* **2005**, *44*. [CrossRef]

Article

Mechatronic Prototype of Parabolic Solar Tracker

Carlos Morón [1,*], Jorge Pablo Díaz [2], Daniel Ferrández [1] and Mari Paz Ramos [2]

[1]   Sensors and Actuators Group, Department of Construction Technology, Polytechnic University of Madrid,
       Madrid 28040, Spain; daniel.ferrandez.vega@alumnos.upm.es
[2]   Professional Institution Salesiana, Salesianos Carabanchel, Madrid 28044, Spain;
       jdiaz@salesianoscarabanchel.com (J.P.D.); pramos@salesianoscarabanchel.com (M.P.R.)
*   Correspondence: carlos.moron@upm.es; Tel.: +34-91-336-7583; Fax: +34-91-336-7637

Academic Editor: Gonzalo Pajares Martinsanz
Received: 31 March 2016; Accepted: 7 June 2016; Published: 15 June 2016

Abstract: In the last 30 years numerous attempts have been made to improve the efficiency of the parabolic collectors in the electric power production, although most of the studies have focused on the industrial production of thermoelectric power. This research focuses on the application of this concentrating solar thermal power in the unexplored field of building construction. To that end, a mechatronic prototype of a hybrid paraboloidal and cylindrical-parabolic tracker based on the Arduido technology has been designed. The prototype is able to measure meteorological data autonomously in order to quantify the energy potential of any location. In this way, it is possible to reliably model real commercial equipment behavior before its deployment in buildings and single family houses.

Keywords: energetic simulation; concentrating solar thermal power; architectural integration; mechatronic solar tracker; parabolic; infrared thermography

1. Introduction

In recent decades, the increasing demand for energy has encouraged the development of new renewable energy production systems with more competitive economic costs and lower environmental impact [1]. Solar power is one of the clean energies able to reduce human need for fossil fuel consumption. One of the main uses of solar power is electricity production based on two existing on the market technologies: photovoltaic panels and concentrating collectors [2,3]. The former are used for the production of energy from both direct and diffuse radiation, working at moderate temperatures often under 100 °C: therefore they are less dependent on exhaustive solar tracking. On the other hand, the latter function requiring concentrating strategies and high temperatures, using in most cases direct solar radiation and requiring complete solar tracking.

This research work is focused on the second capturing technology, based specifically on cylindrical-parabolic and paraboloidal concentrating for electricity production by the Rankine and Stirling thermodynamic cycles, which are considered to be highly efficient and productive methods [4–6]. Currently this type of collectors are widely applied at the industrial level (Figure 1), to produce large amounts of energy in solar thermal power plants [7,8]. That is why the idea to integrate this type of systems in the building sector is remarkable, taking into consideration its high yields.

Some authors have tried to improve architectural integration of these systems [9,10], making them more flexible and esthetic for use of space. All of these authors agree on the need to position the collector correctly in order to achieve performance similar to that obtained under laboratory conditions. It is for this reason that solar tracking systems play a crucial role in the development of solar thermal energy systems, making it possible to state that collector efficiency directly depends on the quality of tracking equipment which can constantly provide the maximum amount of direct solar radiation [11,12].

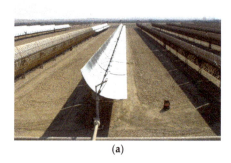
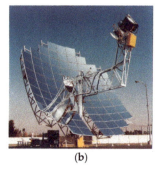

(a)   (b)

Figure 1. Examples of solar thermal plants: (**a**) Parabolic-trough collector technology; (**b**) Paraboloidal solar collector with Stirling engine. Source: Flickr Creative Commons.

Furthermore, the solar tracking mechanism has to be able to track the Sun´s movement with a high level of precision, returning the collector to its original or at-rest position at the end of the day or in unfavorable weather conditions, even performing optimal tracking on cloudy days. In this regard, there are two types of solar trackers: those which use solutions based on astronomical ephemerides with open-loop control, and those which are based on active tracking systems with close-loop approximations [13].

In the first case, the tracking is performed following satellite-based estimates obtained through incoming radiation reflection on the Earth. On occasion, to this source of information the possibility to include historical and empirical data of visible and infrared spectrum obtained using weather stations placed on the Earth's surface is added. Finally, this set of data is calculated mathematically using interpolation algorithms between direct and global radiation [14]. The problem resides in the fact that the resolution obtained using this method (3 km × 3 km) is not always optimal. Moreover, obtained data is horizontal what requires one to introduce inclination angle correction factors. The statistical process in turn gives inevitable errors when working with large database compounding the problem of sensor degradation and saturation.

The trackers which are able to interpret direct solar radiation received through the use of sensors belong to the second type [15]. These photosensitive sensors are used to continually send a signal to data processing equipment, which allows real-time continuous tracking of the Sun during the whole day.

The most commonly used tracking configuration is the one which is able to perform at the same time azimuth and zenith (or polar) movements of the collector [16]. For the azimuth tracking the tracker has to rotate from east to west following parallel to the Earth's surface trajectory. Instead, for the zenith or raising movement solar tracker has to rise following the solar trajectory from sunrise, and descend slowly to sunset varying its height in different seasons of the year. Both movements should be carefully coordinated and performed as accurately as possible. However, there are less precise solar trackers which use only azimuth movement presetting determined inclination angle depending on area latitude.

The solution presented in this research is based on the fabrication of mechatronic solar thermal tracker with hybrid, paraboloidal and cylindrical-parabolic technology, in such a way that it is possible to capture data of different measurable physical magnitudes, and at the same time to simulate two axis tracking of both technologies. For the construction of this prototype, it was necessary to carry out three-dimensional design and simulations using computer software which allowed us to validate a solar tracker prototype. Subsequently, the data was collected *in situ* in order to improve and calibrate the mechatronic tracker. Finally, with a view to improve the architectural integration of this type of energy technologies in the building sector, the work was focused on the paraboloidal disc technology in combination with the Stirling cycle as the most promising solution for possible deployment in buildings.

2. Materials and Methods

In this section the tasks related to design and manufacturing of the CSP tracker are described.

2.1. Design of the 3D Prototype

The mechanical, optical and electronic 3D design was made with a powerful CAD software known as FreeCAD 0.15 [17], which allows one to give parameters to all the components of the drawing, and is compatible with other standard design formats. Figure 2 shows fully designed CSP solar tracker.

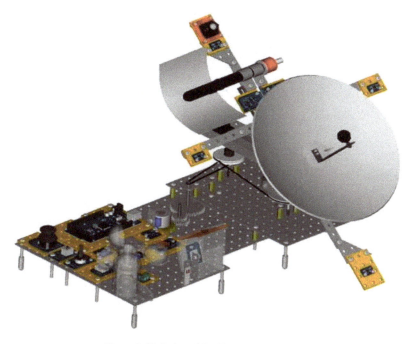

Figure 2. 3D design of the CSP tracker prototype.

Without describing in detail the 3D design process, but focusing rather on the components and devices that define it, it is possible to establish the following classification of subsystems:

- Mechanical subsystem: made up of the aluminium framing, plastic fasteners, stainless steel screws, gears and torque transmission belts, bolts and sleeves.
- Optical subsystem: formed by one paraboloidal collector plus a parabolic trough, being both of them made of highly reflective aluminium.
- Electronic subsystem: constituted with different physical magnitudes sensors (pressure, temperature, humidity, wind, irradiance, presence of water and air quality); positioning variables (accelerometer, compass and camera); driving devices (step by step motor and linear actuator); Arduino programmable electronic controller and many other additional devices [18,19].

The most representative components that define the device are listed in Table 1.

Table 1. Devices used in the construction of the prototype.

Terminology	Function
Arduino Mega 2560 ADK	Programmable electronic controller of the solar tracker, 256 Kb
Pololu Nema 17	Bipolar step by step motor 3.2 kg·cm torque and step angle 1.8°, to azimuthal tracking
Firgelli L12-100-100-6R	Linear actuator up to 17 mm/s and forces higher than 30 N for elevation tracking
SeedStudio	(a) Vibration piezoelectric sensor to measure the impact over the prototype of hail or heavy rain (b) Water detection sensor over the solar tracker (c) Ambiance temperature and relative air humidity sensor (d) Barometric pressure monitoring sensor (e) Quality air sensor in terms of CO_2 concentration (f) Irradiance sensor in the four cardinal positions of the solar tracker, in W/m^2 (g) Electronic compass installed in the brace head (h) Three axis accelerometer to identify movement and orientation of the solar tracker in order to follow its position (i) Digital camera to take pictures in real time
Adafruit Max6675	Temperature sensor (up to 500 °C) and signal amplifier
Todoelectrónica 6710 WIND02	Windmeter (speed of wind 10 km/h = 4 turns/s)
Ventus Ciencia and PHYWE	Paraboloidal and parabolic trough collectors, respectively

2.2. Definitive State of Mechatronic Solar Tracker

After the previous design tasks, the next step was the real construction of the mechatronic prototype. The definitive result can be shown in Figure 3, with all the sensors of the CSP tracker already installed plus the driving devices and solar collectors. This solar tracker prototype can work in real irradiance exposed to outdoors conditions, on the situation that these meteorological conditions allow to do so, because it does not have encapsulation for any waterproof sensor. But it is true that its behaviour can be analyzed by mean on controlled tests in laboratory, or even with computing dynamic simulators that model atmospheric conditions.

Figure 3. Final state of the solar thermal tracker.

2.3. Tracking System Design and Arrangement

In order to show in detail the two axis tracking mechanism of the prototype, it can be useful to have a close look to the linear actuator and stepper motor included. In relation to the first one, as Figure 4a shows, the linear actuator is responsible for the tilt movement of the tracker, since it has an axial rod with a tiny but powerful internal gearbox and a rectangular section to increase its rigidity. Its working principle consists in pulling or pushing a certain load (solar collectors in this case) throughout the length of the rod. The speed of displacement depends on the gearbox mentioned before and the load the actuator is facing.

In the same way, the stepper motor is the device that describes the azimuth movement in the best way: to position in a reliable and precise way the azimuth of the tracker Figure 4b, together with a torque enough to move the load (solar collectors again). Specifically, the shaft of the stepper has to move a gear assembly that moves in its turn another shaft that impels two belt pulleys engaged.

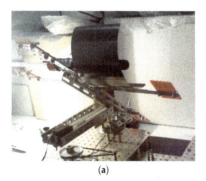
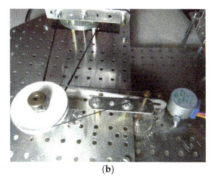

(a)   (b)

Figure 4. Details of the two axis tracking mechanism. (**a**) Linear actuator; (**b**) Stepper motor.

2.4. Definition of the Solar Tracking Algorithm

With an aim of describing better how the prototype tracks the Sun rays, it is also important to indicate how the irradiance sensors work. Firstly, it is necessary to take into account that there are four sensors located in cross in positions North-South-Est-West, which mission is to capture irradiance constantly (Figure 5).

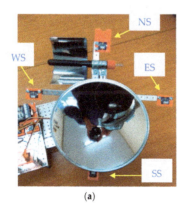

(a)   (b)

Figure 5. Data capture equipment. (**a**) Position of four irradiance sensors; (**b**) Detail of an irradiance sensor.

According to the values of irradiance of each sensor and the algorithm described later (seeking the maximum perpendicularity to sun rays, a mandatory condition to concentrate them in both collectors), the Arduino controller activates linear actuator or stepper motor shown before. Electronically, this kind of sensor is based on an irradiance to digital signal converter with the capability to measure a wide range of radiance spectrum using built-in dual sensibility diodes. In order to face the acquisition of irradiance data, it is possible to select between three detection modes: infrared mode (especially interesting for CSP); full spectrum; and visible spectrum for human being. In terms of irradiance and luminosity (respectively and making a conversion between them) the working range includes from $0 \, \text{W/m}^2$ (0.1 lux) up to $1000 \, \text{W/m}^2$ (40,000 lux).

Subsequently, an algorithm for two axis solar tracking was developed. To this end, it was necessary to identify the most important tasks and to modernize them. There are two operations: azimuthal movement (East-West) and zenithal movement (North-South).

Firstly, it was necessary to identify by abbreviations the names of the irradiance sensors, which intervene in the tracking algorithm, as it is shown in Table 2.

Table 2. Irradiance sensors.

Abbreviation	Sensor
NS	North Sensor, located in the upper part of the CSP solar tracker
SS	South Sensor, located in the lower part of the CSP solar tracker
ES	East Sensor, located in the right part of the CSP solar tracker
WS	West Sensor, located in the left part of the CSP solar tracker

The azimuthal (East-West) tracking works according to the principle shown in the flow chart of Figure 6. At first, the variables are reset, where the azimuthal angle β can be increased in steps of $\Delta\beta = 10°$ (user-adjustable value). Once the controller captures data from ES and WS sensors, if any of them is greater than the other one, this will imply a higher irradiance level in itself, and therefore, it will be necessary to increase or decrease that azimuth β in steps of $\Delta\beta$ searching for the maximum perpendicularity with respect to the Sun, ensuring the highest solar irradiation.

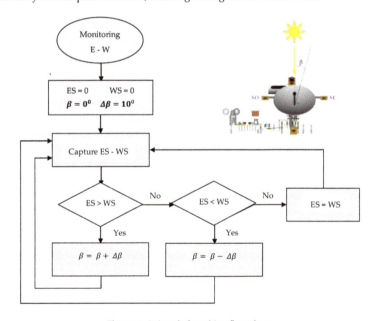

Figure 6. Azimuthal tracking flow chart.

In the same way, the elevation (North-South) tracking can be done, where its working principle is shown with the help of the flow chart of Figure 7. As mentioned before, the variables are reset, where the elevation angle α can be increased in steps of $\Delta\alpha = 10°$ (user-adjustable value too). Once the controller captures data from NS and SS sensors, it detects oscillations in irradiance values, if any of them is greater than the other one. So, the mechatronic device is able to increase or decrease that elevation α in steps of $\Delta\alpha$ searching for the maximum perpendicularity with respect to the sun, achieving the higher solar radiation.

Combining these two movements (azimuthal and elevation), the optimal position that allows to gain the maximum use of solar rays can be obtained, which therefore, permits to obtain a higher performance of the solar prototype. Because of the simplicity of Arduino-compatible devices, the programming of this controller is easy if the sequences detailed are followed as shown in the flow charts.

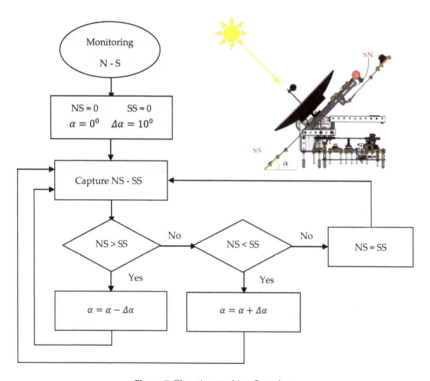

Figure 7. Elevation tracking flow chart.

On the other hand, it can be useful to model the tracking system of the prototype, by means of Laplace Transforms. These automatic control systems can command the variables that characterize any physical process and are built with the following items: input signal–process–output signal. The first one (input signal) is also known as control signal, set point or reference, and is used to command the process itself. And finally, the output signal is nothing but the variable to control.

The open-loop control system (non-feedback controller) computes its input into a system using only the current state and its model of the system. In this sense, a characteristic of this type of controller is that it does not use feedback to determine if its output has achieved the desired goal of the input. In such a way, the system does not observe the output of the processes that it is controlling. Consequently, an open-loop system cannot correct any errors that it could make, and that is why it also may not compensate disturbances in the system, what is not valid for the solar trackers.

On the contrary, the close-loop control system (feedback controller) is capable of using a feedback loop to control the behavior of any system by comparing its output with respect to a desired value with the help of a sensor. This way, is possible to calculate the difference as an error signal to dynamically change the output. Obviously, it is closer to the desired output and it is the strategy used in this paper as it can be seen below.

In effect, once the features of all the techniques related to automatic control are seen, it is interesting to obtain the Transference Function that models the working principles of a solar tracker. To do so, firstly the differential equations of the physical model are considered in terms of variable time (t) according to the model as it is shown in Figure 8.

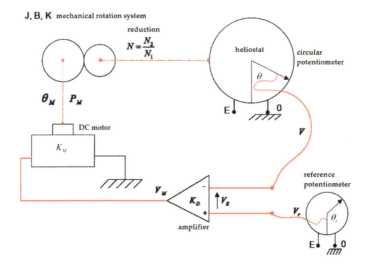

Figure 8. Modelling of the solar tracker in terms of differential equations.

$$v(t) = \frac{E}{2\pi} \cdot \theta(t) \tag{1}$$

$$v_r(t) = \frac{E}{2\pi} \cdot \theta_r(t) \tag{2}$$

$$v_E(t) = v_r(t) - v(t) \tag{3}$$

$$v_M(t) = K_D \cdot v_E(t) \tag{4}$$

$$p_M(t) = K_M \cdot v_M(t) \tag{5}$$

$$p_M(t) = J \cdot \frac{d^2\theta_M(t)}{dt^2} + B \cdot \frac{d\theta_M(t)}{dt} + K \cdot \theta_M(t) \tag{6}$$

$$\frac{N_1}{N_2} = \frac{\theta_2(t)}{\theta_1(t)} \tag{7}$$

where $\theta(t)$ is the angular position of the tracker, $\theta_r(t)$ is the reference angular of the tracker, $v(t)$ is the voltage obtained, $v_r(t)$ is the reference voltage from the potentiometer, between 0 V up to EV, $v_E(t)$ is the diference between $v_r(t)$ and $v(t)$, K_D is the gain of the amplificatory, p_M is the torque of the DC motor, K_M is the proportionality constant of the DC motor, J is the inertia of the load, B is the friction between joints, K is the elastic torque, $\theta_M(t)$ is the angle moved by the tracker and N is the gear relation.

Therefore, once the Laplace Transform is aplied to the expressions mentiond before, making a change of variable in terms of the new variable s = jw, and acording to its mathematical definition, the following set of aditional equations can be obtained:

$$L\ [f(t)] = F(s) = \int_0^\infty e^{-t\cdot s} \cdot f(t) \cdot dt \tag{8}$$

$$V(s) = \frac{E}{2\pi} \cdot \theta(s) \tag{9}$$

$$V_r(s) = \frac{E}{2\pi} \cdot \theta_r(s) \tag{10}$$

$$V_E(s) = V_r(s) - V(s) \tag{11}$$

$$V_M(s) = K_D \cdot V_E(s) \tag{12}$$

$$P_M(s) = K_M \cdot V_M(s) \tag{13}$$

$$P_M(s) = \left(J \cdot s^2 + B \cdot s + K\right) \cdot \theta_M(s) \tag{14}$$

$$\frac{N_1}{N_2} = \frac{\theta_2(s)}{\theta_1(s)} \tag{15}$$

Figure 9 shows the close loop working block diagram of the solar tracker in terms of transfer function.

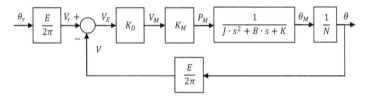

Figure 9. Close-loop block diagram for one axis of the solar tracker.

Even though this mathematical tool is really powerful, trying to simulate 'by hand' a device such as the solar tracker in order to predict its behaviour against wind, increase/decrease of irradiance and so on, is a very tedious task. However, it can be very useful to apply this model with the use of simulation environment software such as Modelica, Wolfram Systemodeller or Simulink.

2.5. Energetic Characterization

Before modelling the previous working tasks of the solar tracker, it was energetically characterized according to real measurements of paraboloidal and parabolic trough collector. Even though this prototype is not going to produce energy, it can be interesting to make this energetic characterization in order to foresee results for commercial collectors. In order to achieve this, Greenius 4.1.1 [17] software with meteorological data from the latest years in Madrid can be used.

In relation to the paraboloidal collector, the equations that model the power generated are:

$$P_{gro} = a \cdot E_{corr} + b \iff E_{DNI} \geqslant E_{DNI,\ min} \tag{16}$$

$$P_{gro} = 0 \iff E_{DNI} < E_{DNI,\ min} \tag{17}$$

where P_{gro} is the gross electrical power generated in W, E_{corr} is the corrected irradiance in W/m^2 that can be obtained in Equation (18), E_{DNI} is the direct normal irradiance in W/m^2, $E_{DNI,\ min}$ is the minimum E_{DNI} to start the production in W/m^2, a and b are two performance constants from the model, in W$_e$/(W/m^2). In this way:

$$E_{corr} = E_{DNI} \cdot f_{ref} \cdot f_{tem} \tag{18}$$

where f_{ref} and f_{tem} are the reflection and temperature correction factors obtained by:

$$f_{ref} = f_{sha} \cdot f_{int} \tag{19}$$

Here f_{sha} is the shadowing factor and f_{int} is the interception factor. So:

$$f_{tem} = \frac{\theta_{nor,amb} + 273,15\,°C}{\theta_{amb} + f_{ref} \cdot E_{DNI} \cdot c_{coo} + 273,15\,°C} \tag{20}$$

where $\theta_{nor,amb}$ is the normalized temperature for the performance of the model in °C, θ_{amb} is the higher temperature reachable in Stirling thermodynamic cycle in °C and c_{coo} is the cooling constant in °C/(W/m^2).

On the other hand we also considered the parasitic power with the help of:

$$P_{par} = P_{ope} \Leftrightarrow E_{DNI} \geqslant E_{DNI,\,min} \tag{21}$$

$$P_{par} = P_{sle} \Leftrightarrow E_{DNI} < E_{DNI,\,min} \tag{22}$$

where P_{par} is the parasitic global power in W, P_{ope} is the operating power in W and P_{sle} is the sleeping power in W.

In case of power injected to net, it can be obtained with the Equation (8):

$$P_{gri} = P_{gro} - P_{par} \tag{23}$$

where P_{gri} is the injected power towards net or consumed from net in W.

The gross profit in energy obtained for our paraboloidal collector using equations shown before, can be seen in Figure 10a. In the same way, for the case of cylindrical-parabolic collector a modelling process was designed according to the following equations that allow to obtain its performance:

$$\eta_{col} = K \cdot \eta_{opt,\,0} \cdot \eta_{cle} - \left(K \cdot b_0 \cdot \Delta T + \frac{b_1 \cdot \Delta T + b_2 \cdot \Delta T^2 + b_3 \cdot \Delta T^3 + b_4 \cdot \Delta T^4}{DNI} \right) \tag{24}$$

where η_{col} is the performance of the cylindrical-parabolic collector, b_i $(i = 0 - 4)$ are the heat losses in vacuum tubes estimated experimentally, $\eta_{opt,\,0}$ is the optical performance for the collector for $\theta = 0°$, θ is the incidence angle for solar rays, ΔT is the difference of temperature according to the Equation (25), K is the dependency factor of $\eta_{opt,\,0}$ over θ as a result of Equation (26) and DNI is the normal incidence irradiance in W/m^2:

$$\Delta T = \frac{T_{SF,\,in} + T_{SF,\,out}}{2} - T_{amb} \tag{25}$$

Here $T_{SF,\,in}$ is the inlet temperature of heat transfer fluid in °C, $T_{SF,\,out}$ is the outlet temperature of heat transfer fluid in °C y T_{amb} is the ambiance temperature in °C, and

$$K = IAM \cdot \cos\theta \tag{26}$$

where IAM is the incidence angle modifier according to Equation (12):

$$IAM = 1 - \frac{a_1 \cdot \theta + a_2 \cdot \theta^2 + a_3 \cdot \theta^3}{\cos\theta} \tag{27}$$

The last equation a_i contains empiric parameters from the manufacturer of the collector. The efficiency obtained for the parabolic trough collector can be seen in Figure 10b. In addition, although it has been said that the main goal of this device is not energy production (its target is monitoring meteorological data), it can be interesting to analyze the power consumption (in order of magnitude) of the linear actuator and stepper motor to make a comparison with the energy generated.

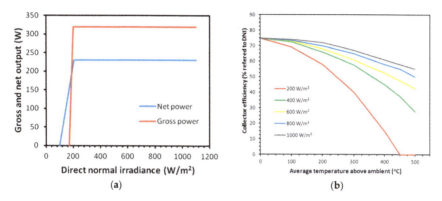

Figure 10. Energetic characterization of collectors. (**a**) Energetic profit obtained by the paraboloidal collector; (**b**) Energetic performance of parabolic trough collector.

In relation to the linear actuator (a Firgelli model L12-100-100-6R) and its features according to its technical datasheet, the electrical working conditions of 450 mA at 6 V can be obtained, so considering Equation (28):

$$P = u \cdot i \tag{28}$$

It is easy to obtain the result of the power substracted to power generated, but only in moments when the linear actuator is working, that is, only when it is acting to extend/contract the solar collectors. It implies a consumption of 2.7 W. Making the hypothesis of a working time of 1 s per movement, and a change of position every 6 min, that implies 10 s/h of working time. Assuming 8 h/day of sun light duration as a medium value in Madrid, this adds up to 80 s/day. In such a way the final quantity of consumed energy is 216 J, according to Equation (29):

$$E = P \cdot t \tag{29}$$

On the other hand, following the same procedure and starting points, the electrical working conditions of the stepper (model 28BYJ48) are: 1 mA at 5 V. Considering again Equation (28), the power substracted is 0.05 W and the energy consumed, according to Equation (29) sums only 4 J.

3. Results

In the first place, measures of irradiance and temperature were taken with the help of the thermal solar tracker throughout winter in 2015 and 2016. The data were collected sequentially with the Arduino device every 30 min from 8.00 a.m. to 18.30 p.m., that is, from sunrise to sunset from December until March in Madrid, except for rainy days. The averages of irradiance obtained are shown in Figure 11. It can be seen how the correct calibration of the device reflects higher values at the times closer to midday. On the other hand, the temperature average for both sensors (paraboloidal and parabolic trough) is shown in Figure 12.

As it can be inferred from Figure 12, the temperature peaks are higher in the paraboloidal transducer, whose geometry and concentration ratio can reach values above the parabolic trough collector for the same conditions of irradiance. As a check, it was decided to carry out an inspection using thermal images that allow to verify the results obtained by the sensor.

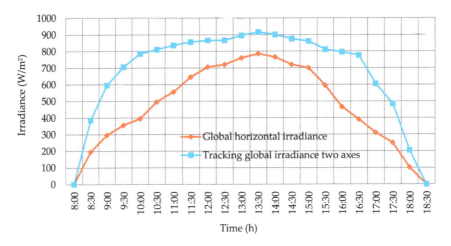

Figure 11. Averages values of irradiance every 30 min.

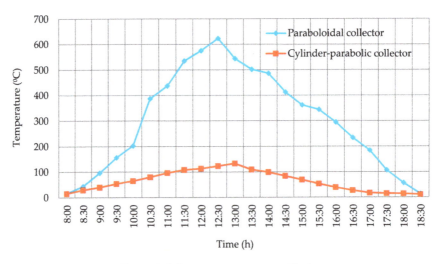

Figure 12. Collectors temperature every 30 min.

In this regard, two representative thermographic images (Figure 13) of both collectors taken at 12:00 p.m. and then proceeded to processing them using the software SmartView 3.12© were taken. In addition, the three-dimensional representation of the temperatures is displayed in function of the geometry of both collectors (Figures 14 and 15), since it identifies the higher ones. Specifically, observing the paraboloidal collector (Figure 14) is appreciated, as expected, a very pronounced peak in its focus and in the parabolic trough a higher temperature along the cylindrical receiver (Figure 15) is also displayed.

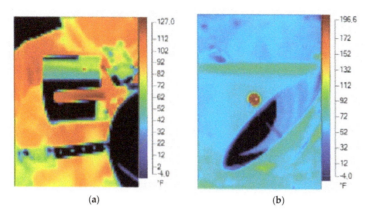

Figure 13. Inspection using thermographic camera. (**a**) Parabolic trough collector thermography; (**b**) Paraboloidal collector thermography.

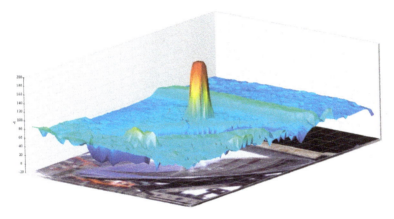

Figure 14. 3D graph that shows Stirling dish collector temperatures.

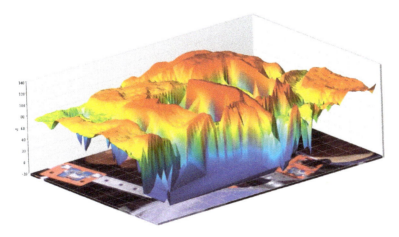

Figure 15. 3D graph that shows the parabolic trough collector temperatures.

4. Conclusions

In this paper it has been possible to design, model, build and calibrate a mechatronic paraboloidal and parabolic trough thermal solar tracker prototype which allows one to collect meteorological and energy data more reliably than using traditional methods based on satellite estimations obtained by the reflection of the incident radiation on the Earth. Furthermore, a scaled simulation of the behaviour of both collectors (paraboloidal and parabolic trough) has been achieved, anticipating the results regarding the energy potential in buildings wherever it takes place.

However, as further research lines and future implementation of this prototype in buildings, it would be appropiate to create a shell which can allow the degrees of freedom of the prototype movement and ensure the protection of the electronic components outdoors. In addition, it could be possible to use paraboloidal collectors whose geometry is less parabolic becoming pseudo-flat and reducing the visual impact, or retractable collectors which in case of a lack of irradiance could fold their surface up in a small sheet form favoring building integration.

Acknowledgments: This work has been partially supported by the Polytechnic University of Madrid and would not have been possible without the collaboration of students of vocational education training by the Salesianos Carabanchel center. Specially the technicians Adrián García Pardo and Víctor Blanco Peña for their effort and dedication in the trials.

Author Contributions: C.M. and J.P.D. created and designed the experiments. J.P.D. and D.F. performed the experiments; C.M. and J.P.D. analyzed the data; D.F. wrote the article. M.P.R. helped to analyze the documents.

Conflicts of Interest: The authors declare no conflict of interest.

References

1. Gil, C.M.; Gil, M.-A.C.; Castro, M.; Santos, C.A.; EIbañez, C.J. *Energía Solar Térmica de Media y Alta Temperatura (Monografías Técnicas de Energías Renovables)*, 1st ed.; PROGENSA-Promotora General de Estudios: Sevilla, Spain, 2001; pp. 1–68.
2. Imadojemu, H.E. Concentrating parabolic collectors: A patent survey. *Energy Convers. Manag.* **1995**, *36*, 225–237. [CrossRef]
3. Feng, C.; Zheng, H.; Wang, R.; Ma, X. Perfomance investigation of a concentrating photovoltaic/termal system with transmissive Fresnel solarconcentrator. *Energy Convers. Manag.* **2016**, *111*, 401–408. [CrossRef]
4. Ahmadi, M.H.; Sayyadi, H.; Dehghani, S.; Hosseinzade, H. Designing a solar powered Stirling heat engine based on multiple criteria: Maximized thermal efficiency and power. *Energy Convers. Manag.* **2013**, *75*, 282–291. [CrossRef]
5. Arora, R.; Kaushik, S.C.; Kumar, R.; Arora, R. Multi-objective thermo-economic optimization of solar parabolic dish Stirling heat engine with regenerative losses using NSGA-II and decision making. *Electr. Power Energy Syst.* **2016**, *74*, 25–35. [CrossRef]
6. Tlili, I.; Timoumi, I.; Nasrallah, S.B. Thermodynamic analysis of the Stirling heat engine with regenerative losses and internal irreversible. *Int. J. Engine Res.* **2008**, *9*, 45–56. [CrossRef]
7. Aldegheri, F.; Baricordi, S.; Bernardoni, P.; Brocato, M.; Calabrese, G.; Guidi, V.; Mondardini, L.; Pozzetti, L.; Tonezzer, M.; Vincenzi, D. Building integrated low concentration solar system for a self-sustainable Mediterranean villa: The Astonyshine house. *Energy Build.* **2014**, *77*, 355–363. [CrossRef]
8. Mao, Q. Recent developments in geometrical configurations of termal energy storage for concentrating solar power plant. *Renew. Sustain. Energy Rev.* **2016**, *59*, 320–327. [CrossRef]
9. Liu, M.; Steven, T.; Bell, S.; Belusko, M.; Jacob, R.; Will, G.; Saman, W.; Bruno, F. Review on concentrating solar power plants and new developmentes in high temperatura termal energy storage technologies. *Renew. Sustain. Energy Rev.* **2016**, *53*, 1411–1432. [CrossRef]
10. Lamnatou, C.; Mondol, J.D.; Chemisana, D.; Maurer, C. Modelling and simulation of Building-Integrated solar thermal systems: Behaviour of the system. *Renew. Sustain. Energy Rev.* **2015**, *45*, 36–51. [CrossRef]
11. Bakos, G.C. Design and construction of a two-axis Sun tracking system for parabolic trough collector (PTC) efficiency improvement. *Renew. Energy* **2006**, *31*, 2411–2421. [CrossRef]

12. Skouri, S.; Ben Haj, A.; Bouadila, S.; Ben Salah, M.; Ben Nasrallah, S. Design and construction of sun tracking systems for solar parabolic concentrator displacement. *Renew. Sustain. Energy Rev.* **2016**, *60*, 1419–1429. [CrossRef]

13. Chong, K.K.; Wong, C.W. General formula for on-axis sun tracking system and its application in improving tracking accuracy of solar collector. *Sol. Energy* **2009**, *83*, 289–305. [CrossRef]

14. Reda, I.; Andreas, A. Solar position algorithm for solar radiation applications. *Sol. Energy* **2004**, *76*, 577–589. [CrossRef]

15. Beltrán, J.A.; González, J.L.S.; García-Beltrán, C.D. Design, manufacturing and performance test of a solar tracker made by an embedded control. In Proceedings of the Electronics Engineering, Robotics and Automotive Mechanics Conference (CERMA 2007), Cuernavaca, Morelos, Mexico, 25–28 September 2007; pp. 129–134.

16. Chen, Y.T.; Lim, B.H.; Lim, C.S. General sun tracking formula for heliostats with arbitrarily oriented axes. *J. Sol. Energy Eng.* **2006**, *121*, 245–250. [CrossRef]

17. Falck, D.Y.; Colle'e, B. *Freecad [How-To]*, 1st ed.; Packt Publishing: London, UK, 2012; pp. 1–70.

18. Warren, J.-D.; Adams, J.Y.; Molle, H. *Arduino Robotics*, 1st ed.; Apress: New York, NY, USA, 2001; pp. 1–180.

19. Salamone, F.; Belussi, L.; Danza, L.; Ghellere, M.Y.; Meroni, I. An open source low-cost wireless control system for a forced circulation solar plant. *Sensors* **2015**, *15*, 27990–28004. [CrossRef] [PubMed]

Article

A Compact Forearm Crutch Based on Force Sensors for Aided Gait: Reliability and Validity

Gema Chamorro-Moriana [1,*], José Luis Sevillano [2] and Carmen Ridao-Fernández [1]

[1] Department of Physiotherapy, University of Seville, Sevilla 41009, Spain; mcrf.2817@gmail.com
[2] Department of Computer Technology and Architecture, University of Seville, Sevilla 41012, Spain; sevi@atc.us.es
*   Correspondence: gchamorro@us.es; Tel.: +34-639-868-009

Academic Editor: Gonzalo Pajares Martinsanz
Received: 31 March 2016; Accepted: 16 June 2016; Published: 21 June 2016

Abstract: Frequently, patients who suffer injuries in some lower member require forearm crutches in order to partially unload weight-bearing. These lesions cause pain in lower limb unloading and their progression should be controlled objectively to avoid significant errors in accuracy and, consequently, complications and after effects in lesions. The design of a new and feasible tool that allows us to control and improve the accuracy of loads exerted on crutches during aided gait is necessary, so as to unburden the lower limbs. In this paper, we describe such a system based on a force sensor, which we have named the GCH System 2.0. Furthermore, we determine the validity and reliability of measurements obtained using this tool via a comparison with the validated AMTI (Advanced Mechanical Technology, Inc., Watertown, MA, USA) OR6-7-2000 Platform. An intra-class correlation coefficient demonstrated excellent agreement between the AMTI Platform and the GCH System. A regression line to determine the predictive ability of the GCH system towards the AMTI Platform was found, which obtained a precision of 99.3%. A detailed statistical analysis is presented for all the measurements and also segregated for several requested loads on the crutches (10%, 25% and 50% of body weight). Our results show that our system, designed for assessing loads exerted by patients on forearm crutches during assisted gait, provides valid and reliable measurements of loads.

Keywords: walking; crutches; instrumentation; validity; reliability

1. Introduction

Gait training is one of the most prominent processes in the physiotherapy area, as gait is one of the main functions of human beings [1–3]. Frequently, recovery of musculoskeletal injuries to an affected lower member involves gait training using forearm crutches for partial unloading thereof [2–4]. In this sense, the current trend is to load the maximum amount of weight depending on the lesion and its evolution on the lower limb. Authors such as Xu *et al.* [5] assert that lower limb unloading damages segmental circulation and decreases muscle tone, which consequently reduces the osteoblastic action and increases osteoclastic action. This result particularly affects the recovery of patients with sequelae of fractures in lower members, even more than in those who suffer from osteopenia or osteoporosis [6]. If we add inhibition of joint and muscle plantar proprioceptive receptors to these aspects, a functional deficit is obtained in the patient that could hamper and delay their recovery [7,8]. On the other hand, an excessive load on the injured lower member can lead to compressions or undue stress of structures on the patient even without regeneration or in the process of recovery, thus causing relapses and sequelae of their original injury [9].

Therefore, it is fundamental that there are feasible measuring instruments in daily practice that objectively monitor or control the unloading that the patients perform on an injured limb. The design of this kind of measuring instrument is an area of active research. As force platforms are usually very expensive and limit user movements, other devices are more convenient. For instance, Gonzalez *et al.* [10] have recently developed a system for gait monitoring based on a wireless sensorized insole. However, although sensorized insoles allow us to identify foot pathologies [11], they have several drawbacks when used during aided gait: for instance, they have to be adapted to each specific user (e.g., his/her foot size) and require using shoes (which may not be used with bandages). On the other hand, crutches can be shared by different users and used in many different situations without any modifications. Furthermore, they permit the comparison of ipsilateral and contralateral loads. One of the authors (Chamorro-Moriana) as part of her PhD Thesis, designed the so-called GCH System 1.0 (GCH is the abbreviation of the name of its inventor) that allows measuring loads applied on the forearm crutches, a prototype patented in 2009 [12], developed and validated in a previous study [13].

Other authors have also recently discussed the use of crutches for gait monitoring, with a prototype and some preliminary results being reported in another study [14]. However, their system overlooks many clinical issues. For instance, it includes a vibrating signal in the crutch grip that is activated only if the patient loads more weight on the crutch than recommended. Clinically, this is wrong feedback since the most important problem for the patient occurs when the patient loads less than the appropriate amount on the crutch though their arm, thus implying an excessive and dangerous load on the affected lower limb. To the best of our knowledge, none of the systems described in literature (including [13,14]) are ready to be used in clinical settings.

The system described in this paper, which we name GCH System 2.0, is an improved version of GCH System 1.0 described in another study [13]. The use of GCH System 1.0 in the laboratory enabled us to verify its effectiveness, after an extensive trial period, and also take into account the feedback provided by real users and the researchers themselves, and a number of improvements were introduced in the original design that led to GCH 2.0. The aim of this paper is to describe all of these improvements, as well as the design, validation and calculation of the reliability of the GCH 2.0 load measurement system. In addition to these technological and methodological contributions, specific clinical features were added that let the new system adapt to different processes of the functional recovery, as the authors show in the following section. The new tool will allow us to objectively know the weight bearing exerted by patients on crutches during the aided gait, and, thus, the unloading on the affected lower limb. In addition, it will enable training based on the patient's biofeedback [15,16]. Finally it will promote monitoring, evaluation and analysis of load progressions to establish clinical protocols [17].

The rest of the paper is organized as follows: Section 2 describes the GCH 2.0 system, with an in-depth comparison to the first GCH prototype. Section 2 also describes the experimental setting of the study of the validity and reliability of measurements via a comparison with the validated AMTI OR6-7-2000 platform. Section 3 presents the results of these experiments, which are discussed in Section 4. Finally, Section 5 presents our Conclusions.

2. Experimental Section

2.1. GCH System 2.0

In this section, the main components of GCH 2.0 are described, emphasizing the main improvements carried out on the previous prototype [13].

The core of GCH 2.0 System consists of the coupling of a miniature force sensor, an Exact Sensor Instrument's EX601D (Shenzhen Exact Sensor Instrument Co., Ltd., Shenzhen, China) [18], within the distal part of the forearm crutch shaft. Compression load cell features include: cylindrical (in the shape of a coin), diameter of 19 mm, stainless steel measuring element, hermetically sealed (fully welded) and easy installation. Safe load limit = 100 kg, rated output = 0.7198 mV/V, linearity $\leqslant 0.05\%$ Full Scale (FS), hysteresis $\leqslant 0.05\%$ FS, repeatability $\leqslant 0.03\%$ FS, Zero balance $\pm 2\%$ FS and operating temperature range = $-20\,^{\circ}\text{C}$ to $60\,^{\circ}\text{C}$.

The said cell is connected to the electronic board and power batteries housed in this area. The data acquisition card has the function of emitting a radiofrequency signal. The communication protocol used is SimpliciTI, a low-power Radio Frequency (RF) protocol from Texas Instruments Inc. (Dallas, Texas, USA) [19] at 898 MHz with a period of 80 samples per second. The digital and amplified signal arrives at an ultra low power MSP430 microcontroller also from Texas Instruments (input voltage 2.4 V DC—Direct Current, battery/autonomy 6000 mAh).

GCH 2.0, as opposed to GCH 1.0 [12], introduces a compact design of the system integrating all these elements inside the crutch shaft. The sensor with its coupling mechanism; the data acquisition and radio cards, incorporated in a single printed circuit board (PCB); and the power supply are integrated in the most distal part of the crutch, an independent and extensible pipe to allow for the height adjustment tube (Figure 1). All external cables are eliminated. Manufacture of the PCB has been conducted with miniature electronic components, as a surface mount device (SMD). This allows us to integrate electronics inside the crutch, unlike the first prototype that wired the sensor installed inside the crutch with an external box that patients had to carry in their belt.

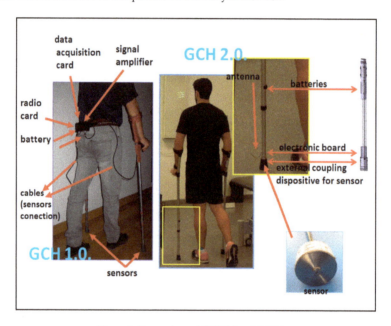

Figure 1. Comparison of GCH 1.0 and GCH 2.0.

It is worth noting that when the GCH 2.0 crutch is started, the offset process is automatically activated. Once the crutch is on, it is placed vertically on the floor for 5 s, thus its weight is not recorded. This method allows us to exchange or add the usual standard components to the crutches (ergonomic handles, casings, beads, *etc.*) without the weight difference affecting measurements.

One of our main objectives with this new system is to provide a compact and ready-to-use device that could be used not only in clinical trials or under direct supervision by the physiotherapist, but also during daily life, thus speeding up the patient's recovery. With this aim in mind, a clock-shaped, portable receiver has been introduced that can be used independently by the patient while performing the aided gait with crutches even outdoors. Thus, not only would they walk on stable ground within a room, but they could also walk on steps, ramps or uneven floors, *i.e.*, the usual outdoor obstacles and difficulties [20]. In addition, thanks to the use of a compact force sensor, GCH 2.0 does not require frequent calibrations.

Regarding the batteries, we decided to replace them with double standard AA that, together with an adapter, were also integrated inside the crutch, with a minimum amperage of 3000 mA each to prolong their use. The batteries are directly rechargeable through a connector or by extracting them and using a standard charger. This latter mode allows the user to replace the batteries while the used ones are being charged so that the crutches are available at all times.

The design of a specific program has been another major improvement in the applicability of GCH 2.0 as a measurement tool. The Crossbow sensor of GCH System 1.0 emitted signals that were monitored and recorded by a generic program called Moteview 2.0 (Moog Crossbow, East Aurora, NY, USA) [21]. This program works at a frequency of 0 to 10 Hz, so that even using the maximum frequency, the curve recorded was not accurate enough. The new GCH 2.0 software (called GCH Control Software 1.0) records data up to 80 Hz, thus obtaining a more optimized curve of weight bearing at each step. The basic application covers one or two modules (one for each crutch), although the system is able to cover more. That is, many patients can be walking with instrumented crutches at the same time. The signal sent by the crutches is detected by a small USB receiver that is connected to a computer ("fixed system") by means of a virtual COM (Communication) port, or by a receptor incorporated into a watch, mobile phone, pendant or substitute (the "portable system" described before). This portable receiver provides autonomy to the patient in order to perform aided gait outdoors.

The fixed system is used during patient care in the clinic or laboratory as a training or research object, respectively. The software translates the millivolts from the sensor signal into units of force (force kilograms—kg), records the data, analyzes them and allows their numerical and graphical display of the loads that the subject carries out on screen in real-time. Besides specific graphs of weight bearing by each crutch, the unification or overlapping of both are obtained. These graphs can be seen by the patient through a canon projector to perform a visual feedback during gait.

This visual feedback is part of what we call a "self-correcting" mechanism. Thanks to this mechanism, new in GCH 2.0, the patient receives acoustic and/or visual information about the load exerted on the crutches. The visual feedback is based on the above-mentioned projection on a screen of the amount of load exerted, which is directly proportional to the injured lower member unloading. This feature is mainly intended for patient's training during physiotherapy sessions. On the other hand, the acoustic feedback allows the system to be used without supervision even outdoors. With this self-correcting mechanism, if the amount of load is wrong, the subject is able to increase or reduce the force to improve the accuracy and exactly manage the ideal load recommended by the physiotherapist.

Finally, the program contains a database specifically designed for patients, in which we can record all their data and the applied sessions with our tool. Likewise, the recordings of the loads applied, number of steps, mistakes made, *etc.*, are designed for their posterior analysis within the care or research area.

The System was registered at the Spanish Patent and Trademark Office with number P201031779, and international expansion has been carried out [22]. A summary of all the improvements is shown in Table 1.

Table 1. The main differences between GCH 1.0 and GCH 2.0.

GCH 1.0.	GCH 2.0.
Distributed system	Compact system
Patients have to carry an electronic box place on their belts.	Electronic component inside the crutch tube.
External cables are necessary to connect the sensors to control box placed on the patient's belt.	Internal cables. Patients do not have any contact with cables.
External electronic components.	Internal miniature electronic components/ surface mount device (SMD).
Weight: 1150 g.	Weight: 720 g.
Non standard battery/rechargeable/700 mA.	Standard battery/AA/rechargeable/6000 mA.
Zero is not automatic.	Offset process is automatically activated.

Table 1. *Cont.*

GCH 1.0.	GCH 2.0.
Only for a patient walking with one or two crutches.	Several patients can use the System simultaneously, with one or two crutches.
Discretized biofeedback. System informs if the load is wrong only with a binary signal.	The physiotherapist/patient can choose between continuous or discretized visual biofeedback. In the continuous mode, the patient receives information throughout the whole process [23].
Moteview 2.0. Generic software that shows: amount of load and a simple linear chart. This is visualized by the researcher. It is not useful for the patient.	GCH Control Software 1.0.: Specific program to control assisted gait. The load could be shown in percentages of the patient's weight-bearing (data of clinic interest). It offers specific charts and data for researchers, physiotherapists and patients. It is adaptable to the kind of patient. (Figure 2).
No database.	Patients' clinical database.
Data sampling frequency ⩽10 Hz	Data sampling frequency ⩽80 Hz
The portable system. The physiotherapist selects the ideal load without percentages. It does not allow for comparisons and research.	The portable system (watch). The physiotherapists or researchers select the ideal load or the percentage of the patient's weight-bearing (data of clinic interest).

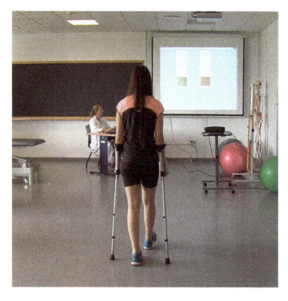

Figure 2. Individual walking while the quantity of the load exerted on the crutches is observed on the screen to improve its accuracy. The image on the board, which is different from the computer screen, is a specific chart for the patient.

Basic Functions of the GCH System 2.0

From an operational point of view, the two main functions of GCH 2.0 are:

- Load control. The objective measurement of the loads applied to the crutches is the basis of the System. It shows the kilograms exerted on the crutches and the percentage of patient body weight (PPBW). This datum is the most relevant in the clinic, which always requires the subject's current weight to be entered. The percentages allow researchers to compare intra-subject and inter-subject tests in order to establish treatment protocols.
- Feedback mechanism. The feedback information includes, individually, the ideal load exerted on the crutches (directly proportional to the unloading on the injured lower member), PPBW,

according to the pathology and the treatment phase, introduced into the software as well as a percentage of error tolerated clinically due to excess or defect load. The fixed system alerts the subject of the mistakes made during gait for immediate self-correction (Figure 2). Acoustic signals, a continuous whistling, will be used if the recommended load is exceeded or discontinuous if it does not reach it; and visual, by using a projector. The portable system only uses an audible feedback. The fixed system shows only the feedback information useful for the patient's training in the projector, and additional personalized clinical information (useful for the physiotherapist/researcher) on the computer screen.

The inclusion of the patient's weight especially benefits the functionality of the program, since it allows us to extrapolate load amounts to PPBW [9], both in the recommendation of the ideal weight as well as the margins of errors allowed. This offers the possibility of performing intra-subject and inter-subject comparisons. Only in this way can we advance towards the creation of protocols of performance based on scientific evidence [24].

2.2. Study Design

The research presented in this section is a concordance study on the equivalence between the values obtained for the same variable (vertical reaction force), under the same conditions and synchronously by two different measurement procedures [25,26]: the force measurement system applied to a crutch, GCH System 2.0, and the already validated AMTI OR6-7-2000 force platform. The following sections, 2.3 to 2.5, present sufficient details to allow reproducibility of the results.

2.3. Study Variables

The variables analyzed were quantitative, since they referred to amounts of load or the difference between various measurements.

- Variable 1: GCH. Vertical reaction force of the Platform on the crutch (Z component) measured using our System. This variable is secreted in GCH_right and GCH_left, right and left crutches, respectively. The kg is used as a unit of measurement for GCH.
- Variable 2: Platform (kg). Vertical reaction force of the crutches on the Platform (Z component) calculated by the AMTI Platform. The unit of measurement was the kg.

We only consider the axial force or Z component because, clinically, we are interested in analyzing the peak load and its maintenance over time, and this peak load occurs when the crutch is perpendicular to the ground. At present, the components X and Y are negligible, being assumed as measurement errors.

2.4. Measurements and Participants

The measurements were made by means of the GCH System 2.0 and the AMTI Platform during the assisted gait carried out by 30 participants, 18 women (66.7%) and 12 men (33.3%), with an age range from 18 to 45 years (mean = 29.87 years; Standard Deviation SD = 7.26).

Different PPBW applied to the crutches were requested in order to obtain heterogeneous measurements during gait. The sampling was considered non-probabilistic due to guidelines provided. We obtained loads between 2.13 kg and 50.60 kg.

For each crutch (2) and participant (30), nine measurements were taken. Of these, three belonged to a requested load on the crutch at 10% of body weight, another three at 25% and three more at 50% or the maximum possible. A total of 540 measurements were taken (270 with each crutch). These values were compared with the 540 measurements undertaken by the force platform. The number of measurements (540) is high enough according to the required sample size when comparing two means in two samples [27].

Finally, participants were also selected in a non-probabilistic and convenience mode.

Inclusion Criteria

- healthy subjects between 18 and 60 years old with previous experience with crutches;
- presenting a normal gait, being asymptomatic on walking at free cadence;
- overcome a simple test of static equilibrium, consisting of keeping one's balance on each foot for 30 s without great bodily movements [26].

Exclusion Criteria

- having an evident disorder of overall coordination and physical skill which could alter the aided gait.

The research protocol was approved by the ethics committee of the University Hospital Virgen Macarena (Seville, Spain). All participants gave written informed consent prior to participation in the study.

2.5. Data Collection

The measurements carried out by the two systems during aided gait were taken simultaneously and under the same conditions. The biomechanics laboratory includes a walking corridor 8.5 m long. The force platform was located halfway along the walk. Laterally, and along this corridor, signals were placed in a straight line that offered: the control of the gait direction and a distracting effect that prevented the subjects from centering their attention on the Platform so as to make the crutches match it (Figure 3).

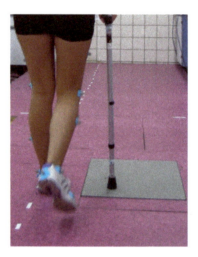

Figure 3. Individual performing aided gait in two stages along the walkway with direction signs and distracting effects so as to avoid him/her focusing on the platform.

To avoid differences among the tests, the laboratory always used the same artificial light (with lowered blinds) and a constant temperature (25 °C). This data is relevant as the subjects were asked to do physical activity and wear special clothes (sports short, short-sleeved T-shirt and sports shoes).

The participants first had a learning and familiarization period followed by them completing an 8.5 m walk, 10 times with each crutch and with each load percentage (10%, 25% and 50% or maximum possible). Aided gait with a partial load was in two stages, with a contra-lateral crutch and simultaneous heel and crutch support. The chosen height of the crutch was associated with an elbow flexion of 20° to 30° [28,29]. The required speed was at free cadence.

Data collection was performed 10 times with the right crutch and 10 more with the left in each load modality and every patient. Of the valid measurements, we took the three central ones to conduct our statistical analysis. GCH Control Software was used with the GCH System 2.0 and Vicon Nexus (Vicon, Oxford, UK) with the AMTI Platform, controlled by two researchers who recorded the load applied when a crutch matched the Platform. Subsequently, the timing and scoring reference of this support was recorded to correlate the data from both pieces of software in time.

Both measurement systems started at offset or zero, so the weight of the implemented crutch, 0.720 kg, was subtracted from the values of the Platform.

2.6. Statistics

The data obtained were organized and analyzed using the IBM SPSS statistical software (Version 22.0; SPSS Inc., Chicago, IL, USA). The descriptive analysis included: mean, standard deviation (SD), minimum, maximum, and percentiles 25, 50 and 75 (P_{25}, P_{50} and P_{75}). The inferential analysis considered a confidence level of 95%, so that the experimental *p*-value was compared to a significance level of 5%.

To determine the most appropriate test according to data behaviour, we performed the Kolmogorov–Smirnov normality test. According to whether these normality criteria are met or not, the following parametric tests are considered appropriate [30]:

- *t*-test for related samples [30]: it compares the mean values of related samples when the values of the variables meet the normality criteria. This test was used to determine whether the two measurements can be considered similar. In addition, the study was performed based on the different ranges of weight loaded onto the crutches.
- Wilcoxon signed-rank test [30]: it compares the related sample distribution when the values of the variables do not meet the normality criteria.

The intra-class correlation coefficient [31] was used to carry out the concordance analysis between the GCH System and AMTI Platform.

3. Results

The descriptive analysis of the variables *Platform* and *GCH* is shown in Table 2 for each crutch and different requested loads (10%, 25% and 50% of body weight). Note that three repeated measures on six different conditions (arm by percentage of load) are taken for every subject.

Table 2. Descriptive analysis of the Platform and GCH for each crutch and different loads.

| | | Load | N | Mean* | SD | Minimum | Maximum | Percentiles | | |
								25	50	75
Platform	Right crutch	10	90	7.33	4.00	2.24	20.09	4.38	6.37	9.00
		25	90	16.03	4.86	9.07	29.89	12.44	15.04	19.00
		50	90	25.66	8.45	12.64	50.60	19.33	24.23	30.63
	Left crutch	10	90	7.76	4.30	2.13	23.15	4.58	6.73	9.33
		25	90	15.79	5.07	6.72	30.50	12.64	14.89	17.42
		50	90	26.48	8.28	9.58	46.52	20.12	25.45	32.59
GCH	Right crutch	10	90	7.31	3.97	2.20	19.65	4.40	6.30	9.07
		25	90	15.93	4.82	8.99	29.73	12.42	14.95	18.90
		50	90	25.46	8.38	12.70	50.06	19.02	24.38	30.40
	Left crutch	10	90	7.73	4.26	2.14	22.92	4.57	6.76	9.25
		25	90	15.69	5.02	6.70	30.19	12.56	14.77	17.32
		50	90	26.27	8.22	9.58	46.02	19.88	25.23	32.17

* Values are presented in Kg.

Table 3 shows the intra-class correlation coefficients (ICCs) for the different values of loads and crutch. Note that ICCs are always between 0.99937 and 0.99995 with $p < 0.001$.

Table 3. Intra-class correlation coefficients between Platform and GCH.

	Load	Intra-Class Correlation	Confidence Interval (95%)		*p*-Value
			Lower Bound	Upper Bound	
	10	0.99964	0.99946	0.99976	<0.001
Right crutch	25	0.99937	0.99904	0.99958	<0.001
	50	0.99985	0.99977	0.99990	<0.001
	10	0.99990	0.99985	0.99994	<0.001
Left crutch	25	0.99993	0.99990	0.99996	<0.001
	50	0.99995	0.99992	0.99996	<0.001
Global		**0.99992**	**0.99990**	**0.99993**	**<0.001**

The differences between the average values of AMTI and the right crutch and AMTI and the left crutch were both ⩽0.11 kg. In addition, 78.1% of the results (422/540) showed higher values in the Platform than in GCH; in 20.2% (109/540), values were lower and 1.7% (9/540) recorded exactly the same measurement. Although the differences were significant ($p < 0.01$), the global effect-size is 0.028 (and, in all cases, it is always lower than 0.1), so the differences are not relevant from a statistical point of view.

Figure 4 shows the similarities between the GCH and Platform measurements.

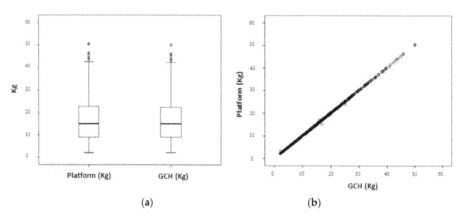

(a) (b)

Figure 4. Box plot representing percentiles 25, 50 and 75 (**a**) and scatter plot (**b**) of the relationship between platform and GCH measurements.

The Limits of Agreement (LOA) are shown using a Bland-Altman graphical analysis [32] in Figure 5. The differences between (platform minus crutch) and GCH, versus the means between (platform minus crutch) and GCH, are represented. The study is complemented by Table 4, which determined that the higher the load, the less the number of tolerable measurements. Even so, all of the values were within a 93% tolerance level.

Table 4. Frequencies and load percentages of adjustment for the tolerance levels according to the Bland–Altman method. (*n* = number of measurements).

Requested Load	Tolerance Level	*n*	%
10%	Tolerable	178	98.9
	Not tolerable	2	1.1
25%	Tolerable	173	96.1
	Not tolerable	7	3.9
50%	Tolerable	169	93.9
	Not tolerable	11	6.1
Global	Tolerable	520	96.3
	Not tolerable	20	3.7

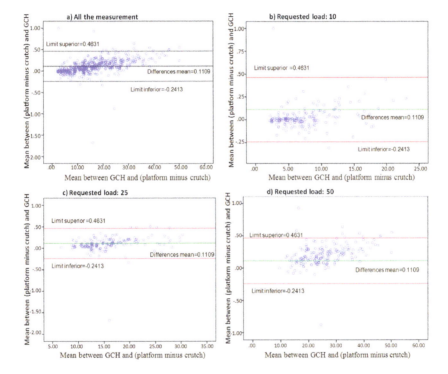

Figure 5. Bland-Altman Method representing the values differences (platform minus crutch) and GCH, versus the means (platform minus crutch) and GCH; for all the measurements (**a**), measurement requesting load on the crutches at 10% of body weight (**b**), 25% (**c**) and 50% (**d**).

ANOVA, with $p < 0.001$, determined the existence of a linear relationship between both variables. The regression line obtained 99.3% accuracy since the value of the adjusted squared R was 0.993. In the contrast to coefficients, a $p < 0.05$ was obtained, so that both were valid in the regression line. The regression equation obtained was:

Measurements on Platform = −0.383 + 1.036· *Measurements on GCH*. Confidence intervals at 95% for the constant: (−0.519 to −0.247), and for the slope: (1.029 to 1.044).

The breakdown of the regression analysis for each crutch and the different loads is presented in Table 5. Note that the adjusted squared R is in all cases higher than 0.986.

Table 5. Regression analysis for each crutch and the different loads.

	Right Crutch			Left Crutch			Global
Load	10	25	50	10	25	50	
Constant	−0.072	−0.023	−0.016	−0.369	−0.045	0.009	−0.383
Constant Lower Bound	−0.167	−0.198	−0.146	−0.598	−0.093	−0.066	−0.519
Constant Upper Bound	0.024	0.153	0.115	−0.140	0.003	0.083	−0.247
Constant *p*-value	0.138	0.799	0.809	0.002	0.067	0.813	**<0.001**
Slope	1.014	1.008	1.009	1.062	1.009	1.008	**1.036**
Slope Lower Bound	1.003	0.997	1.004	1.036	1.006	1.005	**1.029**
Slope Upper Bound	1.026	1.018	1.013	1.089	1.012	1.010	**1.044**
Slope *p*-value	<0.001	<0.001	<0.001	<0.001	<0.001	<0.001	**<0.001**
Adjusted Squared R	0.997	0.998	0.999	0.986	>0.999	>0.999	**0.993**
Regression *p*-value	<0.001	<0.001	<0.001	<0.001	<0.001	<0.001	**<0.001**

Figure 6 shows a comparison between measurements of the Platform and the values obtained through the regression line. The difference between the actual value and the predicted value of the measurements on the Platform was quantified, obtaining an average of 0 (SD = 0.80) and the following percentiles: p25 = −0.12, p50 = 0.03 and p75 = 0.24.

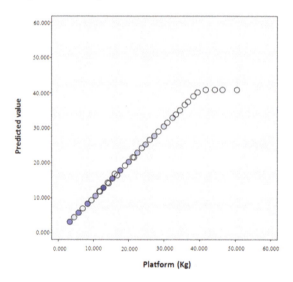

Figure 6. Dispersion graph. Ordenate axis: predicted values by the regression line. Abcissa axis: values recorded by the platform. Color increase in the representation indicates that there are higher values in those in which the variables coincide (0, 10, 20, 30, 40, 50, 60, 70).

4. Discussion

Validity and Reliability

Intra-class correlation coefficients (ICCs) for the different values of loads and crutches assess the reliability of GCH 2.0 measurements. To validate GCH 2.0, a regression line [33] to determine the predictive ability of the GCH system towards the AMTI Platform was found. In this way, a suitable description of the relationship between both quantitative variables is obtained. The R-squared measure of goodness of fit and the ANOVA of this regression [34] were included in order to check the regression line accuracy and its veracity. Since the force platform is a validated and an effective method to determine the accuracy of the loads, this predictive method is useful as it allows us to use the platform as a reference to validate our system and its measurements. Note that the platform is not a feasible system in gait recovery due to its complexity, difficult handling and high cost [35]. The GCH 2.0 solves these issues for providing effectiveness and efficiency: it is cheaper, requires a minimum space, offers information about load along the way and not only for a single step, the handling is easy, *etc.* Consequently, the instrumented crutch is a better option for gait recovery in patients using forearm crutches.

Regarding the data obtained, the GCH System 2.0 is a valid and reliable tool for assessing the amount of load exerted on the forearm crutches. Consequently, it allows us to quantify said force, providing, at the same time, an objective control of the unloading exerted on the affected limb [13]. Thus, the statistics showed minimum differences between the measurements of the Platform and GCH, interpreting them as non significant. In addition, these non-significant differences were corroborated by conducting comparisons between the Platform and the right crutch as well as between the Platform and the left crutch. Therefore, we conclude that the measurements of both prototypes are equally effective and reproducible.

Secondly, a detailed analysis of these slight disagreements announced that the Platform measurements had a tendency to be higher than those of GCH. This result was expected and assumed by the authors from the beginning of the study, since there are slight mechanical frictions of the outer sleeve that slip over the distal end of the crutch shaft to press the sensor each time a weight is exerted upon it. This friction reduces the GCH values regarding the Platform.

Furthermore, the Bland-Altman plots (Figure 5) show that the greater the load, the greater the differences between measurements. Depending on the different ranges of weight bearing, the difference between measurements of the Platform and GCH increases as weight bearing increases, which is once again expected due to the increase of mechanical friction.

The exhaustive study of the minimum errors due to a mechanical component confirms that they are neither statistically significant nor clinically relevant since a tolerance of up to a kg of error is established [13]. Despite this, the design is more demanding and is aimed at not exceeding 0.5 kg.

This study showed the predictive ability of GCH towards AMTI using a regression line [34]. The mean of the differences between the actual and the predicted value of measurements of the Platform was 0 (SD = 0.80). By comparing the predicted values with the actual values, an almost perfect diagonal between the two variables was achieved, so that it can be confirmed that the predictive model achieved is optimal for research. It is as from 40 kg on the Platform where the predictive model ceases to have a growing tendency, although it is not decisive, as there are very few cases that occur in this range. As noted in Section 3, the differences between the Platform and GCH 2.0 are always too small to be statistically relevant. In all cases, these differences are definitely not clinically relevant.

To sum up, the GCH System 2.0 offers effectiveness in quantifying partial weight-bearing loads applied to the crutches in gait training according to each patient´s particular conditions. Consequently, it is a useful tool to control assisted gait, to obtain accuracy of loads exerted on the crutches and unloadings on the injuried lower limb, to carry out objective progressions [17,36] and to establish new treatment protocols, involving higher quality in the functional gait recovery [37]. Eliminating the subjectivity present in the treatment and raising the accuracy implies a shorter duration and reduction in relapses and sequelae.

As a final comment, according to our experience with healthy subjects with excellent coordinative skills, experience with the use of crutches and training of the exerted loads on the crutches based on feedback, we found that they were unable to control an exact load and maintain it for some time. This conviction is corroborated by Isakov [38], Kaplan [39] and Fu *et al.* [16]. Therefore margins of error of body weight are necessary in order to avoid constant warnings of error by the feedback system. Further work of this paper is the establishment of the ideal margins. In addition, a new study with patients without experience with the use of crutches should be carried out, this being a limitation of this research.

5. Conclusions

The GCH System 2.0 constitutes a reliable and valid instrument for measuring weight bearing on implemented forearm crutches during aided gait. The concordance study with the validated AMTI OR6-7-2000 force platform, carried out in assisted gait dynamic conditions, determined the effectiveness of the new prototype in association with the GCH Control Software 1.0 specific program, which allows us to numerically and graphically display loads applied to the crutches through the upper limbs during gait, record them and later analyze them. The System incorporates an optimized acoustic and visual feedback mechanism that increases and maintains the accuracy of the recommended loads for each patient.

As a result, and as opposed to other systems described in literature, our paper presents a compact and ready-to-use device that focuses on accurately monitoring the force applied to crutches during gait, providing personalized feedback to both the patient and the physiotherapist/researcher in order to improve gait training and evaluate the process of functional recovery.

Acknowledgments: We would like to thank the *Telefónica* Chair "Intelligence in Networks" of the University of Seville for funding our study. Likewise, we are grateful for the cooperation of the Department of Mechanics at the University of Seville and Antonia Sáez of Axioma Comunicaciones.

Author Contributions: G.C.-M. conceived and designed the study, acquired the data, interpreted the data, drafted the article and approved the final version to be published. J.L.S. designed the study, analyzed the data, and approved the final version to be published. C.R.-F. acquired the data, interpreted the data, drafted the article and approved the final version to be published.

Conflicts of Interest: The authors declare no conflict of interest.

Abbreviations

The following abbreviations are used in this manuscript:

ANOVA	Analysis of Variance
FS	Full Scale
PCB	Printed Circuit Board
PPBW	Percentage of Patient Body Weight
SD	Standard Deviation
SMD	Surface Mount Device

References

1. Van Kammen, K.; Boonstra, A.; Reinders-Messelink, H.; den Otter, R. The combined effects of body weight support and gait speed on gait related muscle activity: A comparison between walking in the Lokomat exoskeleton and regular treadmill walking. *PLoS ONE* **2014**, *9*, e107323.
2. Chamorro-Moriana, G.; Ridao-Fernández, C.; Ojeda, J.; Benítez-Lugo, M.; Sevillano, J.L. Reliability and validity study of the Chamorro Assisted Gait Scale for people with sprained ankles, walking with forearm crutches. *PLoS ONE* **2016**, *11*, e0155225. [CrossRef] [PubMed]
3. Fischer, J.; Nuesch, C.; Gopfert, B.; Mundermann, A.; Valderrabano, V.; Hugle, T. Forearm pressure distribution during ambulation with elbow crutches: A cross-sectional study. *J. Neuroeng. Rehabil.* **2014**, *11*, 61. [CrossRef] [PubMed]
4. Gates, D.H.; Darter, B.J.; Dingwell, J.B.; Wilken, J.M. Comparison of walking overground and in a Computer Assisted Rehabilitation Environment (CAREN) in individuals with and without transtibial amputation. *J. Neuroeng. Rehabil.* **2012**, *9*, 81. [CrossRef] [PubMed]
5. Xu, P.T.; Li, Q.; Sheng, J.J.; Chang, H.; Song, Z.; Yu, Z.B. Passive stretch reduces calpain activity through nitric oxide pathway in unloaded soleus muscles. *Mol. Cell. Biochem.* **2012**, *367*, 113–124. [CrossRef] [PubMed]
6. Moreira, L.D.F.; de Oliveira, M.L.; Lirani-Galvao, A.P.; Marin-Mio, R.V.; dos Santos, R.N.; Lazaretti-Castro, M. Physical exercise and osteoporosis: Effects of different types of exercises on bone and physical function of postmenopausal women. *Arq. Bras Endocrinol. Metabol.* **2014**, *58*, 514–522. [CrossRef] [PubMed]
7. Robroy, L.M.; Davenport, D.E.; Paulseth, S.; Wukich, D.K.; Godges, J. Ankle stability and movement coordination impairments: Ankle ligament sprains. *J. Orthop. Sports Phys. Ther.* **2013**, *43*, A1–A40.

8. Domingo, A.; Lam, T. Reliability and validity of using the Lokomat to assess lower limb joint position sense in people with incomplete spinal cord injury. *J. Neuroeng. Rehabil.* **2014**, *11*, 167. [CrossRef] [PubMed]
9. Gusinde, J.; Pauser, J.; Swoboda, B.; Gelse, K.; Carl, H. Foot loading characteristics of different graduations of partial weight bearing. *Int. J. Rehabil. Res.* **2011**, *34*, 261–264. [CrossRef] [PubMed]
10. González, I.; Fontecha, J.; Hervás, R.; Bravo, J. An Ambulatory system for gait monitoring based on wireless sensorized insoles. *Sensors* **2015**, *15*, 16589–16613. [CrossRef] [PubMed]
11. Wafai, L.; Zayegh, A.; Woulfe, J.; Aziz, S.M.; Begg, R. Identification of foot pathologies based on plantar pressure asymmetry. *Sensors* **2015**, *15*, 20392–20408. [CrossRef] [PubMed]
12. Chamorro-Moriana, G. Sistema de Medición de Cargas en Bastones de Antebrazo. Spanish Patent P200901942, 2 October 2009. (In Spanish)
13. Chamorro-Moriana, G.; Rebollo-Roldán, J.; Jiménez-Rejano, J.J.; Chillón-Martínez, R.; Suárez-Serrano, C. Design and validation of GCH System 1.0 which measures the weight-bearing exerted on forearm crutches during aided gait. *Gait Posture* **2013**, *37*, 564–569. [CrossRef] [PubMed]
14. Sardini, E.; Serpelloni, M.; Lancini, M. Wireless instrumented crutches for force and movement measurements for gait monitoring. *IEEE Trans. Instrum. Meas.* **2015**, *64*, 3369–3379. [CrossRef]
15. Tveit, M.; Karrholm, J. Low effectiveness of prescribed partial weight bearing. Continuous recording of vertical loads using a new pressure-sensitive insole. *J. Rehabil. Med.* **2001**, *33*, 42–46. [PubMed]
16. Fu, M.C.; DeLuke, L.; Buerba, R.A.; Fan, R.E.; Zheng, Y.J.; Leslie, M.P.; Baumgaertner, M.R.; Grauer, J.N. Haptic biofeedback for improving compliance with lower-extremity partial weight bearing. *Orthopedics* **2014**, *37*, e993–e998. [CrossRef] [PubMed]
17. Clark, B.C.; Manini, T.M.; Ordway, N.R.; Ploutz-Snyder, L.L. Leg muscle activity during walking with assistive devices at varying levels of weight bearing. *Arch. Phys. Med. Rehabil.* **2004**, *85*, 1555–1560. [CrossRef] [PubMed]
18. Shenzhen Exact Sensor Instrument Co., Ltd. Available online: http://shenzhen-exact.en.ywsp.com/ (accessed on 11 May 2016).
19. Friedman, L. *SimpliciTI: Simple Modular RF Network Specification*; Texas Instruments, Inc.: San Diego, CA, USA, 2009.
20. Richardson, J.K.; Thies, S.; Ashton-Miller, J.A. An exploration of step time variability on smooth and irregular surfaces in older persons with neuropathy. *Clin. Biomech.* **2008**, *23*, 349–356. [CrossRef] [PubMed]
21. Chaiwatpongsakorn, C.; Lu, M.; Keener, T.C.; Khang, S.J. The deployment of carbon monoxide wireless sensor network (CO-WSN) for ambient air monitoring. *Int. J. Environ. Res. Public Health* **2014**, *11*, 6246–6264. [CrossRef] [PubMed]
22. Chamorro-Moriana, G. Sistema de Medición de Cargas en Bastones de Antebrazo. International Patent PCT/ES2011/000340, 2 February 2012.
23. D'Anna, C.; Schmid, M.; Bibbo, D.; Bertollo, M.; Comani, S.; Conforto, S. The Effect of continuous and discretized presentations of concurrent augmented visual biofeedback on postural control in quiet stance. *PLoS ONE* **2015**, *10*, e0132711. [CrossRef] [PubMed]
24. Hol, A.; van Grinsven, S.; Lucas, C.; van Susante, J.; van Loon, C. Partial *versus* unrestricted weight bearing after an uncemented femoral stem in total hip arthroplasty: Recommendation of a concise rehabilitation protocol from a systematic review of the literature. *Arch. Orthop. Trauma Surg.* **2010**, *130*, 547–555. [CrossRef] [PubMed]
25. Audette, I.; Dumas, J.P.; Cote, J.N.; De Serres, S.J. Validity and between-day reliability of the cervical range of motion (CROM) device. *J. Orthop. Sports Phys. Ther.* **2010**, *40*, 318–323. [CrossRef] [PubMed]
26. Ruiz-Morales, A.; Morillo-Zárate, L.E. *Epidemiología Clínica: Investigación Clínica Aplicada*; Editorial Médica Panamericana: Bogotá, Columbia, 2006.
27. Martínez-González, M.A.; Sánchez-Villegas, A.; Faulin-Fajardo, J. *Bioestadística Amigable*, 2nd ed.; Díaz De Santos: Madrid, Spain, 2008.
28. Osaki, Y.; Kunin, M.; Cohen, B.; Raphan, T. Relative contribution of walking velocity and stepping frequency to the neural control of locomotion. *Exp. Brain Res.* **2008**, *185*, 121–135. [CrossRef] [PubMed]
29. Jones, A.; Alves, A.; de Oliveira, L.; Saad, M.; Natour, J. Energy expenditure during cane-assisted gait in patients with knee osteoarthritis. *Clinics* **2008**, *63*, 197–200. [CrossRef] [PubMed]
30. Altman, D.G. *Practical Statistics for Medical Research*, 1st ed.; Chapman and Hall: London, UK, 1991.

31. McGraw, K.O.; Wong, S.P. Forming inferences about some intraclass correlation coefficients. *Psychol. Methods* **1996**, *1*, 30–46. [CrossRef]
32. Buffa, R.; Mereu, E.; Lussu, P.; Succa, V.; Pisanu, T.; Buffa, F.; Marini, E. A new, effective and low-cost three-dimensional approach for the estimation of upper-limb volume. *Sensors* **2015**, *15*, 12342–12357. [CrossRef] [PubMed]
33. Greenland, S. *Introduction to Regression Models*, 2nd ed.; Lippincott-Raven: Philadelphia, PA, USA, 1998; pp. 359–399.
34. Dagnino, S.J. Regresión lineal. *Rev. Chil. Anest.* **2014**, *43*, 143–149.
35. Hausdorff, J.M.; Ladin, Z.; Wei, J.Y. Footswitch system for measurement of the temporal parameters of gait. *J. Biomech.* **1995**, *28*, 347–351. [CrossRef]
36. Bateni, H.; Maki, B.E. Assistive devices for balance and mobility: Benefits, demands, and adverse consequences. *Arch. Phys. Med. Rehabil.* **2005**, *86*, 134–145. [CrossRef] [PubMed]
37. Terjesen, T.; Lofterod, B.; Skaaret, I. Gait improvement surgery in ambulatory children with diplegic cerebral palsy. *Acta Orthop.* **2015**, *86*, 511–517. [CrossRef] [PubMed]
38. Isakov, E. Gait rehabilitation: A new biofeedback device for monitoring and enhancing weight-bearing over the affected lower limb. *Eura Medicophys.* **2007**, *43*, 21–26. [PubMed]
39. Kaplan, Y. The use of a new biofeedback insole weight-bearing measuring device in the assessment and rehabilitation of soccer players: A case study review. *J. Sport Sci. Med.* **2007**, Suppl 10, S30–S34.

Article

Robust Decentralized Nonlinear Control for a Twin Rotor MIMO System

Lidia María Belmonte [1], Rafael Morales [1,*], Antonio Fernández-Caballero [1] and José Andrés Somolinos [2]

[1] Escuela de Ingenieros Industriales de Albacete, Universidad de Castilla-La Mancha, 02071 Albacete, Spain; LidiaMaria.Belmonte@uclm.es (L.M.B.); Antonio.Fdez@uclm.es (A.F.-C.)
[2] Escuela Técnica Superior de Ingenieros Navales, Universidad Politécnica de Madrid, 28040 Madrid, Spain; joseandres.somolinos@upm.es
*   Correspondence: Rafael.Morales@uclm.es; Tel.: +34-967-599-200 (ext. 2542); Fax: +34-967-599-224

Academic Editor: Gonzalo Pajares Martinsanz
Received: 10 June 2016; Accepted: 19 July 2016; Published: 27 July 2016

Abstract: This article presents the design of a novel decentralized nonlinear multivariate control scheme for an underactuated, nonlinear and multivariate laboratory helicopter denominated the twin rotor MIMO system (TRMS). The TRMS is characterized by a coupling effect between rotor dynamics and the body of the model, which is due to the action-reaction principle originated in the acceleration and deceleration of the motor-propeller groups. The proposed controller is composed of two nested loops that are utilized to achieve stabilization and precise trajectory tracking tasks for the controlled position of the generalized coordinates of the TRMS. The nonlinear internal loop is used to control the electrical dynamics of the platform, and the nonlinear external loop allows the platform to be perfectly stabilized and positioned in space. Finally, we illustrate the theoretical control developments with a set of experiments in order to verify the effectiveness of the proposed nonlinear decentralized feedback controller, in which a comparative study with other controllers is performed, illustrating the excellent performance of the proposed robust decentralized control scheme in both stabilization and trajectory tracking tasks.

Keywords: decentralized control; nonlinear control; time-scale model; Euler–Lagrange model; TRMS

---

1. Introduction

In the last few years, there has been an increased interest from researchers in developing control algorithms for unmanned aerial vehicles (UAVs) [1–6], due to the multiple applications and uses of this type of vehicle. This has motivated the use of new laboratory platforms capable of simulating the operation of the UAVs. This way, it is possible to perform experimental tests for evaluating the different designs developed. We can highlight the three-DOF hover system [7], the three-DOF helicopter system [8] and the twin rotor MIMO system (TRMS) [9], which is the platform used in this research.

The TRMS is a nonlinear and multivariate laboratory helicopter specifically designed to test and evaluate control algorithms by means of the MATLAB/Simulink® software environment. The dynamic behavior of the system is similar to a real helicopter, but with some differences due to the construction of the model that greatly hinder the modeling and design of control algorithms for this platform. As can be seen in Figure 1, the TRMS is formed by a base attached to a tower, at which end is a two-dimensional pivot that allows the mobile structure to rotate freely. The mobile part is composed of two metal beams: the horizontal beam in which ends the main and tail rotors with the corresponding propellers are positioned in perpendicular planes and the counterbalance beam affixed to the horizontal beam at the pivot to move the equilibrium point of the system.

Figure 1. Twin rotor MIMO system (TRMS).

The electrical part of the TRMS is mainly composed of two DC motors that drive the propellers of both rotors and the interface circuit, an internal electrical circuit that adapts the input control voltages, applied in MATLAB/Simulink®, to the actual voltage value applied to each DC motor. Thus, a change in the control voltages produces a variation in the supply voltages of the motors, which results in a variation of the rotational speed of each propeller, measured by a tachometer. This way, a change in propulsive forces finally results in the movement of the platform. The movement, in the vertical and horizontal planes, is measured by two encoders that determine the pitch and yaw angles, respectively.

The movement of the TRMS presents not only a high cross-coupling between the two rotors as in a real helicopter, but also a coupling effect between rotor dynamics and the body of the model. This is due to the action-reaction principle originated in the acceleration and deceleration of the motor-propeller groups. Therefore, the control system of the TRMS generates a significant difference with regard to a real helicopter by varying the voltages applied to the rotors, which greatly complicates the system dynamics. On the other hand, the TRMS is also an underactuated system as a result of fewer control actions, which are the voltages applied to the respective rotors, compared to the four degrees of freedom of the system, which are: the pitch and yaw angles and the angular velocities of the propellers. Moreover, there are many physical parameters that cannot be measured exactly, and some of the parameters supplied by the manufacturer are changed by time, such as the friction coefficients. All of this makes the modeling and control of the system a difficult task to achieve.

There are many research works that have addressed this challenging experimental platform. In fact, the dynamic modeling of the TRMS has been studied from different approaches. Rahideh et al. define the dynamic model of the TRMS using Newtonian and Lagrangian methods and also by means of two models based on neuronal networks using Levenberg–Marquardt (LM) and gradient descent (GD) algorithms [10]. Toha et al. develop a parametric model for the TRMS based on dynamic spread factor particle swarm optimization [11]. A linear parameter varying (LPV) method of identification, by taking a local approach, is considered in order to derive an LPV model for TRMS by means of interpolation and approximation in the work of Tanaka et al. [12]. The Euler–Lagrange method is employed in the research of Tastermirov [13] to obtain a complete dynamic model of the TRMS, which is tuned and validated experimentally. More recently, a model based on first-principle modeling and later improved by gray box modeling, has been presented in [14]. The design of control algorithms for the TRMS platform has been also investigated via several approaches and control methods. Among the different contributions in this area, we can cite the following works. Juang and colleagues present a comparative study [15], by means of numerical simulations, between classical control schemes, based on the Ziegler–Nichols proportional-integral-derivative (PID) rule, the gain margin and phase margin rule, the pole placement method and novel controllers based on fuzzy logic and genetic algorithms. In the research of Wen et al. [16], the dynamic model for the TMRS is decoupled into two single input single output (SISO) systems in order to apply a PID-based robust deadbeat control scheme for each of them, thus achieving the control of the platform. The design and experimental validation of a multi-step Newton-type model predictive control (MPC) to control the TRMS is presented in [17] where a nonlinear dynamic model of the platform is also developed. An adaptive fuzzy controller to stabilize the TRMS in a desired position or to track a specified trajectory is discussed in [18]. The work of Pandey et al. [19] in which two conventional PID controllers, improved by the use of derivative filter coefficients, are employed to the control of the pitch and yaw angles, the work of Belmonte et al. based on active disturbance rejection control (ADRC) [20] and the research of Alagoz et al. [21] about a reference model-based optimization approach for the online auto-tuning of PIDs using the stochastic multi-parameters divergence optimization (SMDO) method are other interesting investigations that are focused in the design of control schemes for the TRMS.

In the particular case of this research, we present the design of a novel decentralized nonlinear controller for the TRMS, composed of two control loops in a cascade scheme. The development of the proposed control scheme has been separated into two independent stages: the design of the inner loop or electrical controller, which is used to control the angular velocity of each propeller, and the design of the outer loop or mechanical controller, which is employed to determine the necessary velocities to control the space position of the TRMS. The effectiveness of the proposed scheme is validated by means of the experiments performed in the laboratory platform in which the proposed nonlinear controller shows an excellent performance for both stabilization and tracking tasks.

The rest of the article is organized as follows: Section 2 introduces the dynamic model of the TRMS, showing the modeling of the electrical part formed by the interface circuit and the DC motors, and the modeling of the mechanical part composed by the equations of motion of the system. Next, the design of the proposed decentralized control scheme is detailed in Section 3. The experiments carried out in order to verify the efficiency of the proposed control algorithm are presented in Section 4, where we detail the experimental setup and the obtained results, which include a comparative study with other classical controllers. Finally, some conclusions are provided in Section 5.

2. Dynamic Model

This section describes the dynamic modeling of the TRMS, and according to [10], it has been divided into the following two stages. In the first place, the electrical part of the platform is modeled, including the interface circuit, the DC motors and the propulsive forces produced by these motors. Then, a Lagrangian-based model is employed for the remaining mechanical structure. Next, each part of the dynamic model is dealt with in the next subsections.

2.1. Dynamics of the Electrical Part

The main and tail rotors (denominated as m and t, respectively) are assumed to be identical with different mechanical loads. The mathematical expressions governing the main and tail rotors are the following:

- Main rotor:

$$L_m \frac{di_m}{dt} = v_m - k_{v_m}\omega_m - R_m i_m \tag{1}$$

$$I_{m_1}\dot{\omega}_m = k_{t_m}i_m - f_{v_m}\omega_m - C_{Q_m}\omega_m|\omega_m| \tag{2}$$

- Tail rotor:

$$L_t \frac{di_t}{dt} = v_t - k_{v_t}\omega_t - R_t i_t \tag{3}$$

$$I_{t_1}\dot{\omega}_t = k_{t_t}i_t - f_{v_t}\omega_t - C_{Q_t}\omega_t|\omega_t| \tag{4}$$

where i_m and i_t are the main and tail motor currents, respectively, L_m and L_t represent the motor inductances, R_m and R_t denote the motor resistances, k_{v_m} and k_{v_t} express the motor back electromotive force (EMF) constants, ω_m and ω_t are the angular velocities of the propellers and v_m and v_t represent the input voltage of the DC motors. I_{m_1} and I_{t_1} define the moment of inertia of the rotors; the terms $k_{t_m} i_m$ and $k_{t_t} i_t$ express the main and tail electromechanical torques generated by the DC motors; $C_{Q_m} \omega_m |\omega_m|$ and $C_{Q_t} \omega_t |\omega_t|$ illustrate the aerodynamic torques; and $f_{v_m} \omega_m$ and $f_{v_t} \omega_t$ denote the friction torques. Following a similar argument as [13], the dynamics of the current of the motors defined in Expressions (1) and (3) is ignored due to the higher value of the DC motor mechanical time constants against the electrical ones. In fact, the DC motor mechanical constants (c_{m_m} and c_{m_t}) are in the order of 10^3-times higher than the DC motor electrical constants (c_{e_m} and c_{e_t}), as you may observe in Table 1, which shows the parameters of both rotors. Thereby, for the DC motor circuits, the following algebraic equations are obtained:

$$v_m - k_{v_m} \omega_m - R_m i_m = 0 \tag{5}$$

$$v_t - k_{v_t} \omega_t - R_t i_t = 0 \tag{6}$$

Table 1. Dynamic model of the TRMS: electrical parameters.

Symbol	Parameter	Value	Units
	Parameters of the Main Rotor		
k_{v_m}	Motor velocity constant	0.0202	$V \cdot rad^{-1} \cdot s$
R_m	Motor armature resistance	8	Ω
L_m	Motor armature inductance	0.86×10^{-3}	H
k_{t_m}	Electromagnetic constant torque motor	0.0202	$N \cdot m \cdot A^{-1}$
k_{u_m}	Coefficient linear relationship interface circuit	8.5	—
$C_{Q_m}^+$	Load factor ($\omega_m \geq 0$)	2.695×10^{-7}	$N \cdot m \cdot s^2 \cdot rad^{-2}$
$C_{Q_m}^-$	Load factor ($\omega_m < 0$)	2.46×10^{-7}	$N \cdot m \cdot s^2 \cdot rad^{-2}$
f_{v_m}	Viscous friction coefficient	3.89×10^{-6}	$N \cdot m \cdot rad^{-1} \cdot s$
I_{m1}	Moment of inertia about the axis of rotation	1.05×10^{-4}	$kg \cdot m^2$
c_{e_m}	Electrical time constant (L_m / R_m)	1.075×10^{-4}	s
c_{m_m}	Mechanical time constant ($I_{m1} R_m / k_{t_m} k_{v_m}$)	2.058	s
	Parameters of the Tail Rotor		
k_{v_t}	Motor velocity constant	0.0202	$V \cdot rad^{-1} s$
R_t	Motor armature resistance	8	Ω
L_t	Motor armature inductance	0.86×10^{-3}	H
k_{t_t}	Electromagnetic constant torque motor	0.0202	$N \cdot m \cdot A^{-1}$
k_{u_t}	Coefficient linear relationship interface circuit	6.5	—
C_{Q_t}	Load factor	1.164×10^{-8}	$N \cdot m \cdot s^2 \cdot rad^{-2}$
f_{v_t}	Viscous friction coefficient	1.715×10^{-6}	$N \cdot m \cdot rad^{-1} \cdot s$
I_{t1}	Moment of inertia about the axis of rotation	2.1×10^{-5}	$kg \cdot m^2$
c_{e_t}	Electrical time constant (L_t / R_t)	1.075×10^{-4}	s
c_{m_t}	Mechanical time constant ($I_{t1} R_t / k_{t_t} k_{v_t}$)	0.4117	s

It should be noted that the magnitude input voltages of the main and tail rotors in the MATLAB/Simulink® environment, defined as u_m and u_t, respectively, and the motor terminal voltages, defined as v_m and v_t, respectively, are nonlinear (the signals pass through a circuit interface), as was demonstrated in [10]. In our developments, it is assumed that the relationship between the control signals and the motor voltages is linear and that the differences will be canceled at the

controller stage. Therefore, the relationships between the control signals and the MATLAB/Simulink®
environment are the following:

$$v_m = k_{u_m} u_m \tag{7}$$

$$v_t = k_{u_t} u_t \tag{8}$$

in which k_{u_m} and k_{u_t} are defined as constant gains. Upon operating with Equations (1)–(8) and
rearranging terms, the following two equations are yielded for the main and rail rotors of the TRMS:

$$\dot{\omega}_m = \frac{k_{t_m} k_{u_m}}{I_{m_1} R_m} u_m - \left(\frac{k_{t_m} k_{v_m}}{R_m} + f_{v_m} \right) \frac{\omega_m}{I_{m_1}} - \frac{C_{Q_m}}{I_{m_1}} \omega_m |\omega_m| \tag{9}$$

$$\dot{\omega}_t = \frac{k_{t_t} k_{u_t}}{I_{t_1} R_t} u_t - \left(\frac{k_{t_t} k_{v_t}}{R_t} + f_{v_t} \right) \frac{\omega_t}{I_{t_1}} - \frac{C_{Q_t}}{I_{t_1}} \omega_t |\omega_t| \tag{10}$$

in which the value and units of each parameter of the main and tail rotors are detailed in Table 1.
Finally, if we use matrix notation, the dynamic model of the electrical part of the TRMS can be expressed
by means of the following expression:

$$\dot{\omega}(t) = \mathbf{N}\mathbf{u}(t) + \mathbf{\Gamma}(\omega(t)) \tag{11}$$

where $\omega(t) = [\omega_m(t), \omega_t(t)]^T$ is the angular velocity vector, $\mathbf{u}(t) = [u_m(t), u_t(t)]^T$ is the input
control voltage vector and, finally, the diagonal positive matrix $\mathbf{N} = diag(n_m, n_t)$ and the vector
$\mathbf{\Gamma}(\omega(t)) = [\Gamma_m(t), \Gamma_t(t)]^T$ are given by:

$$\mathbf{N} = \begin{bmatrix} n_m & 0 \\ 0 & n_t \end{bmatrix} = \begin{bmatrix} \frac{k_{t_m} k_{u_m}}{I_{m_1} R_m} & 0 \\ 0 & \frac{k_{t_t} k_{u_t}}{I_{t_1} R_t} \end{bmatrix} \tag{12}$$

$$\mathbf{\Gamma}(\omega(t)) = \begin{bmatrix} \Gamma_m(t) \\ \Gamma_t(t) \end{bmatrix} = \begin{bmatrix} -\left(\frac{k_{t_m} k_{v_m}}{R_m} + f_{v_m} \right) \frac{\omega_m}{I_{m_1}} - \frac{C_{Q_m}}{I_{m_1}} \omega_m |\omega_m| \\ -\left(\frac{k_{t_t} k_{v_t}}{R_t} + f_{v_t} \right) \frac{\omega_t}{I_{t_1}} - \frac{C_{Q_t}}{I_{t_1}} \omega_t |\omega_t| \end{bmatrix} \tag{13}$$

2.2. Dynamics of the Mechanical Part

If the developments reported in [10] are used as a basis, the dynamics of the TRMS can be derived
using Lagrange's formulation:

$$\frac{d}{dt} \left(\frac{\partial L}{\partial \dot{q}} \right) - \frac{\partial L}{\partial q} = \mathbf{Q} \tag{14}$$

where $L = K - V$ is the Lagrangian function, K and V are the kinetic and potential energies of the
TRMS, $\mathbf{q}(t) = [\psi(t), \phi(t)]^T$ is a vector of generalized coordinates and $\mathbf{Q}(t) = [Q_\psi(t), Q_\phi(t)]^T$ denotes
the vector of generalized forces in the TRMS. All of the necessary terms of (14) are obtained in the
following subsections.

2.2.1. Evaluation of the Kinetic Energy

In order to calculate the energy of the TRMS, we consider the platform as divided into the
following three subsystems: (1) the subsystem composed of the free-free beam (tail and main beam),
tail rotor, main rotor, tail shield and main shield; (2) the counterbalance beam with the counterweight;
and (3) the pivoted beam (see Figures 2–4). The positions of the subsystems can be expressed as the
position of a point for each one, P_1, P_2, P_3, parametrized by the distance between it and the point

where the subsystem can rotate, as can be observed in the following expressions (where $S_\psi \equiv \sin \psi$, $C_\psi \equiv \cos \psi$, $S_\phi \equiv \sin \phi$ and $C_\phi \equiv \cos \phi$):

$$\mathbf{P_1}\,(R_1) \;=\; \begin{bmatrix} P_{1_x} & P_{1_y} & P_{1_z} \end{bmatrix}^T \;=\; \begin{bmatrix} -R_1 S_\phi C_\psi + h C_\phi & R_1 C_\phi C_\psi + h S_\phi & R_1 S_\psi \end{bmatrix}^T \tag{15}$$

$$\mathbf{P_2}\,(R_2) \;=\; \begin{bmatrix} P_{2_x} & P_{2_y} & P_{2_z} \end{bmatrix}^T \;=\; \begin{bmatrix} -R_2 S_\phi S_\psi + h C_\phi & R_2 C_\phi S_\psi + h S_\phi & -R_2 C_\psi \end{bmatrix}^T \tag{16}$$

$$\mathbf{P_3}\,(R_3) \;=\; \begin{bmatrix} P_{3_x} & P_{3_y} & P_{3_z} \end{bmatrix}^T \;=\; \begin{bmatrix} R_3 C_\phi & R_3 S_\phi & 0 \end{bmatrix}^T \tag{17}$$

where R_1 and R_2 are the distances from point O_1 to P_1 and P_2, respectively, and R_3 is the distance from P_3 to the center of the reference system, that is the point O.

In this way, the total amount of kinetic energy consists of the sum of the following three terms:

$$K = \sum_{i=1}^{3} K_i = \frac{1}{2}\sum_{i=1}^{3} \int v_i^2(R_i)\,dm(R_i) \tag{18}$$

where K_i denotes the kinetic energy of each subsystem and $v_i(R_i)$ is the velocity of each subsystem parameterized by R_i, which represents the distances R_1, R_2 and R_3 that have been defined above. The calculations of these energies are the following:

$$K_1 \;=\; \frac{1}{2}J_1\left(C_\psi^2\dot\phi^2 + \dot\psi^2\right) + \frac{1}{2}h^2 m_{T_1}\dot\phi^2 - h S_\psi l_{T_1} m_{T_1}\dot\phi\dot\psi \tag{19}$$

$$K_2 \;=\; \frac{1}{2}J_2\left(S_\psi^2\dot\phi^2 + \dot\psi^2\right) + \frac{1}{2}h^2 m_{T_2}\dot\phi^2 + h C_\psi l_{T_2} m_{T_2}\dot\phi\dot\psi \tag{20}$$

$$K_3 \;=\; \frac{1}{2}J_3\dot\phi^2 \tag{21}$$

where:

$$J_1 \;=\; m_{ts}r_{ts}^2 + \frac{1}{2}m_{ms}r_{ms}^2 + \left(\frac{1}{3}m_t + m_{tr} + m_{ts}\right)l_t^2 + \left(\frac{1}{3}m_m + m_{mr} + m_{ms}\right)l_m^2$$

$$m_{T_1} \;=\; m_m + m_{mr} + m_{ms} + m_t + m_{tr} + m_{ts}$$

$$l_{T_1} \;=\; \frac{\left(\frac{m_t}{2} + m_{tr} + m_{ts}\right)l_t - \left(\frac{m_m}{2} + m_{mr} + m_{ms}\right)l_m}{m_{T_1}}$$

$$J_2 \;=\; \frac{1}{3}m_b l_b^2 + m_{cb} l_{cb}^2$$

$$m_{T_2} \;=\; m_b + m_{cb}$$

$$l_{T_2} \;=\; \frac{m_b \frac{l_b}{2} + m_{cb} l_{cb}}{m_{T_2}}$$

$$J_3 \;=\; \frac{1}{3}m_h l_h^2$$

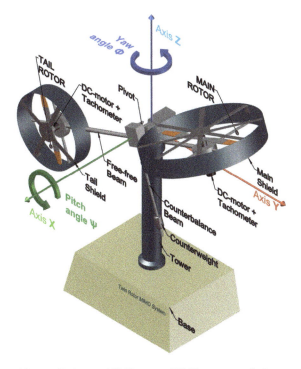

Figure 2. Twin rotor MIMO system (TRMS) prototype platform.

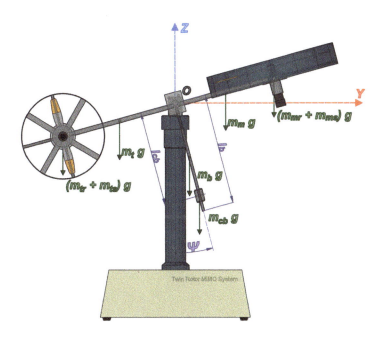

Figure 3. View of the TRMS on the vertical plane.

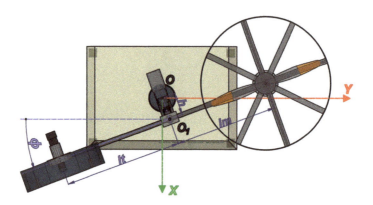

Figure 4. View of the TRMS on the horizontal plane.

2.2.2. Evaluation of the Potential Energy

The total potential energy consists of the sum of the following three terms:

$$V = \sum_{i=1}^{3} V_i = g \sum_{i=1}^{3} \int r_{zi}(R_i) dm(R_i) \tag{22}$$

where g denotes the gravity constant, V_i represents the potential energy of each one of the three subsystems in which we have divided the platform and $r_{zi}(R_i)$ is the coordinate on the z-axis of the position of each subsystem (P_{i_z}). The calculation of these energies is as follows:

$$V_1 = gS_\psi l_{T_1} m_{T_1} \tag{23}$$
$$V_2 = -gC_\psi l_{T_2} m_{T_2} \tag{24}$$
$$V_3 = 0 \tag{25}$$

2.2.3. Lagrangian

After substituting Expressions (18)–(25) in the Lagrangian expression, we obtain:

$$L = K - V = \frac{1}{2} \left(J_1 C_\psi^2 + J_2 S_\psi^2 + J_3 + h^2 \left(m_{T_1} + m_{T_2} \right) \right) \dot{\phi}^2 + \frac{1}{2} \left(J_1 + J_2 \right) \dot{\psi}^2$$
$$+ h \left(l_{T_2} m_{T_2} C_\psi - l_{T_1} m_{T_1} S_\psi \right) \dot{\phi} \dot{\psi} - g \left(l_{T_1} m_{T_1} S_\psi - l_{T_2} m_{T_2} C_\psi \right) \tag{26}$$

2.2.4. Generalized Forces

The external forces in the mechanical system are owing to the following four physical effects: (a) aerodynamic forces created by the propellers; (b) the electromechanical forces generated by the propellers; (c) the viscous forces that model the dissipative effects that are present in the system and; (d) the elastic force created by the cable. After grouping the effect of these forces for each generalized coordinate, the following result is achieved for $\mathbf{Q}(t) = [Q_\psi(t), Q_\phi(t)]^T$:

$$Q_\psi(t) = C_{T_m} \omega_m |\omega_m| l_m - C_{R_t} \omega_t |\omega_t| - \left(f_{v_\psi} \dot{\psi} + f_{c_\psi} sign \left(\dot{\psi} \right) \right) + k_t \dot{\omega}_t \tag{27}$$

$$Q_\phi(t) = C_{T_t} \omega_t |\omega_t| l_t C_\psi - C_{R_m} \omega_m |\omega_m| C_\psi - \left(f_{v_\phi} \dot{\phi} + f_{c_\phi} sign \left(\dot{\phi} \right) \right) - C_c \left(\phi - \phi_0 \right)$$
$$+ k_m \dot{\omega}_m C_\psi \tag{28}$$

where $C_{T_m}\omega_m\,|\omega_m|\,l_m$ and $C_{T_t}\omega_t\,|\omega_t|\,l_t C_\psi$ represent the aerodynamic thrust torques acting along the ψ and ϕ angles, respectively; $C_{R_t}\omega_t\,|\omega_t|$ and $C_{R_m}\omega_m\,|\omega_m|\,C_\psi$ denote the aerodynamic cross-couplings effects generated by the propeller; the terms $\left(f_{v_\psi}\dot{\psi} + f_{c_\psi}sign\,(\dot{\psi})\right)$ and $\left(f_{v_\phi}\dot{\phi} + f_{c_\phi}sign\,(\dot{\phi})\right)$ define the magnitudes of friction torques for each generalized coordinate; $k_t\dot{\omega}_t$ and $k_m\dot{\omega}_m C_\psi$ express the inertial counter torques that are owing to the reaction produced by a change in the rotational speed of the rotor propellers; and the term $C_c\,(\phi - \phi_0)$ is the magnitude of the torque exerted by the cable (it has a certain stiffness that allows us to model it as a spring) on the ϕ angle.

Finally, it should be noted that the works of Tastermirov et al. [13] and Mullhaupt et al. [22] provide more details about the external forces in the TRMS and other laboratory platforms with similar dynamics.

2.2.5. Equations of Motion

Upon substituting Expressions (26)–(28) in Equation (14) and after some straightforward manipulations, we obtain the following equations of motion:

$$(J_1 + J_2)\,\ddot{\psi} + h\left(l_{T_2}m_{T_2}C_\psi - l_{T_1}m_{T_1}S_\psi\right)\ddot{\phi} + \left(\frac{(J_1 - J_2)}{2}S_{2\psi}\right)\dot{\phi}^2 + g\left(l_{T_1}m_{T_1}C_\psi + l_{T_2}m_{T_2}S_\psi\right) =$$

$$= C_{T_m}\omega_m\,|\omega_m|\,l_m - C_{R_t}\omega_t\,|\omega_t| - \left(f_{v_\psi}\dot{\psi} + f_{c_\psi}sign\,(\dot{\psi})\right) + k_t\dot{\omega}_t \qquad (29)$$

$$h\left(l_{T_2}m_{T_2}C_\psi - l_{T_1}m_{T_1}S_\psi\right)\ddot{\psi} + \left(J_1 C_\psi^2 + J_2 S_\psi^2 + J_3 + h^2\left(m_{T_1} + m_{T_2}\right)\right)\ddot{\phi}$$

$$-h\left(l_{T_1}m_{T_1}C_\psi + l_{T_2}m_{T_2}S_\psi\right)\dot{\psi}^2 + \left((J_2 - J_1)\,S_{2\psi}\right)\dot{\phi}\dot{\psi} =$$

$$= C_{T_t}\omega_t\,|\omega_t|\,l_t C_\psi - C_{R_m}\omega_m\,|\omega_m|\,C_\psi - \left(f_{v_\phi}\dot{\phi} + f_{c_\phi}sign\,(\phi)\right) - C_c\,(\phi - \phi_0) + k_m\dot{\omega}_m C_\psi \qquad (30)$$

in which the value and units of all of the mechanical parameters are illustrated in Tables 2 and 3, respectively. Finally, we can express the motion equations of the system in a compact form by means of matrix notation, thus obtaining the complete dynamic model of the mechanical part of the TRMS in the following expression:

$$\mathbf{M}(\mathbf{q}(t))\ddot{\mathbf{q}}(t) + \mathbf{C}(\mathbf{q}(t), \dot{\mathbf{q}}(t))\dot{\mathbf{q}}(t) + \mathbf{G}(\mathbf{q}(t)) + \mathbf{F}(\dot{\mathbf{q}}(t)) + \mathbf{T}(\mathbf{q}(t), \dot{\omega}(t)) = \mathbf{E}(\mathbf{q}(t))\mathbf{\Omega}(t) \qquad (31)$$

where:

$$\mathbf{M}(\mathbf{q}(t)) = \begin{bmatrix} J_1 + J_2 & h\left(l_{T_2}m_{T_2}C_\psi - l_{T_1}m_{T_1}S_\psi\right) \\ h\left(l_{T_2}m_{T_2}C_\psi - l_{T_1}m_{T_1}S_\psi\right) & J_1 C_\psi^2 + J_2 S_\psi^2 + J_3 + h^2\left(m_{T_1} + m_{T_2}\right) \end{bmatrix} \qquad (32)$$

$$\mathbf{C}(\mathbf{q}(t), \dot{\mathbf{q}}(t)) = \begin{bmatrix} 0 & \frac{1}{2}(J_1 - J_2)\,S_{2\psi}\dot{\phi} \\ -h\left(l_{T_1}m_{T_1}C_\psi + l_{T_2}m_{T_2}S_\psi\right)\dot{\psi} & (J_2 - J_1)\,S_{2\psi}\dot{\psi} \end{bmatrix} \qquad (33)$$

$$\mathbf{G}(\mathbf{q}(t)) = \begin{bmatrix} g\left(l_{T_1}m_{T_1}C_\psi + l_{T_2}m_{T_2}S_\psi\right) \\ 0 \end{bmatrix} \qquad (34)$$

$$\mathbf{F}(\dot{\mathbf{q}}(t)) = \underbrace{\begin{bmatrix} f_{v_\psi} & 0 \\ 0 & f_{v_\phi} \end{bmatrix}}_{\mathbf{F_v}}\dot{\mathbf{q}}(t) + \underbrace{\begin{bmatrix} f_{c_\psi}sgn\,(\dot{\psi}) \\ f_{c_\phi}sgn\,(\dot{\phi}) \end{bmatrix}}_{\mathbf{F_c}(\dot{\mathbf{q}}(t))} = \begin{bmatrix} f_{v_\psi}\dot{\psi} + f_{c_\psi}sgn\,(\dot{\psi}) \\ f_{v_\phi}\dot{\phi} + f_{c_\phi}sgn\,(\dot{\phi}) \end{bmatrix} \qquad (35)$$

$$\mathbf{T}(\mathbf{q}(t), \dot{\omega}(t)) = \underbrace{\begin{bmatrix} 0 \\ C_c\,(\phi - \phi_0) \end{bmatrix}}_{\mathbf{M_c}(\mathbf{q}(t))} - \underbrace{\begin{bmatrix} 0 & k_t \\ k_m C_\psi & 0 \end{bmatrix}}_{\mathbf{M_i}(\mathbf{q}(t))}\dot{\omega}(t) = \begin{bmatrix} -k_t\dot{\omega}_t \\ C_c\,(\phi - \phi_0) - k_m\dot{\omega}_m C_\psi \end{bmatrix} \qquad (36)$$

$$E(\mathbf{q}(t)) = \begin{bmatrix} C_{T_m}l_m & -C_{R_t} \\ -C_{R_m}C_\psi & C_{T_t}l_tC_\psi \end{bmatrix} \tag{37}$$

$$\Omega(t) = \begin{bmatrix} \omega_m \,|\omega_m| \\ \omega_t \,|\omega_t| \end{bmatrix} \tag{38}$$

Table 2. Dynamic model of the TRMS: mechanical parameters.

Symbol	Parameter	Value	Units
l_t	Length of the tail part of the free-free beam	0.282	m
l_m	Length of the main part of the free-free beam	0.246	m
l_b	Length of the counterbalance beam	0.290	m
l_{cb}	Distance between the counterweight and the joint	0.276	m
r_{ms}	Radius of the main shield	0.155	m
r_{ts}	Radius of the tail shield	0.1	m
h	Length of the pivoted beam	0.06	m
m_{tr}	Mass of the tail DC motor and tail rotor	0.221	kg
m_{mr}	Mass of the main DC motor and main rotor	0.236	kg
m_{cb}	Mass of the counterweight	0.068	kg
m_t	Mass of the tail part of the free-free beam	0.015	kg
m_m	Mass of the main part of the free-free beam	0.014	kg
m_b	Mass of the counterbalance beam	0.022	kg
m_{ts}	Mass of the tail shield	0.119	kg
m_{ms}	Mass of the main shield	0.219	kg
m_h	Mass of the pivoted beam	0.01	kg

Table 3. Dynamic model of the TRMS: parameters of the pitch and yaw movements.

Symbol	Parameter	Value	Units
Parameters of the Pitch movement			
$C_{T_m}^+$	Thrust torque coefficient of the main rotor ($\omega_m \geq 0$)	1.53×10^{-5}	$N \cdot s^2 \cdot rad^{-2}$
$C_{T_m}^-$	Thrust torque coefficient of the main rotor ($\omega_m < 0$)	8.8×10^{-6}	$N \cdot s^2 \cdot rad^{-2}$
C_{R_t}	Load torque coefficient of the tail rotor	9.7×10^{-8}	$N \cdot m \cdot s^2 \cdot rad^{-2}$
f_{v_φ}	Viscous friction coefficient	0.0024	$N \cdot m \cdot s \cdot rad^{-1}$
f_{c_φ}	Coulomb friction coefficient	5.69×10^{-4}	$N \cdot m$
k_t	Coefficient of the inertial counter torque due to change in ω_t	2.6×10^{-5}	$N \cdot m \cdot s^2 \cdot rad^{-1}$
Parameters of the Yaw movement			
$C_{T_t}^+$	Thrust torque coefficient of the tail rotor ($\omega_t \geq 0$)	3.25×10^{-6}	$N \cdot s^2 \cdot rad^{-2}$
$C_{T_t}^-$	Thrust torque coefficient of the tail rotor ($\omega_t < 0$)	1.72×10^{-6}	$N \cdot s^2 \cdot rad^{-2}$
$C_{R_m}^+$	Load torque coefficient of the main rotor ($\omega_m \geq 0$)	4.9×10^{-7}	$N \cdot m \cdot s^2 \cdot rad^{-2}$
$C_{R_m}^-$	Load torque coefficient of the main rotor ($\omega_m < 0$)	4.1×10^{-7}	$N \cdot m \cdot s^2 \cdot rad^{-2}$
f_{v_φ}	Viscous friction coefficient	0.03	$N \cdot m \cdot s \cdot rad^{-1}$
f_{c_φ}	Coulomb friction coefficient	3×10^{-4}	$N \cdot m$
c_c	Coefficient of the elastic force torque created by the cable	0.016	$N \cdot m \cdot rad^{-1}$
ϕ_0	Constant for the calculation of the torque of the cable	0	rad
k_m	Coefficient of the inertial counter torque due to change in ω_m	2×10^{-4}	$N \cdot m \cdot s^2 \cdot rad^{-1}$

To conclude, the dynamic model of the mechanical part of the TRMS (31) can be summarized in a simplified form if we consider that the movement of the platform is sufficiently smooth. In this way, the terms of the inertial counter torques, $k_t \dot{\omega}_t$ and $k_m \dot{\omega}_m C_\psi$, can be considered negligible in comparison with the other terms. Thereby, the dynamic model of the TRMS can be rewritten as:

$$\mathbf{M}(\mathbf{q}(t))\ddot{\mathbf{q}}(t) + \mathbf{D}(\mathbf{q}(t), \dot{\mathbf{q}}(t)) = \mathbf{E}(\mathbf{q}(t))\mathbf{\Omega}(t) \tag{39}$$

where the matrices $\mathbf{M}(\mathbf{q}(t))$, $\mathbf{E}(\mathbf{q}(t))$ and $\mathbf{\Omega}(t)$ have been defined in Equations (32), (37) and (38), respectively, and the new matrix $\mathbf{D}(\mathbf{q}(t), \dot{\mathbf{q}}(t)) = [D_\psi(t), D_\phi(t)]^T$ is given by:

$$D_\psi(t) = \frac{1}{2}(J_1 - J_2)S_{2\psi}\dot{\phi}^2 + g\left(l_{T_1}m_{T_1}C_\psi + l_{T_2}m_{T_2}S_\psi\right) + \left(f_{v_\psi}\dot{\psi} + f_{c_\psi}sgn\left(\dot{\psi}\right)\right) \tag{40}$$

$$D_\phi(t) = -h\left(l_{T_1}m_{T_1}C_\psi + l_{T_2}m_{T_2}S_\psi\right)\dot{\psi}^2 + ((J_2 - J_1)S_{2\psi})\dot{\phi}\dot{\psi} + \left(f_{v_\phi}\dot{\phi} + f_{c_\phi}sgn\left(\dot{\phi}\right)\right) + C_c\left(\phi - \phi_0\right) \tag{41}$$

3. Design of the Control System

The proposed decentralized nonlinear control scheme is based on decoupling the electrical dynamics from the mechanical dynamics. Once these dynamics have been decoupled, a nonlinear multivariate inner loop is closed in order to control the vector of the angular velocities of the propellers, $\omega(t) = [\omega_m(t), \omega_t(t)]^T$, and then, a nonlinear multivariate outer loop is closed to control the vector of the generalized coordinates of the system, $\mathbf{q}(t) = [\psi(t), \phi(t)]^T$, in order to achieve stabilization and precise trajectory tracking tasks for the controlled position of the generalized coordinates of the TRMS. If we make the dynamics of the inner loop control much faster than the mechanical dynamics of the TRMS in Equation (39), the dynamics of the inner loop can be therefore made approximately equal to $\mathbf{I}^{2 \times 2}$, (i.e., $\omega^*(t) \approx \omega(t)$), and the outer loop can be designed independently [23].

Among the advantages of this control scheme are: (a) the robust nonlinear controller design procedure is simplified to a great extent, since it allows one to design the multivariate inner loop in an independent manner from the multivariate outer loop, thus dividing the control design process into two much simpler design processes; (b) this scheme can be more easily and safely implemented than the standard controllers used in the control of the TRMS platform, which involve closing a single loop, because the nested control loops proposed in this work are sequentially implemented, first closing the inner loop, which exhibits a very high relative stability in the presence of system uncertainties, external disturbances and noisy corruptions, and later closing the outer loop, which is more prone to becoming unstable, but for which the risk of exhibiting unstable motions has been significantly reduced by previously having closed the inner loop; (c) the disturbances affecting the secondary or inner loop are effectively compensated before they affect the main process output, thereby improving the stability of the system; (d) the closing of the control loop around the secondary part of the process reduces the phase lag seen by the primary or outer controller, resulting in increased speed of response; (e) the cascade control scheme is not strongly sensitive to modeling errors, although large errors could lead to oscillations or instability in one of the feedback controllers; (f) any variation in the static gain of the secondary part of the process is compensated by its own tie; (g) the use of this scheme can dramatically improve the performance of control strategies, reducing both the maximum deviation and the integral error for disturbance responses. In the scheme shown in Figure 5, the outer loop controller generates an auxiliary command reference vector $\omega^*(t) = [\omega_m^*(t), \omega_t^*(t)]^T$ for the velocities of the propellers on the basis of the tracking objective for the vector of generalized coordinates: $\mathbf{q}(t) = [\psi(t), \phi(t)]^T$. The inner loop controller takes the command vector signal generated by the outer loop $\omega^*(t)$ as its reference for the inner loop propeller velocity control system. The different parts of the proposed control scheme are explained next.

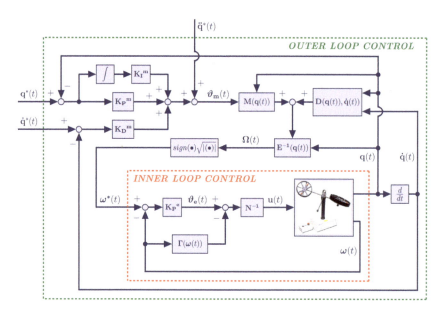

Figure 5. Robust decentralized nonlinear control scheme for the TRMS.

3.1. Inner Loop Control

The inner loop control is designed to calculate the required values for the input control voltages of the motors in the MATLAB/Simulink® environment, $\mathbf{u}(t) = [u_m(t), u_t(t)]^T$, in order to reduce and eliminate the difference between the vector of angular velocities of the propellers of the TRMS, $\boldsymbol{\omega}(t) = [\omega_m(t), \omega_t(t)]^T$, and the reference vector of these angular velocities, $\boldsymbol{\omega}^*(t) = [\omega_m^*(t), \omega_t^*(t)]^T$, which is the output of the outer loop. In this sense, the feedback multivariate control input, $\mathbf{u}(t) = [u_m(t), u_t(t)]^T$, is synthesized as a nonlinear input transformation and classical proportional controller with a nonlinear cancellation vector:

$$\mathbf{u}(t) = \mathbf{N}^{-1}[\boldsymbol{\vartheta_e}(t) - \boldsymbol{\Gamma}(\boldsymbol{\omega}(t))] \tag{42}$$

in which $\mathbf{N}$ and $\boldsymbol{\Gamma}(\boldsymbol{\omega}(t))$ are defined in Equations (12) and (13), respectively, and $\boldsymbol{\vartheta_e}(t) = [\vartheta_m(t), \vartheta_t(t)]^T$ represents a vector of auxiliary control inputs, given by:

$$\boldsymbol{\vartheta_e}(t) = \dot{\boldsymbol{\omega}}(t) = -\mathbf{K_P^e}[\boldsymbol{\omega}(t) - \boldsymbol{\omega}^*(t)] \tag{43}$$

where $\mathbf{K_P^e} \in \mathbb{R}^{2 \times 2}$ is a constant diagonal positive definitive matrix that represents the design elements of a vector-valued classical proportional controller.

The closed loop tracking error vector, $\mathbf{e}_\omega(t) = \boldsymbol{\omega}(t) - \boldsymbol{\omega}^*(t)$, for the electrical part is obtained after substituting Expression (42) in the dynamic model of the electrical part of the system in Equation (11), yielding the following expression:

$$\dot{\boldsymbol{\omega}}(t) + \mathbf{K_P^e}\mathbf{e}_\omega(t) = 0 \tag{44}$$

The controller design matrix $\mathbf{K_P^e}$ is designed so as to render the following 2×2 complex-valued diagonal matrix, $\mathbf{p_c^e}(s)$, defined as:

$$\mathbf{p_c^e}(s) = \mathbf{I}^{2 \times 2}s + \mathbf{K_P^e} \tag{45}$$

as first degree Hurwitz polynomials with the desired roots located in the left half of the complex plane in order to achieve the convergence of the tracking error dynamics to a small vicinity around the origin of the error phase space. In particular, the constant controller gain matrix $\mathbf{K_P^e}$ of the closed loop characteristic polynomial is determined by means of a term by term comparison with the following desired Hurtwitz 2×2 diagonal matrix:

$$\mathbf{p_{c_d}^e}(s) = \mathbf{I}^{2 \times 2} s + \mathbf{p_c^e} \tag{46}$$

where $\mathbf{p_c^e} \in \mathbb{R}^{2 \times 2}$ is a diagonal positive definite matrix, which represents the desired position of the poles in closed loop. Therefore, the design controller gain is given by:

$$\mathbf{K_P^e} = \mathbf{p_c^e} \tag{47}$$

Finally, to conclude the description of the inner loop control, we highlight again that the design parameters are selected for the sake of making the dynamics of the inner loop control much faster than the outer loop dynamics, this way ensuring the functioning of the cascade controller [24]. The secondary controller must be relatively quick so that it attenuates a disturbance before the disturbance affects the primary controlled variable. A general guideline is that the secondary one should be three-times faster than the primary [25]. It should be noted that the cascade strategy has to be tuned in a sequential manner. In this procedure, the inner loop control should be tuned first, because the secondary controller or inner loop affects the open-loop dynamics of the primary or outer loop. Thereby, and in order to tune the parameters in the inner loop control, which are the gains of the proportional controller defined in matrix $\mathbf{K_P^e}$, the primary controller will be disconnected, i.e., the cascade should be open, and then, the electrical controller will be tuned in a conventional manner, which involves a plant experiment, initial tuning calculation and fine-tuning based on a closed-loop dynamic response.

3.2. Outer Loop Control

The objective of the outer loop control is to determine the required values for the angular velocities of the main and tail rotors, i.e., the reference vector for the angular velocities, which is the reference input of the inner loop, $\omega^*(t) = [\omega_m^*(t), \omega_t^*(t)]^T$, in order to eliminate the difference between the generalized coordinates of the TRMS, $\mathbf{q}(t) = [\psi(t), \phi(t)]^T$, and the reference trajectories for these coordinates, $\mathbf{q}^*(t) = [\psi^*(t), \phi^*(t)]^T$. To achieve this goal, the following multivariate nonlinear feedback control input vector, $\mathbf{\Omega}(t)$, is synthesized as a nonlinear input transformation and a proportional-integral-derivative (PID) controller with a nonlinear cancellation vector:

$$\mathbf{\Omega}(t) = \mathbf{E}^{-1}(\mathbf{q}(t))[\mathbf{M}(\mathbf{q}(t))\boldsymbol{\vartheta_m}(t) + \mathbf{D}(\mathbf{q}(t), \dot{\mathbf{q}}(t))] \tag{48}$$

where $\boldsymbol{\vartheta_m}(t) = [\vartheta_\psi(t), \vartheta_\phi(t)]^T$ represents a vector of auxiliary control variables, given by:

$$\boldsymbol{\vartheta_m}(t) = \ddot{\mathbf{q}}(t) = \ddot{\mathbf{q}}^*(t) - \mathbf{K_D^m}(\dot{\mathbf{q}}(t) - \dot{\mathbf{q}}^*(t)) - \mathbf{K_P^m}(\mathbf{q}(t) - \mathbf{q}^*(t)) - \mathbf{K_I^m}\int(\mathbf{q}(t) - \mathbf{q}^*(t)) \tag{49}$$

where $\mathbf{K_D^m}$, $\mathbf{K_P^m}$ and $\mathbf{K_I^m} \in \mathbb{R}^{2 \times 2}$ are diagonal positive definitive matrices that represent the design elements of a vector-valued classical proportional-integral-derivative multivariate controller.

The closed loop tracking error vector, $\mathbf{e_q}(t) = \mathbf{q}(t) - \mathbf{q}^*(t)$, for the mechanical part is obtained after substituting Expression (48) in the simplified model of the mechanical part in Equation (39), yielding the following expression:

$$\mathbf{e_q}^{(3)}(t) + \mathbf{K_D^m}\ddot{\mathbf{e}}_\mathbf{q}(t) + \mathbf{K_P^m}\dot{\mathbf{e}}_\mathbf{q}(t) + \mathbf{K_I^m}\mathbf{e_q}(t) = 0 \tag{50}$$

In order to achieve the convergence of the tracking error dynamics to a small vicinity around the origin of the tracking error phase space, the controller design matrices $\mathbf{K_D^m}$, $\mathbf{K_P^m}$ and $\mathbf{K_I^m}$ are chosen in such a manner that all non-zero components of the 2×2 complex valued diagonal matrix, $\mathbf{p_c^m}(s)$, defined as,

$$\mathbf{p_c^m}(s) = \mathbf{I}^{2 \times 2} s^3 + \mathbf{K_D^m} s^2 + \mathbf{K_P^m} s + \mathbf{K_I^m} \tag{51}$$

are all third degree Hurwitz polynomials whose roots are located sufficiently far into the left half on the complex plane. The stability of Expression (51) can be studied by using the Routh–Hurwitz criterion. Bearing in mind that the set of design matrices $\mathbf{K_P^m}$, $\mathbf{K_D^m}$ and $\mathbf{K_I^m}$ are diagonal, the stability of each error variable $\mathbf{e_q}(t) = [e_\psi(t); e_\phi(t)]^T = [\psi(t) - \psi^*(t); \phi(t) - \phi^*(t)]^T$ can be studied in an independent manner. After applying the Routh–Hurwitz criterion, one obtains the following stability conditions: (i) $K_{D_i}^m, K_{P_i}^m > 0$; and (ii) $0 < K_{I_i}^m < K_{D_i}^m \cdot K_{P_i}^m$ for $i = \psi, \phi$. After considering the previous stability restrictions, the constant controller gains $\mathbf{K_D^m}$, $\mathbf{K_P^m}$ and $\mathbf{K_I^m}$ of the closed loop characteristic polynomial are determined by using a term by term comparison with the following desired Hurtwitz 2×2 complex-valued diagonal matrix:

$$\mathbf{p_{c_d}^m}(s) = \left(\mathbf{I}^{2 \times 2} s + \mathbf{p_c^m} \right) \left(\mathbf{I}^{2 \times 2} s^2 + 2\zeta_c^m \omega_c^m s + (\omega_c^m)^2 \right) \tag{52}$$

where $\mathbf{p_c^m}$, ζ_c^m and $\omega_c^m \in \mathbb{R}^{2 \times 2}$ are diagonal positive definite matrices. Therefore, the design controller gains are given by:

$$\mathbf{K_D^m} = 2\zeta_c^m \omega_c^m + \mathbf{p_c^m} \tag{53}$$

$$\mathbf{K_P^m} = (\omega_c^m)^2 + 2\zeta_c^m \omega_c^m \mathbf{p_c^m} \tag{54}$$

$$\mathbf{K_I^m} = \mathbf{p_c^m} (\omega_c^m)^2 \tag{55}$$

Finally, the necessary angular velocity vector values, $\omega^*(t) = [\omega_m^*(t), \omega_t^*(t)]^T$, are obtained from the input control vector, $\Omega(t) = [\omega_m \, |\omega_m|, \omega_t \, |\omega_t|]^T$, by performing the following operation:

$$\omega^*(t) = \begin{bmatrix} \omega_m^*(t) \\ \omega_t^*(t) \end{bmatrix} = \begin{bmatrix} sign\,(\omega_m \, |\omega_m|) \, \sqrt{|\omega_m \, |\omega_m||} \\ sign\,(\omega_t \, |\omega_t|) \, \sqrt{|\omega_t \, |\omega_t||} \end{bmatrix} \tag{56}$$

4. Experimental Section

This section describes the experiments carried out to verify the effectiveness of the proposed control algorithm. In the following subsections, we briefly explain the experimental platform and the software tools, and after that, we illustrate the experimental results on the real platform, including a comparison with other control algorithms in terms of both stabilization and trajectory tracking task performance.

4.1. Experimental Setup

The implementation of the designed robust decentralized controller is carried out by using the following equipment:

- A twin rotor MIMO system provided by Feedback Instruments® (see Figure 1 and [9]).
- A PC operating in a Windows® environment using software tools from *MathWorks*® *Inc* (MATLAB®, Simulink, Control Toolbox, Real Time Workshop® (RTW), Real Time Windows Target® (RTWT)) and Visual C++ Professional®.
- The real TRMS is connected to the computer by means of an Advantech® PCI1711 card, which is accessible in the MATLAB/Simulink® environment through the Real-Time Toolbox®.
- The control signals in the MATLAB/Simulink® environment consist of two input voltages (in the range $[-2.5, 2.5]$ V) for the two DC motors A-max 26 provided by Maxon Motor®.
- The vector of generalized coordinates, $\mathbf{q}(t) = [\psi(t), \phi(t)]^T$, are measured by using two HCTL 2016 digital encoders provided by Agilent Technologies®, and the angular velocity vector $\omega(t) = [\omega_m(t), \omega_t(t)]^T$ is measured by using two DC-Tacho DCT 22 provided by Maxon Motor®.

- The sampling rate for the controlled system is 0.002 s.

On the other hand, the executable file for the proposed control scheme is achieved by performing the following steps (see Figure 6): MATLAB® acts as the application host environment, in which the other MathWorks® products run, and Simulink® provides a well-structured graphical interface for the implementation of the proposed nonlinear control scheme. Real Time Workshop® automatically builds a C++ source program from the Simulink Model. The C++ Compiler® compiles and links the code created by Real Time Workshop® to produce an executable program. Real Time Windows Target® communicates with the executable program acting as the control program and interfaces with the TRMS through the PCI1711 card. Real Time Windows Target® controls the two-way data, or signal flow, to and from the model (which is now an executable program), and to and from the PCI1711 card. The advantage of this approach is that the designer only needs to model the process, using the graphical tools available in Simulink®, without having to worry about the mechanics of communication to and from the TRMS.

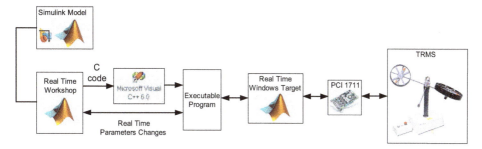

Figure 6. Control system development flow diagram.

4.2. Experimental Results

This subsection discusses the experiments carried out to verify the efficiency of the control strategy proposed in Section 3 in the following aspects: (a) robustness with regard to large initial errors; (b) quick convergence of the tracking errors to a small neighborhood of zero; (c) smooth transient responses; (d) low control effort; (e) robustness with regard to modeling errors. In the trials, the desired reference trajectories for the pitch (ψ) and the yaw (ϕ) angles have been selected in order to obtain a complex and challenging trajectory for the TRMS, which allows one to show the main characteristics and excellent performance of the proposed control scheme, but avoiding at the same time the saturation of the actuators of the laboratory platform. This reference trajectory is defined by the following expression:

$$\mathbf{q}^*(t) = \begin{bmatrix} \psi^*(t) \\ \phi^*(t) \end{bmatrix} = \begin{bmatrix} A_{0_\psi} + A_{1_\psi}\left(2sin(\omega_{1_\psi}t) + sin(\omega_{2_\psi}t)\right) \\ A_{1_\phi}sin(\omega_{1_\phi}t) + A_{2_\phi}\left(sin(\omega_{2_\phi}t) + sin(\omega_{3_\phi}t)\right) \end{bmatrix} \tag{57}$$

where $\mathbf{q}^*(t) = [\psi^*(t), \phi^*(t)]^T$ is the reference trajectory vector of the generalized coordinates, and the values of the above constants are given by:

$$A_{0_\psi} = 0.4 \ rad; \qquad \omega_{1_\psi} = 0.0785 \ rad/s;$$
$$A_{1_\psi} = 0.1 \ rad; \qquad \omega_{2_\psi} = 0.0157 \ rad/s;$$
$$A_{1_\phi} = 0.8 \ rad; \qquad \omega_{1_\phi} = 0.157 \ rad/s;$$
$$A_{2_\phi} = 0.3 \ rad; \qquad \omega_{2_\phi} = 0.0785 \ rad/s;$$
$$\omega_{3_\phi} = 0.0157 \ rad/s; \tag{58}$$

   In order to demonstrate the exponential convergence of the desired trajectories, and the robustness with respect to large initial errors, the initial position of the TRMS is defined as $\mathbf{q}_0(t) = [0,0]^T$, which represents a different value than the initial position of the reference trajectory vector $\mathbf{q}^*(t)$. With regard to the parameters of the plant used in the experimentation, the values of which are presented in Tables 1–3, we have to highlight that the discrepancies in the model due to modeling errors are around 5%, as a consequence of the difficulty involved in adequately modeling all of the dynamics terms. The small errors observed in the dynamics identification trials, which have been performed in our research, are compensated by the action of the proposed control scheme. With the use of an integral action on the outer loop, eliminating the possible steady state errors is achieved.

   On the other hand, the design of the proposed nonlinear control scheme and the choice of the values of the gain vectors, which are tuned according to the procedure explained in Section 3, have been done in order to achieve a control as fast as possible, but avoiding possible saturations of the input voltages of the motors in the MATLAB/Simulink® environment, which occur at ±2.5 V. The summary of the procedure carried out to tune the designer parameters is explained next. Firstly, the inner loop control has been tuned using the model of the electrical part of the TRMS by means of numerical simulations. In this first stage, the parameters of the proportional controller have been tuned in order to achieve the fast dynamics of the inner loop. In other words, the aim is to achieve a quick convergence of the closed loop tracking error vector, $\mathbf{e}_\omega(t)$, to a small vicinity around the origin of the tracking error phase space. Secondly, we have assumed the dynamics of the inner loop to be equal to $\mathbf{I}^{2\times2}$, and then, we have tuned, again by means of numerical simulations, the parameters of the PID controller in the outer loop. Finally, the values obtained in the simulations have been slightly adjusted in the experimental trials with the laboratory platform. Thereby, for the inner loop controller, the values of the desired Hurtwitz 2×2 complex diagonal matrix for the controller are $\mathbf{p}_c^e(s) = diag(12.0, 9.0)$, and for the outer loop controller, the values of the matrices of the desired Hurtwitz polynomial vector for the feedback controller are $\mathbf{p}_c^m = diag(1.0, 1.0)$, $\boldsymbol{\zeta}_c^m = diag(1.5, 1.5)$ and $\boldsymbol{\omega}_c^m = diag(2.0, 1.8)$. More details about how to tune controllers based on a cascade scheme can be consulted in some reference works [25–28].

   In the following lines, we discuss the performance of the proposed decentralized control scheme, which is shown in the next graphs (Figures 7–11), where, in order to show the improvements of this design, we shall also compare the experimental results obtained using the proposed control (denoted in the graphs as decentralized nonlinear control (DEC NON)), a standard PID control [29] (denoted in the graphs as PID CLASSIC) and a PID control with a derivative filter coefficient [19] (denoted in the graphs as PID DFC). Figure 7 illustrates a comparison between the desired trajectory, $\mathbf{q}^*(t) = [\psi^*(t), \phi^*(t)]^T$, and the real trajectory of the TRMS, $\mathbf{q}(t) = [\psi(t), \phi(t)]^T$. This graph shows that the three algorithms are robust with regard to large initial errors. However, the proposed decentralized control scheme has the smoothest transient response and the best performance in trajectory tracking, as can also be observed in Figure 8, which shows, for each control, the closed loop tracking error vector, $\mathbf{e}_q(t) = \mathbf{q}(t) - \mathbf{q}^*(t) = [\psi(t) - \psi^*(t), \phi(t) - \phi^*(t)]^T$. The proposed decentralized controller has a closed loop tracking error vector that remains bounded within a vicinity of radius $[0.04, 0.10]^T$ *rad*, while the standard PID and the PID with derivative filter have error vectors bounded in $[0.02, 0.35]^T$ *rad* and $[0.02, 0.33]^T$ *rad*, respectively. Therefore, although the three control algorithms achieve a quick convergence of the tracking error to a small neighborhood of zero, the proposed control scheme presents the smallest error.

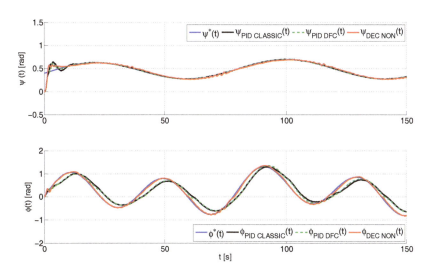

Figure 7. Real and desired evolution trajectories of the vector of the generalized coordinates of the TRMS, $\mathbf{q}(t) = [\psi(t), \phi(t)]^T$.

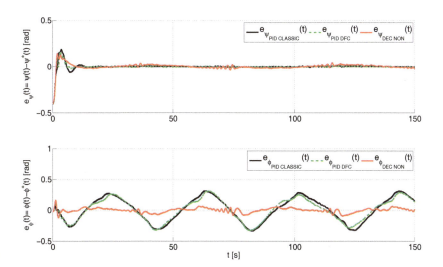

Figure 8. Evolution of the error vector of generalized coordinates of the TRMS, $\mathbf{e_q}(t) = \mathbf{q}(t) - \mathbf{q}^*(t) = [\psi(t) - \psi^*(t), \phi(t) - \phi^*(t)]^T$.

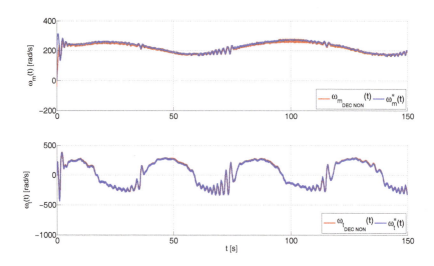

Figure 9. Real and desired evolution trajectories of the angular velocity vector, $\omega^*(t) = [\omega_m^*(t), \omega_t^*(t)]^T$ and $\omega(t) = [\omega_m(t), \omega_t(t)]^T$.

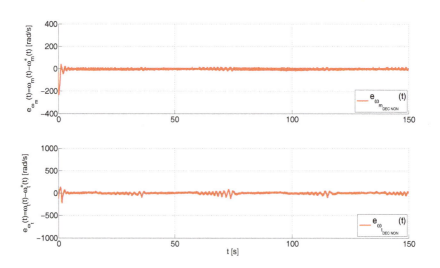

Figure 10. Evolution of the angular velocity error vector, $\mathbf{e}_\omega(t) = \omega(t) - \omega^*(t) = [\omega_m(t) - \omega_m^*(t), \omega_t(t) - \omega_t^*(t)]^T$.

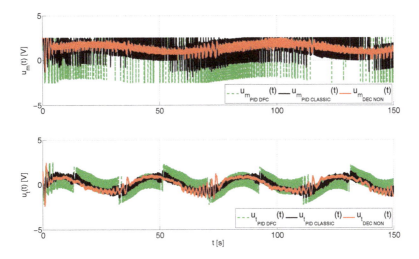

Figure 11. Evolution of the input voltage vector, $\mathbf{u}(t) = [u_m(t), u_t(t)]^T$, in the MATLAB/Simulink® environment.

On the other hand, the performance of the inner control loop is shown in Figure 9, which illustrates a comparison between the angular velocity vector, $\boldsymbol{\omega}^*(t) = [\omega_m^*(t), \omega_t^*(t)]^T$, obtained from the output of the outer loop, and the real magnitudes of the angular velocity vector, $\boldsymbol{\omega}(t) = [\omega_m(t), \omega_t(t)]^T$. Again, the proposed control has a smooth transient response and a fast convergence of the tracking error to a neighborhood near to zero as evidenced in Figure 10, which shows, for each control, that the angular velocity error vector, $\mathbf{e}_\omega(t) = \boldsymbol{\omega}(t) - \boldsymbol{\omega}^*(t) = [\omega_m(t) - \omega_m^*(t), \omega_t(t) - \omega_t^*(t)]^T$, remains bounded in $[20, 120]^T$ rad/s. Finally, the input voltage vectors in the MATLAB/Simulink® environment, $\mathbf{u}(t) = [u_m(t), u_t(t)]^T$, are shown in Figure 11. This graph illustrates that the smallest control input effort is provided by the proposed control scheme, which furthermore presents a smooth evolution of the input voltage vector without saturations unlike both PID controls, the standard PID and the PID with derivative filter coefficient. As you may observe at the top of this figure, both PID controls cause the saturation of the control signal of the main rotor, which occurs at ± 2.5 V, for long periods of time during the trials. These saturations cause a worse performance of each one of these controllers in comparison with the proposed control scheme.

Additionally, the performances of the control methods have been measured in terms of the integral squared tracking error, $ISE = \int_{t_A}^{t_B} \mathbf{e_q}(t)^T \mathbf{e_q}(t) dt = \int_{t_A}^{t_B} (e_\psi(t)^2 + e_\phi(t)^2) dt$, the integral absolute tracking error, $IAE = \int_{t_A}^{t_B} (|e_\psi(t)| + |e_\phi(t)|) dt$, and the integral time absolute tracking error, $ITAE = \int_{t_A}^{t_B} t(|e_\psi(t)| + |e_\phi(t)|) dt$, where $t_A = 0$ s and $t_B = 150$ s denote the initial and final time of the simulation, and $\mathbf{e_q}(t) = [e_\psi(t), e_\phi(t)]^T = [\psi(t) - \psi^*(t), \phi(t) - \phi^*(t)]^T$ is the closed loop tracking error vector. The *ISE* and the *IAE* criteria will treat all of the tracking errors in a uniform manner. However, the *ITAE* criterion, as time appears as a factor, will heavily penalize errors that occur late in time, but ignore errors that occur early in time. The results achieved are illustrated in Table 4, showing the best performance of the proposed decentralized control scheme (DEC NON) in comparison to the other conventional controls (PID CLASSIC and PID DFC). Both PID controls show a similar behavior and have a worse performance when they are compared to the proposed control method.

Table 4. Performance of the control methods.

Control Method	ISE	IAE	ITAE
Robust Decentralized Nonlinear Control (DEC NON)	0.3956	6.6579	435.7
Standard PID control (PID CLASSIC)	5.8275	26.7591	2002.4
PID control with the derivative filter coefficient (PID DFC)	5.0814	24.8175	1834.4

To sum up, the experimental results show a better performance of the proposed decentralized control scheme against the other control laws. The proposed control law illustrates a better performance in the following aspects: (1) robustness in relation to large initial errors with a smooth transient response; (2) better tracking of the reference trajectories; (3) quick convergence of the tracking errors to the smallest neighborhood of zero; (4) less control effort; and (5) the absence of saturations in the input control voltages.

5. Conclusions

In this study, we have successfully designed a novel robust nonlinear multivariate decentralized control scheme for the underactuated and nonlinear twin rotor MIMO system (TRMS) laboratory platform. This control system is based on decoupling the electrical from the mechanical dynamics and the use of two nested nonlinear multivariate loops. The inner loop is designed as a nonlinear input transformation and classical proportional controller with a nonlinear cancellation vector and is responsible for the stabilization and tracking of the vector of angular velocities of the propellers of the TRMS. The outer loop control is designed as a nonlinear input transformation, a proportional-integral-derivative (PID) linear action and nonlinear compensation vector, which determines the required values for the reference velocities in order to achieve the elimination of the difference between the generalized coordinates of the TRMS and the reference trajectories for these. This independence in the design of the control loops is possible thanks to having made the dynamics of the inner loop much faster than the dynamics of the mechanical part. This control system is very simple and allows the platform to be perfectly stabilized and positioned in space. Additional advantages of this control approach are: (a) simplification of the control design procedure due to the design of two much simpler dynamics, which are controlled separately; (b) this scheme can be more easily and safely implemented than the standard controllers used in the control of the TRMS platform, which involve closing a single loop, because the nested control loops proposed in this work are sequentially implemented, first by closing the inner loop, which exhibits a very high relative stability in the presence of system uncertainties, external disturbances and noisy corruptions, and later through closing the outer loop, which is more prone to becoming unstable, but whose risk of exhibiting unstable motions has been significantly reduced by having previously closed the inner loop. The experimental tests carried out, in order to verify the performance of the proposed decentralized controller, show not only the accurate tracking of the reference trajectories, but also the better performance of the proposed control compared to the other two conventional controllers. The robustness in regards to large initial errors and possible modeling errors, the quick convergence to a small neighborhood of zero and the smooth transient response with a low control effort are the main features of the proposed design.

Acknowledgments: This work has been partially supported by the Spanish Ministerio de Economía y Competitividad/FEDER under the TEC2016-80986-R, DPI2016-80894-R, TIN2013-47074-C2-1-R and DPI2014-53499-R grants. Lidia M. Belmonte holds an Formación del Profesorado Universitario (FPU) Scholarship (FPU014/05283) from the Spanish Government.

Author Contributions: L.M.B., R.M, A. F.-C and J.A.S conceived, designed and performed the experiments. Additionally, L.M.B., R.M, A. F.-C and J.A.S analyzed the data and participated in writing the paper.

Conflicts of Interest: The authors declare no conflicts of interest.

References

1. Nonami, K.; Kendoul, F.; Suzuki, S.T. *Autonomous Flying Robots—Unmanned Aerial Vehicles and Micro Aerial Vehicles*; Springer: Heidelberg, Germany, 2010.
2. Espizona, T.; Dzul, A.; Llama, M. Linear and Nonlinear Controllers Applied to Fixed-Wing UAV. *Int. J. Adv. Robot. Syst.* **2013**, *10*, 33, doi:10.5772/53616.
3. Fernández-Caballero, A.; Belmonte, L.M.; Morales, R.; Somolinos, J.A. Generalized Proportional Integral Control for an Unmanned Quadrotor System. *Int. J. Adv. Robot. Syst.* **2015**, *12*, 85, doi:10.5772/60833.
4. Alvarenga, J.; Vitzilaios, N.I.; Valavanis, K.P.; Rutherford, M.J. Survey of Unmanned Helicopter Model-Based Navigation and Control Techniques. *J. Intell. Robot. Syst.* **2015**, *80*, 87–138.
5. Ali, Z.A.; Wang, D.; Aamir, M. Fuzzy-Based Hybrid Control Algorithm for the Stabilization of a Tri-Rotor UAV. *Sensors* **2016**, *16*, 652, doi:10.3390/s16050652.
6. Cabecinhas, D.; Naldi, R.; Silvestre, C.; Cunha, R.; Marconi, L. Robust Landing and Sliding Maneuver Hybrid Controller for a Quadrotor Vehicle. *IEEE Trans. Control Syst. Technol.* **2016**, *4*, 400–412.
7. Chen, F.; Wu, Q.; Jiang, B.; Tao, G. A Reconfiguration Scheme for Quadrotor Helicopter via Simple Adaptive Control and Quantum Logic. *IEEE Trans. Ind. Electron.* **2015**, *62*, 4328–4335.
8. Zheng, B.; Zhong, Y. Robust Attitude Regulation of a 3-DOF Helicopter Benchmark: Theory and Experiments. *IEEE Trans. Ind. Electron.* **2011**, *58*, 660–670.
9. Feedback Co. *Twin Rotor MIMO System 33-220 User Manual*; Feedback Co.: Crowborough, UK, 1998.
10. Rahideh, A.; Shaheed, M.H.; Huigberts, H.J.C. Dynamic Modelling of a TRMS Using Analytical and Empirical Approaches. *Control Eng. Pract.* **2008**, *16*, 241–259.
11. Toha, S.F.; Latiff, I.A.; Mohamad, M.; Tokhi, M.O. Parametric modelling of a TRMS using dynamic spread factor particle swarm optimisation. In Proceedings of the UKSim 2009: 11th International Conference on Computer Modelling and Simulation, Cambridge, UK, 25–27 March 2009; pp. 95–100.
12. Tanaka, H.; Ohta, Y.; Okimura, Y. A local approach to LPV-identification of a Twin Rotor MIMO System. *IFAC Proc. Vol.* **2011**, *44*, 7749–7754.
13. Tastemirov, A.; Lecchini-Visintini, A.; Morales, R.M. Complete Dynamic Model of the TWIN Rotor MIMO System (TRMS) with Experimental Validation. In Proceedings of the 39th European Rotorcraft Forum 2013 (ERF 2013), Moscow, Russia, 3–6 September 2013.
14. Chalupa, P.; Prikryl, J.; Novák, J. Modelling of Twin Rotor MIMO System. *Procedia Eng.* **2015**, *100*, 249–258.
15. Juang, J.-G.; Lin, R.-W.; Liu, W.-K. Comparison of classical control and intelligent control for a MIMO system. *Appl. Math. Comput.* **2008**, *25*, 778–791.
16. Wen, P.; Lu, T.W. Decoupling control of a twin rotor mimo system using robust deadbeat control technique. *IET Control Theory Appl.* **2008**, *2*, 999–1007.
17. Rahideh, A.; Shaheed, M.H. Constrained output feedback model predictive control for nonlinear systems. *Control Eng. Pract.* **2012**, *20*, 431–443.
18. Jahed, M.; Farrokhi, M. Robust adaptive fuzzy control of twin rotor MIMO system. *Soft Comput.* **2013**, *17*, 1847–1860.
19. Kumar-Pandey, S.; Laxmi, V. Control of Twin Rotor MIMO System using PID controller with derivative filter coefficient. In Proceedings of the 2014 IEEE Students' Conference on Electrical, Electronics and Computer Science, Bhopal, India, 1–2 March 2014.
20. Belmonte, L.M.; Morales, R.; Fernández-Caballero, A.; Somolinos, J.A. A Tandem Active Disturbance Rejection Control for a Laboratory Helicopter with Variable Speed Rotors. *IEEE Trans. Ind. Electron.* **2016**, doi:10.1109/TIE.2016.2587238.
21. Alagoz, B.B.; Ates, A.; Yeroglu, C. Auto-tuning of PID controller according to fractional-order reference model approximation for DC rotor control. *Mechatronics* **2013**, *23*, 789–797.
22. Mullhaupt, P.; Srinivasan, B.; Levine, J; Bonvin, D. Control of the Toycopter Using a Flat Approximation. *IEEE Trans. Control Syst. Technol.* **2008**, *16*, 882–896.
23. Morales, R.; Feliu, V.; Jaramillo, V. Position control of very lightweight single-link flexible arms with large payload variations by using disturbance observers. *Robot. Auton. Syst.* **2012**, *60*, 532–547.
24. Son, Y.I.; Kim, I.H.; Choi, D.S; Shim, D. Robust Cascade Control of Electric Motor Drives Using Dual Reduced-Order PI Observers. *IEEE Trans. Ind. Electron.* **2015**, *62*, 3672–3682.

25. Marlin, T.E.; *Process Control, Designing Processes and Control Systems for Dynamic Performance*, 2nd ed.; McGraw-Hill: New York, NY, USA, 2000.
26. Arrieta, O., Vilanova, R., Balaguer, P. Procedure for cascade control systems design: Choice of suitable PID tunings. *Int. J. Comput. Commun. Control* **2008**, *3*, 235–248.
27. Alfaro, V.M.; Vilanova, R.; Arrieta, O. Robust tuning of Two-Degree-of-Freedom (2-DoF) PI/PID based cascade control systems. *J. Process Control* **2009**, *19*, 1658–1670.
28. Veronesi, M., Visioli, A. Simultaneous closed-loop automatic tuning method for cascade controllers. *IET Control Theory Appl.* **2011**, *5*, 263–270.
29. Feedback Co. *Twin rotor MIMO system. Control Experiments*; Manual 33-949S Ed01; Feedback Co.: Crowborough, UK, 2008.

MDPI AG

St. Alban-Anlage 66

4052 Basel, Switzerland

Tel. +41 61 683 77 34

Fax +41 61 302 89 18

http://www.mdpi.com

Sensors Editorial Office

E-mail: sensors@mdpi.com

http://www.mdpi.com/journal/sensors

www.ingramcontent.com/pod-product-compliance
Lightning Source LLC
LaVergne TN
LVHW071357070326
832902LV00028B/4629